Studies in Neuroscience, Consciousness and Spirituality

Volume 8

Neuroscience has become one of the major drivers of the scientific progress recently. At the same time studies of spiritual practices such as meditation reveal that these practices can have clinical and individual benefits, but can also tell us a lot about how consciousness functions. The study of consciousness from the perspective of spirituality might help us to understand how a satisfying scientific approach to consciousness would need to be crafted. This series brings together scholarship and science that combines these views. It will publish the outcome of scientific workshops, that allow experts in their fields to discuss the general issues around these topics in a trans-disciplinary spirit. The series will also publish research monographs and edited collections that gravitate around this new topical intersection of science, consciousness studies, and spirituality or religious studies. The guiding idea behind this series is that science is not complete if it leaves out experiential areas that are crucial for humans, and that such experiences might also be relevant for science. At the same time science, being one of the major peaceful enterprises of humankind and the driver for a global enlightenment movement, might be able to inform consciousness and religious studies. The editors of this series believe in the relevance of this dialogue which needs to be radically open in both directions.

All books to be published in this series will be fully peer-reviewed before final acceptance.

Prem Saran Satsangi
Anna Margaretha Horatschek
Anand Srivastav
Editors

Consciousness Studies in Sciences and Humanities: Eastern and Western Perspectives

Springer

Editors
Prem Saran Satsangi
Dayalbagh Educational Institute
Deemed to be University
Agra, India

Anna Margaretha Horatschek
Department of English and American Literatures, Cultures, and Media
Kiel University
Kiel, Germany

Anand Srivastav
Department of Mathematics
Kiel University
Kiel, Germany

ISSN 2211-8918 ISSN 2211-8926 (electronic)
Studies in Neuroscience, Consciousness and Spirituality
ISBN 978-3-031-13919-2 ISBN 978-3-031-13920-8 (eBook)
https://doi.org/10.1007/978-3-031-13920-8

This Springer imprint is published by the registered company Springer Nature Switzerland AG
The registered company address is: Gewerbestrasse 11, 6330 Cham, Switzerland

Paper in this product is recyclable.

Preface

> I would in fact say, see, hear, sense, feel, perceive, such that reality in the real meaning would be open [...] as full consciousness. (Prof. Prem Saran Satsangi, 2016)

This anthology comprises essays that are mostly based on lectures contributed to the Dayalbagh Science of Consciousness (DSC) Conferences from 2019 to 2021. The book gathers Eastern – namely Indian – and Western – namely German, British, American and Canadian – perspectives and approaches to Consciousness Studies from a range of disciplines in the natural sciences and humanities. The essays present ongoing research in the specific disciplines and thus offer a sound introduction to and overview of the broad field of Consciousness Studies. However, the structure of the book deliberately will not stick to disciplinary compartmentalization, but adheres to central aspects of the umbrella term 'consciousness' that are addressed with various disciplinary methodologies. This form of presentation will illustrate that what is meant by the term consciousness is highly determined by the cultural and disciplinary context in which the word is used and highlight the often unacknowledged pre-theoretical assumptions of the respective disciplinary research, as they reciprocally comment on each other. By capturing the internal refractions of the term 'consciousness' along these multiple axes, the book in its very structure thus presents a model of consciousness that facilitates interdisciplinary, transdisciplinary, and intercultural exchange and dialogue – an increasingly stipulated necessity to produce adequate results in the field of Consciousness Studies.

The central challenge of any inter- and trans-disciplinary cooperation consists in the difficulty to establish a shared conceptual language concerning foundational terms. This is also true for Consciousness Studies. Questions about consciousness have been addressed by sages, philosophers, and rishis for thousands of years. In the East, there are long traditions of exploring matters of consciousness in theory and practice, for example with regard to techniques, benefits, and ultimate aims of meditation; in the West also, theologians, philosophers, and artists have dealt with questions of consciousness for ages. In the 1990s, the natural sciences started to seriously study consciousness, and since then evidence-based scientific research has played a dominant role in Consciousness Studies. Most scientific experiments are based on the premise that consciousness – generally – is the result of physical conditions, more specifically of interrelated brain activities. But this model is not uncontested. Consequently, there has been an explosion of work on consciousness from philosophers, psychologists, neurologists, etc. Accordingly, the term 'consciousness' has acquired very different meanings, depending on whether the person comes from the natural sciences or the humanities, from the West or from the East.

A landmark in inter- and transdisciplinary Consciousness Studies worldwide is certainly the Dayalbagh Educational Institute (DEI) (Deemed University) in Agra, India, where research on Consciousness Studies forms an essential aspect of its academic profile. The ultimate objective of former honorary director (CEO) of DEI Prof. Dr. Prem Saran Satsangi, the Eighth Sant Satguru of Radhasoami Faith, Dayalbagh, of an 'Evolutionary Art, Science and Engineering (TEASE) of Consciousness to aim at a certain degree of compatibility and convergence to the extent feasible to raise it to higher and higher status with the passage of time' (Satsangi 2020), has been honoured by the recently instituted Alumni Eminence Award (as distinct from Distinguished Alumni Award), by Banaras Hindu University (BHU-IT) conferred on 1. Prof. Dr. P.S. Satsangi; and 2. Dr. Rajneesh Dube, Additional Chief Secretary, Govt. of Uttar Pradesh; nominated as the incoming Vice-Chancellor of BHU-IT. BHU-IT espouses a holistic approach to education and is tagged as the Institution of Eminence by MHRD (Ministry of Human Resource Development, renamed as Ministry of Higher Education in 2020), Government of India.

In view of this public recognition and the outstanding distinction of Prof. Dr. Prem Saran Satsangi as the instigator of a holistic approach to Consciousness Studies,[1] the 'Comments from the Editor' render verbatim his observations on two *Scientific American* articles by Mark Fischetti (2022) and Katie Worth (2022). Fischetti highlights social and educational developments from 1872 to 1972, which he considers crucial for the Western hemisphere. The respective Comment points out similarly momentous developments in the Eastern hemisphere, and focuses on Radhasoami Faith and the lacto-vegetarian way of life in Dayalbagh, Agra, India. Worth discusses interventions from oil and gas industries to manipulate textbooks from kindergarten to 12th grade, and criticizes the tactic of belabouring from a cost-benefit perspective the drawbacks of the renewable power energy strategies in Germany. The Comment advances a wider perspective as argumentative counter-strategy. Additionally, annotations of Prof. Prem Saran Satsangi on specific essays are included in several cases. Some of them form part of discussions with the authors on potential future research projects and testify to the vivid impact of this book already during its genesis.

Prof. Satsangi's observations provide an emic perspective for the topic of this volume, which might envision levels of consciousness, spirituality, and cosmos that lie at the margins of intellectual comprehensibility – today and beyond.[2]

---

[1] See also the publication of Prof. Dr. Makund Behari Lal, the Seventh Revered Leader of Radhasoami Faith, Dayalbagh Hqs. Zoology in Lucknow. *Nature* 176, 1955, pp. 907–908. https://doi.org/10.1038/176907e0

[2] For a comprehensive explication of tenets of Radhasoami Faith, see Maharaj Sahabji Pandit Brahm Sankar Misra, M.A. *Discourses on Radhasoami Faith*, with the Supplement by Prem Saran Satsangi, and a foreword by Prem Kumar. Radhasoami Satsang Sabha: Dayalbagh (Agra), India. 2004, repr. 2009. A rudimentary modelling framework in the form of a topological graph is given there, which delineates the macrocosm and spiritual domains with corresponding counterparts in the human brain that enable each human being to reach higher levels of consciousness and spirituality by the meditational practice of Surat Shabdh Yoga. This topological graph is discussed by Srivastav and Narayan in the present volume.

Comments from the Editor

Rev. Prof. Prem Saran Satsangi
Eighth Spiritual Leader of the Radhasoami Faith
Chairman, Advisory Committee on Education, Dayalbagh Educational Institute
(Deemed to be University)
Member Radhasoami Satsang Sabha, Dayalbagh, Agra – 282005, INDIA

Comments on Two Articles Published in *Scientific American*, July 2022

A. Article

Mark Fischetti, **50, 100 and 150 years ago.** *Scientific American*. July 2022. n.p.

Comment:

'Radhasoami (RS) Faith ≥ 200 Years
Headquarters Dayalbagh ≥ 100 Years

Lacto-Vegetarianism/ लैक्टो-शाकाहार practiced by Estimated Stabilized approximately 1 Billion Population, holds the capacity of feeding requisite healthy diet to 11 Billion people on Planet Earth in Solar System based Milkyway Galaxy; and promising Imperishable Existence in Purely Spiritual Domain (Articulative Ra-Dha-Soa-Aah-Mi Faith) via Unity-with-Duality-at-Will: thus Ushering-in the Fifth Industrial Revolution (through Continuing Pursuit of Digitisation/ Discretization); while the World at Large, still talks about Coping-Strategy for Inevitable Merger with the nearest Orbiting Marsian Galaxy {even preparing for dumping International Space Station Debris at Point Nemo for surviving Dissolution & Great Dissolution predicted by the Modern Scientific community; without the Benefit of Escalative/ Escalatory 'Fail-Safe' Options through Ultra-Transcendental Meditational Practices of Radhasoami Faith (inherent in Microcosm of Every Living Human Brain) as duly Identified by Modern Neuro-Science; for transiting to the Purely Spiritual Domain Beyond Nature (Macrocosm); endowed with Infinite Degrees of Freedom in Physical & Astral Domains; Ruled separately, as well as jointly, by *MAYA & KAAL*; in collusion, as it were}'

B. Article

Katie Worth. **Climate Miseducation: How oil and gas representatives manipulate the standards for courses and textbooks, from kindergarten to 12th grade.** *Scientific American*. July 2022. Here 47:

'"Inactivism" doesn't deny human-caused climate change but downplays it, deflects blame for it and seeks to delay action on it'.

Comment:

'While retaining Focus on Cost-Benefit Analysis, it is necessary to emphasize equally, if not more, Renewable energy Sources for Survival of Living Beings {Particularly Human Beings, who are endowed with a Perfect Microcosm in their Brains to Cope with the Impending Ravages of Nature (Macrocosm)}.'

Acknowledgements

The book editors are sincerely grateful to many people in India and in Germany, who have contributed to the preparation of this book, and want to thank at least some of them here individually.

From the Dayalbagh Educational Institute (DEI), Deemed to be University, Agra, India, we thank Prof. Prem Kumar Kalra, Director of DEI, for his support, and our colleagues Profs. K. Soami Daya, Gur Saran, and C. M. Markan, as well as DEI PhD students Dheeraj Sharma, Hardik Chaddah, and Amit Kumar for the selective compilation of a pre-print version of the book in very short time. We also thank Mr. Arsh Dhir from Dayalbagh for his reliable commitment to ensure technical assistance and the smooth communication between the editors based in India and Germany.

At Kiel University in Germany, Mrs. Annette Gottwald-Mueller, Dr. Volkmar Sauerland, Dr. Mourad El Ouali, the PhD students Wiete Keller, Mathias Sowa, and student assistant Fiene Lange helped in all aspects of editorial subtleties and formal refinements for many weeks, and far beyond their regular professional occupations and duties. Their extraordinary flexibility and cooperation were essential to master various crises, great and small: Thanks again to all of you for your unfailing support through some quite turbulent times.

We are especially grateful to Profs. Stefan Schmidt and Harald Walach, editors of *Studies in Neuroscience, Consciousness and Spirituality* at Springer Publishing, for their encouragement and helpful discussions in the earliest stages of this book project. We also thank Venkatesh Sarvasiddhi, Managing Director of Springer Nature India Limited, and Nagarajan Paramasivam, Project Coordinator, for the efficient promotion and technical support of the book during its production. In the final stages of preparing the printing process, the openness and commitment of Christopher T. Coughlin, Senior Editor of Springer Nature, New York, provided invaluable help and encouragement.

Our deepest thanks go to the contributors of this volume for their willingness to transform their oral presentations of the conferences into written form, in most cases adding substantial new material to back up their theses, and widening the scope and depth of their analysis considerably. Their patience and cooperation – sometimes at very short notice – deserve special mention. Seeing now so many cutting-edge analyses from different disciplinary angles and culturally induced conceptualizations gathered in this anthology leaves no doubt that it was worth the effort.

Last but not least, we thank our two academic institutions, the Dayalbagh Educational Institute (DEI) and Kiel University, for generous support of this project through staff and financial resources, an exemplary illustration of institutional cooperation under the Memorandum of Understanding (MoU) entered into by both institutions in 2016.

The prolific research in Consciousness Studies conducted at the Dayalbagh Educational Institute (DEI) (Deemed University) is connected to the foundational work of 2020 Nobel Laureate Sir Roger Penrose, who participated in the conference QANSAS 2013 hosted at the DEI. The institution would like to acknowledge this most productive inspiration by making known to the public the congratulatory felicitations extended to Sir Roger Penrose on the occasion of the 2020 Nobel Prize in Physics in full text in this volume.

Congratulatory Note of Prem Prashant, President DEI, for Sir Roger Penrose, 2020 Nobel Laureate in Physics

DAYALBAGH EDUCATIONAL INSTITUTE
(Deemed to be University)
DAYALBAGH, AGRA 282 005, INDIA

Prem Prashant (*I.A.S. Retd.*), President
(Former Chief Secretary, Government of Haryana)

October 8, 2020

Sir Roger Penrose
Oxford University, UK
(email : penroad@wadh.ox.ac.uk)

Subject : Felicitations on receiving the Nobel Prize in Physics, 2020 – from DEI (Deemed University), Agra, India

Dear Prof. Penrose

All of us at Dayalbagh Educational Institute and the global Dayalbagh community are absolutely thrilled that the Nobel Prize in Physics 2020 recognizes the unique blend of physicist, mathematician, philosopher and best-selling author that you are, and we are sure that your 'strange feeling of elation' mentioned in the interview soon after the announcement by the Royal Swedish Academy has now changed to jubilation. Please accept our congratulations.

We have fond memories of your visit to Dayalbagh Educational Institute (Deemed to be University) in December 2013, when you delivered the Valedictory Address at QANSAS 2013 Winter School and also conducted a Workshop on Orch-OR Theory of Consciousness with an intensive Panel Discussion, which I had the opportunity to Chair and where we had Prof. Stuart Hameroff and Prof. Anirban Bandyopadhyay on the panel as well. We appreciate your kind words after the event that you got a great deal out of the Agra meetings. A memorable picture of your warm handshake with Revered Prof. P.S. Satsangi, Chairman, Advisory Committee on Education, Dayalbagh Educational Institutions is enclosed to relive the golden moment.

We are glad that your work and approach on black holes involving mathematical abstraction dovetails very well with the Dayalbagh Science of Consciousness approach of unlocking the deepest secrets known to mankind through scientific techniques of '*Surat-Shabda-Yoga*' or ultra-transcendental meditation as described in *Discourses on Radhasoami Faith* (by Maharaj Sahab, our third Revered Leader) in early 20th century, in the scientific language of the time, which has been updated with an Appendix employing modern scientific terminology at the turn of the 21st century by Revered Prof. Satsangi. We whole-heartedly endorse your concluding ideas put forth in the seminal work '*Fashion, Faith and Fantasy*' that scientists might be missing out in the pursuit of ultimate truth without realizing that this state of ultimate truth or fantasy may be a mathematical abstraction in dynamic equilibrium.

The global media spotlight is once again focused on you. Though adulation from the worldwide scientific and research community has been, and will always be headed your way, but for reasons of deep understanding of cosmic realities, the communities of *homo spiritualis*, like Dayalbagh, will feel closer to you.

With warm regards and best wishes

(Prem Prashant)

Sir Roger Penrose with Revered Prof. Prem Saran Satsangi, Chairman Advisory Committee on Education (Non-Statutory Body) & the Present Eighth "Acclaimed" Revered Leader of Radhasoami Faith, Dayalbagh Hqs., Agra- 282005, India (Since May 18, 2003).

Tuesday, 16-08-2022, 05:53 AM

Contents

About the Editors

Prem Saran Satsangi

Prem Saran Satsangi is Chairman of the Advisory Committee on Education (ACE), a non-statutory body, serving as a think-tank, for building consensus among various stakeholders, Dayalbagh Educational Institute, Dayalbagh, Agra, India, and since 2003 Acclaimed Eighth Spiritual Leader of the Radhasoami Faith, Dayalbagh, Agra, India.

Revered Prof. Prem Saran Satsangi, the father of systems movement in India, having graduated with a BSc in Electrical Engineering from the Banaras Hindu University (presently Indian Institute of Technology, BHU), pursued an MS from the Department of Electrical Engineering at Michigan State University, USA, and earned his PhD in the field of Socio-economic Systems from the University of Waterloo, Canada. Rev. Prof. Satsangi took voluntary retirement from the Indian Institute of Technology, Delhi, to serve as Honorary Director (CEO), Dayalbagh Educational Institute (Deemed to be University). Prof. Satsangi has generalized the application of physical systems theory to a variety of 'conceptual' socio-economic-environmental systems by extending physical systems theory as a rudimentary modelling framework for complete 'creational' systems, including not only physical but also 'esoteric', mental, and spiritual processes. On May 18, 2003, a gathering in Dayalbagh of about 25,000 representatives of the approximately half a million members (followers) of Radhasoami Faith ('Radhasoami Satsang' Dayalbagh) from all over India and abroad proclaimed Prof. Prem Saran Satsangi as the Eighth spiritual leader of the Radhasoami Satsang Community for his lifetime. Prof. Prem Saran Satsangi, the Acclaimed Eighth Spiritual Leader of the Radhasoami Faith, Dayalbagh, is the Chairman of the Advisory Committee on Education (ACE), Dayalbagh Educational Institute, a Non-statuatory Body serving as a "Think-Tank" for DEI. He is the first awardee of the newly instituted 'Alumni of Eminence Award' (Beyond Distinguished Alumni Award) by BHU-IIT with the motto 'Work is Worship' (voluntarily accepted on an Honorary Basis, as was the case for the earlier Award too).

Anna Margaretha Horatschek
Department of English and American Literatures, Cultures, and Media, Kiel University, Kiel, Germany.

Anna Margaretha Horatschek is a full member of the German Academy of Sciences and Humanities in Hamburg, and from 2016 to 2021 was Vice President of the institution. From 2000 to 2018, she held the chair as professor for English Literature at Kiel University, Germany. She studied English Literature, Philosophy, and German Literature in Germany and the USA. She received her BA from UC Berkeley, USA, and her PhD from Freiburg University, Germany, with a dissertation on the epistemological significance of self-reflexive language experiments in the novels of US author Richard Brautigan. She habilitated with a monograph on the epistemological, political, and ethical implications of identity and alterity constructs (nation, gender, 'race') in novels by Joseph Conrad, E.M. Forster, and D.H. Lawrence in 1995, funded by the German Research Foundation (DFG). In 1998, she spent a year as Visiting Professor at the University of Maryland, USA. She has published widely on knowledge formation, consciousness studies, identity and alterity constructs ('race', class, gender), intermedial representation in English and American Literature, body-politics, and on transcultural poetics and gender issues in Indian Literatures in English. Her books include *Competing Knowledges – Wissen im Widerstreit*, De Gruyter, 2020 (ed.), *Identitäten im Prozess: Region, Nation, Staat, Individuum*, De Gruyter, 2015 (ed. with Anja Pistor-Hatam), and *Navigating Cultural Spaces: Maritime Places*, Rodopi, 2014 (ed. with Yvonne Rosenberg and Daniel Schäbler).

Anand Srivastav
Department of Mathematics, Kiel University, Kiel, Germany.

Anand Srivastav has been a Professor in the Department of Mathematics at Kiel University since 1997. He was born in Dayalbagh, Agra. After primary education in Dayalbagh and school education in Germany, he studied Mathematics and Physics at the University of Münster, Germany, where he received a master's degree in Mathematics as well as in Physics. In 1988 he received the doctoral degree Dr.rer. nat. from the University of Münster with a thesis in Functional Analysis. From 1988 to 1993 he was an Assistant Professor at the Research Institute for Discrete Mathematics, University of Bonn, Germany, and from 1993 to 1994 a Visiting Professor at the University of Minnesota, New York University, and Yale University. In the years 1994–1996 he wrote his Habilitation Thesis in the area of Combinatorial Optimization at the Free University and the Humboldt University of Berlin. Since 1997 he has been a professor and chair for Discrete Optimization at Kiel University. His research interests are not only Combinatorial Optimization, Combinatorial Games, Discrete Harmonic Analysis and Discrepancy Theory, and Randomized and Derandomized Algorithms, but also applications of optimization in Marine and Life Sciences. He was awarded the stipend for habilitation work by the German Research Foundation in 1995, a research stipend by the Japan Society of Science in 1997, the Indo-German Guest professorship in 2013 at IIT Delhi by the Max Planck Society, and the DEI Distinguished Alumni Award of the Dayalbagh Educational Institute (Deemed to be University), Agra, India, in 2019. He has published more than 80 papers in peer-reviewed journals and conference proceedings, and accumulated a third-party funding for his research of approx. 5 Million Euros.

About the Contributors

Reinhard Blutner
Institute for Logic, Language and Computation, University of Amsterdam, Amsterdam, The Netherlands.

Reinhard Blutner completed his PhD in Theoretical High Energy Physics at the Karl Marx University of Leipzig (1975). Furthermore, he has a habilitation thesis in Cognitive Science at the Humboldt University of Berlin (Faculty of Philosophy, 1995). Blutner started his scientific career at the East German Academy of Sciences. Later, he taught linguistics at the Humboldt University of Berlin and artificial intelligence at the University of Amsterdam. In Amsterdam, he also gave courses on the philosophy of mind, intensional logic, and quantum cognition. Further, he gave courses at the University of Krakow, Stanford University, the University of Bloomington, the University of Oslo, and the Universities of Stockholm, Zadar, and Ljubljana. Blutner has written several books and about a hundred publications. His research interests concern the semantics and pragmatics of natural language, bi-directional optimality theory, lexical pragmatics, symbolic-connectionist integration, quantum cognition, and reasoning with uncertainty. Since his retirement in summer 2013, his work has concentrated on quantum cognition, including applications in cognitive musicology.

Andrew Davies
Department of Theology and Religion, Edward Cadbury Centre, University of Birmingham, UK.

Andrew Davies is Professor of Public Religion at the University of Birmingham, where he founded and directs the Edward Cadbury Centre for the Public Understanding of Religion. A biblical scholar by training, his interests are primarily in the impact of sacred texts and religious belief upon policy, politics, culture, and society. He is a specialist in civic engagement and public education, and an experienced academic project manager, who led the AHRC-funded 'Megachurches and Social Engagement in London' project from 2013 to 2016, which investigated the social engagement activities of the UK's largest churches, developed and managed the UK government's pilot project for the professional development of religious leaders from 2019 to 2020, and continues to work with UK government departments to promote the understanding of religion in policy-related contexts. He also directs the University of Birmingham's Master of Public Administration in Faith-Based Leadership and MA in Religion, Politics and Society programmes.

Andrea Diem-Lane
Department of Philosophy, Mt. San Antonio College, Walnut, CA, USA.

Andrea Diem-Lane is Professor of Philosophy and currently the Chair of the Department at Mt. San Antonio College, CA, USA. She received her PhD and MA from UCSB and her BA in Psychology from UCSD. Dr. Diem-Lane is the author of over 10 books, including *The Cerebral Mirage*, *How to Study the Sacred*, *The Jain Path*, *The Gnostic Mystery*, *Spooky Physics*, etc. Some of this material was presented at international conferences held in Agra, India, where she was a plenary speaker. Her latest publication is an annotated bibliography for Oxford University Press. She is currently working on a book for Cambridge University titled *Memetic Theory and New Religions*. On a more personal note, she is married to Professor David Lane with whom she has two boys, Shaun and Kelly. Besides being a strict vegetarian for over 40 years, Dr. Diem-Lane has been an avid surfer for 35 years.

Wolfgang J. Duschl
Institute for Theoretical Physics and Astrophysics, Kiel University, Kiel, Germany.

Wolfgang J. Duschl, born in 1958 in Munich, Germany, is Professor of Astrophysics at Christian Albrechts University at Kiel; director of the Institute of Theoretical Physics and Astrophysics; an affiliate member of the research faculty of Steward Observatory, The University of Arizona, Tucson, USA; and Prof. h.c. of Irkutsk State University, Russia. His main research interests include evolution of supermassive black holes in galactic centres, and atmospheres of exoplanets.

Rocco J. Gennaro
Political Science, Public Administration and Philosophy, University of Southern Indiana, Evansville, IN, USA.

Rocco J. Gennaro is Professor of Philosophy at the University of Southern Indiana. He received his PhD in philosophy from Syracuse University in 1991. Dr. Gennaro's primary research and teaching interests are in philosophy of mind/cognitive science (especially consciousness), metaphysics, early modern history of philosophy, and neuro-ethics. He has published 12 books (as either sole author or editor) and over 50 articles and book chapters in these areas. For example, he published the book *The Consciousness Paradox: Consciousness, Concepts, and Higher-Order Thoughts* (MIT Press, 2012) and edited an anthology entitled *Disturbed Consciousness: New Essays on Psychopathologies and Theories of Consciousness* (MIT Press, 2015).

More recently, he has edited *The Routledge Handbook of Consciousness* (Routledge, 2018), and published *Consciousness* (Routledge, 2017) as well as the second edition of his introductory dialogue *Mind and Brain: A Dialogue on the Mind-Body Problem* (Hackett Publishing Company, 2020), which is substantially revised and updated.

Philip Goff
Department of Philosophy, Durham University, Durham, UK.

Philip Goff is Associate Professor of Philosophy at Durham University, UK. His research focuses on how to integrate consciousness into our scientific worldview. He has authored an academic book with Oxford University Press – *Consciousness and Fundamental Reality* – and a book aimed at a general audience – *Galileo's Error: Foundations for a New Science of Consciousness*. Dr. Goff has published over 40 academic articles and written extensively for newspapers and magazines, including *Scientific American*, *The Guardian*, and the *Times Literary Supplement*. An interview with Dr. Goff by Pulitzer Prize-winning journalist Gareth Cook was one of the most viewed articles in *Scientific American* of 2020. Dr. Goff has been interviewed on over 50 podcasts and radio shows, including BBC radio's flagship news show the *Today* programme, and podcasts by Russell Brand and Sean Carroll.

Mark Juergensmeyer
Department of Global Studies, University of California, Santa Barbara, CA, USA.

Mark Juergensmeyer is Distinguished Professor of Sociology and Global Studies at the University of California, Santa Barbara, where he was Founding Director of the Orfalea Center for Global and International Studies. He has taught at the Berkeley, Santa Cruz, and Santa Barbara campuses of the University of California and served as Founding Dean of Asian and Pacific Studies at the University of Hawaii. He is the recipient of Guggenheim, American Council of Learned Societies, and other fellowships, and has two honorary doctorates. He has served as President of the American Academy of Religion, and is the author or editor of 30 books, including *The Oxford Handbook of Global Religion*, *The Encyclopedia of Global Religion*, *God in the Tumult of the Global Square*, *Radhasoami Reality*, *Gandhi's Way*, and the award-winning *Terror in the Mind of God*. He received his PhD in political science from the University of California at Berkeley and a Master of Divinity from Union Theological Seminary, New York.

Subhash Kak
School of Electrical and Computer Engineering, Chapman University and Oklahoma State University, Stillwater, OK, USA.

Subhash Kak born in Srinagar, Kashmir, was educated at various institutions in Jammu and Kashmir. He received his PhD in Electrical Engineering from Indian Institute of Technology, Delhi. Between 1975 and 1976, he was a visiting faculty at

Imperial College, London, and a guest researcher at Bell Laboratories, Murray Hill. In 1977, he was a visiting researcher at Tata Institute of Fundamental Research, Bombay. Between 1979 and 2007, he was with Louisiana State University, Baton Rouge, where he served as Donald C. and Elaine T. Delaune Distinguished Professor of Electrical and Computer Engineering. Kak's research has spanned the fields of information theory, cryptography, neural networks, and quantum information. He is also an achaeoastronomer and Vedic scholar. In 2008–2009, he was appointed one of the eight principal editors for the ICOMOS project of UNESCO for identification of world heritage sites. He is the author of 20 books which include *The Nature of Physical Reality, The Architecture of Knowledge, and Mind and Self.* He is also the author of six books of verse.

Martin Krieger
History Department, Kiel University, Kiel, Germany.

Martin Krieger is Professor of Northern European History at Kiel University (Germany). His major focus of research is the cultural history of the Baltic Sea area and knowledge-exchange between Northern Europe and India. He is former Vice Dean of the Humanities Faculty and was Speaker of the Center for Asian and African Studies (ZAAS) at Kiel University. He has lived in India for some time, has conducted research on the former Danish trading-settlement Tranquebar in today's Tamil Nadu, and extensively published on Indo-European relations. Books published in India comprise *Water and State in Europe and Asia* (with Peter Borschberg, New Delhi: Manohar 2008), *European Cemeteries in South India* (New Delhi: Manohar 2013), and *Nathaniel Wallich. Botanist and Founder of the Indian Museum* (New Delhi, 2022).

Ami Kumar
Irving Medical Center Department of Neurology, Columbia University, New York, NY, USA.

Ami Kumar completed her MSc from Dayalbagh Educational Institute (Deemed University), and her PhD in Neuroscience from Kiel University, Germany. During her PhD, she worked on a multidisciplinary project, associated with the analysis of neuronal networks underlying absence epilepsy seizures, linked with the presence and absence of consciousness in children. Currently she is a postdoctoral research fellow at the Department of Neurology at Columbia University, New York. Her current research involves developing new therapeutic pipelines for neuromodulation in movement disorders such as ataxia and essential tremor. She is using the applications of electroencephalography and non-invasive brain stimulation techniques such as transcranial alternating current stimulation and transcranial direct current stimulation to assess the underlying pathophysiology of tremor in patients.

David Christopher Lane
Department of Philosophy, Mt. San Antonio College, Walnut, CA, USA.

David Christopher Lane is Professor of Philosophy and Sociology at Mt. San Antonio College. Dr. Lane is the author of many books, including *The Making of a Spiritual Movement*; *The Radhasoami Tradition*; *Exposing Cults*; *The Unknowing Sage*; and *The Virtual Reality of Consciousness*, among other titles. He is the founder of the MSAC Philosophy Group, which is responsible for publishing 530 books, 260 audiobooks, and over 150 mini films. Professor Lane has a PhD and an MA in the Sociology of Knowledge from the University of California at San Diego, where he was also a recipient of a Regents Fellowship. Additionally, he has an MA in the History and Phenomenology of Religion from the Graduate Theological Union in Berkeley, California. He has previously taught at the University of California, San Diego; California State University, Long Beach; and the University of London in England. Dr. Lane's most recent book, *The Sound Current Tradition: A Historical Overview*, was published by Cambridge University Press. On a more personal note, Dr. Lane is an avid surfer and bodysurfer, having won the World Bodysurfing Contest in 1999 and the International Bodysurfing Contest eight times (1997–2016). Professor Lane has pioneered virtual reality technology in education, establishing the very first virtual reality course in philosophy at the undergraduate level. He has created two textbooks for the field: *Digital Teleportation: A Philosophic Journey in Virtuality*; and *Virtual Philosophy: Adventures in Digital Reality.*

Christine Mann
Author, Theologue, Psychologist, Cologne, Germany.

Christine Mann, née Heisenberg, is the sixth child of 1932 Nobel Laureate Werner Heisenberg and his wife Elisabeth Heisenberg. She studied Theology and Pedagogics in Munich (1964–1968), Germany, and Psychology at the University of Münster (1974–1978), Germany, and in 1986 obtained her PhD in the field of Educational Science at the University of Göttingen. She has worked at various primary schools in Bavaria and North Rhine-Westphalia, and as head of a school psychological counselling centre in Worms. In 1966, Dr. Mann married Prof. Frido Mann, grandson of 1929 Nobel Laureate Thomas Mann. Since 2001, Dr. Mann has turned to quantum physics. In 2011 Frido and Christine Mann initiated an interdisciplinary discussion group, in which they met regularly with Prof. Dr. Thomas Görnitz and others, to think about questions of the relationship between mind and matter in quantum physics. After numerous publications in the pedagogical-didactic field, in 2017 Frido and Christine Mann jointly published the monograph *Es werde Licht. Die Einheit von Geist und Materie in der Quantenphysik* (*Let there be Light. The Unity of Spirit and Matter in Quantum Physics*), with Fischer-Verlag. The anthology *Im Lichte der Quanten. Konsequenzen eines neuen Weltbildes* (*In the Light of Quanta. Implications of a New World View*), edited by Frido and Christine Mann, has been published by the Wissenschaftliche Buchgesellschaft (WBG) Thesis in 2021.

Andreas Müller
Faculty of Theology, Professor for the History of Church and Religion of the First Millenium, Kiel University, Germany.

Andreas Müller was born in 1966 in Bochum. He studied Protestant Theology in Bethel/Bielefeld, Bern (CH), Heidelberg, and Thessaloniki (GR), obtained his Doctorate in 1998 in Heidelberg, and habilitated 2003 in Munich with a thesis on spiritual obedience in Johannes Sinaites, called John of the Ladder. From 2003–2009 he worked as pastor of the Protestant Church of Westphalia with a church-historical research project and parallel professorships in Jena, Kiel, and Berlin. Since 2009 he has been professor for church and religious history of the first millennium at Kiel University, since 2011 he is Vice-Chairman and Chairman of the Church History Section of the Scientific Society for Theology (Wissenschaftliche Gesellschaft für Theologie), and since 2012 Chairman of the Society of Friends of Christian Mysticism.

Apurva Narayan
Department of Computer Science and Department of Electrical and Computer Engineering, Western University, London, ON, Canada; Systems Design Engineering Department, University of Waterloo, Waterloo, ON, Canada; Department of Computer Science, the University of British Columbia, Vancouver, BC, Canada.

Apurva Narayan is an Assistant Professor at Western University, Department of Computer Science and Department of Electrical and Computer Engineering, Canada. He also holds an appointment as Affiliate Assistant Professor in the Department of Computer Science at the University of British Columbia, Canada, and as Adjunct Assistant Professor in Systems Design Engineering at the University of Waterloo, Canada. He obtained his bachelor's degree in Electrical Engineering from Dayalbagh Educational Institute in 2008 and PhD from the Department of Systems Design Engineering, University of Waterloo, in 2015. Dr. Narayan's research interests lie at the interface of data science, safety-critical systems, systems theory, machine learning, artificial intelligence, and quantum computing. He has authored more than 40 peer-reviewed articles in ACM and IEEE conferences and journals. He has been invited to deliver guest lectures at premier universities around the world. He is currently the director of the Intelligent Data Science Lab jointly at UWO, UW, and UBC. He was the lead organizer of the Landmark First International Conference on Dayalbagh Science of Consciousness, DSC 2019, jointly organized by DEI and University of Waterloo under their MoU for joint research in September 2019. He was conferred the Young Systems Scientists Award by Systems Society of India at QANSAS 2017 and Varshney Award at QANSAS 2019.

Ulrich Stephani
University Clinic of Kiel, Neuropediatrics; former Dean of the Medical Faculty, Kiel University, Kiel, Germany.

Ulrich Stephani, born in 1954, completed his study of medicine at the Universities of Hannover, Würzburg, and Berlin. He received his licence to practise medicine in 1977, and was certified as a pediatrician in 1989, followed by 'Habilitation' and Venia Legendi in 1990. From 1992 to 2018 he was Director of the Dept. of Neuropediatrics at Kiel University, as well as Director of the Northern German Epilepsy Centre for children and adolescents. Between 1998 and 2007 Stephani was the first secretary of the German chapter of the ILAE, and in 1999 and 2000 was President of the German-speaking Society for Neuropediatrics. In 2013 he was appointed Dean of the Medical Faculty at Kiel University until 2020, and from 2018 he was board member of the University Clinics of Schleswig Holstein (UKSH). Retired since 2020, he is active in several functions at Kiel University. He has completed several third-party-funded research projects and published more than 250 articles on neuropediatrics and pediatric epileptology.

Annette Wilke
Institute for the Study of Religion, Westfälische Wilhelms-Universität Münster, Münster, Germany.

Annette Wilke is Professor of the Cultural Study of Religion (emeritus) at the Westfalian Wilhelms University of Muenster, Faculty of Catholic Theology, Germany (1998–2019). She has been head of the Institute for the Study of Religion and founding member of the Centre for Religious Studies (CRS) at the University of Muenster. Until her retirement she was also part of the Cluster of Excellence 'Religion and Politics' with a project on global Hinduism. Academic training in the History of Religions, Philosophy, and Theology at the University of Fribourg (CH), and in Indology in the USA, Zurich, and Varanasi. She holds a PhD in the History of Religions with a dissertation on comparative mysticism (Berne 1994). Wilke combines in her work textual studies and field work and has widely published on her research areas: aesthetics of religion, mysticism, ritual studies, goddess worship, Advaita-Vedānta, Tantra, temple Hinduism in the diaspora, sensory awareness, and the role of sound in Hindu traditions past and present. Her publications include (with Oliver Moebus) *Sound and Communication. An Aesthetic Cultural History of Sanskrit Hinduism* (2011).

Ralph Yarrow
School of Literature, Drama and Creative Writing, University of East Anglia, Norwich, UK.

Ralph Yarrow is Emeritus Professor of Drama and Comparative Literature, University of East Anglia (UEA), School of Literature, Drama, and Creative Writing, Norwich, UK. He is a teacher, theatre director, performer, actor trainer, writer, editor, translator, and project leader. He made a founding international contribution to research into consciousness and the arts, and developed further specializations in improvisation, Indian and Asian theatre, and Applied Theatre, centred on close co-operation with India's leading Forum Theatre/Theatre of the Oppressed organization, Jana Sanskriti. His books include *Improvisation in Drama, Theatre and Performance* (with Anthony Frost, 3rd edition, 2015); *Indian Theatre: Theatre of Origin, Theatre of Freedom*; and the co-written *Sacred Theatre* (with Carl Lavery et al.); translations (from German) of Birgit Fritz's *In Exact Art: The Autopoietic Theatre of Augusto Boal*, and her *The Courage to Be: Augusto Boal's Revolutionary Politics of the Body* (2016). He has directed in the UK, Germany, India, and South Africa. His current research focuses on Theatre of the Oppressed/Applied Theatre practice; theatre and embodiment; Indian and South African practice and contemporary theatre; and theatre and ecology.

Part I
Introduction

Consciousness Studies: Which Consciousness? What Studies?

Anna Margaretha Horatschek

Abstract The essay supplies a – necessarily selective – overview of the state of Consciousness Studies. It outlines various concepts of 'consciousness' in East – mainly India – and West, and in different academic disciplines, and focuses on methodological problems concerning key-terms like experience, experiment, first-person reports, representation, and interpretation. Addressing the problems ensuing from the fact that Western sciences presuppose a hierarchy between various knowledge formations depending on the conditions of their production, Karen Barad's model of 'diffractive' knowledge production might open up a possibility to acknowledge seemingly incompatible knowledge traditions for the benefit of comprehensive Consciousness Studies.

Keywords Consciousness studies · Knowledge production · First-person report · Karen Barad · Experience · Experiment

This book is aimed at specialists and students in the field of Consciousness Studies (CS), but several essays will also be of great interest to the general public. Therefore, this introduction will give a – necessarily selective – overview of the field by focusing on central questions and issues of CS.

The essay will show that the specific methodologies of research in CS and the conceptualisation of the research subject 'consciousness' cannot be separated as neatly as the title of the introduction suggests. For heuristic reasons, however, I will stick to the differentiation between the object of investigation and issues of research methodologies, and structure the text accordingly. Thus, the first part "Which Consciousness?" will focus on the host of competing and often incompatible

---

A. M. Horatschek (✉)
Department of English and American Literatures, Cultures, and Media, Kiel University, Kiel, Germany
e-mail: horatschek@anglistik.uni-kiel.de

P. S. Satsangi et al. (eds.), *Consciousness Studies in Sciences and Humanities: Eastern and Western Perspectives*, Studies in Neuroscience, Consciousness and Spirituality 8, https://doi.org/10.1007/978-3-031-13920-8_1

concepts of consciousness in various disciplines as well as in East and West, and highlight the intricate entanglement of methodology and concepts of consciousness; the second part "What Methods?" will address issues that lie at the heart of CS, and that are approached with widely divergent disciplinary strategies; the third part "What to Do?" opens up vistas of how to answer the salient problems of conceptualisation and methodology in CS.

The titles of all three parts end with a question mark as a typographical marker for the fact that "our empirical knowledge about consciousness is so fragmentary and incomplete, at this early stage of scientific inquiry" (Kriegel 2020: 1) – with the *Cambridge Handbook of Consciousness*, the first of its kind, appearing only in 2007 – that uncertainties and problems outrun by far any plausible answers to "the *How?, Where?,* and *What?* questions" (Papineau 2020: 14) concerning consciousness.

1 Which Consciousness?

Writing about consciousness is a risky undertaking if you are not stationed in the Neurology, Quantum Physics, or Neurobiology Department, or a Philosophy Department specialised in Quantum Physics and IT. Although evidence based scientific research of the natural sciences in matters of consciousness started only in the 1990s, it has dominated the field since then, notwithstanding that philosophers and *rishis* have addressed questions about consciousness for thousands of years in East and West. Along with this discursive dominance, a corresponding model of what consciousness is has gained prevalence in the public understanding, claiming that – despite a host of incompatible views on details – consciousness is the result of physical conditions, more specifically: of interrelated brain activities.

This book takes up the latest developments in empirical research of Neurology, Quantum Physics, Computer Science, and Mathematics in CS, but it does not succumb to the exclusivist and universalising validity claims of the natural sciences that are voiced – sometimes with an astounding absence of methodological self-reflection – by some acknowledged experts of the scientific field. Instead, the respective findings are presented alongside investigations on consciousness in the Humanities, the Social Sciences, and (comparative) Religious Studies, as well as spiritual models mostly from the Indian subcontinent.

This form of presentation highlights the central problem of talking about consciousness, namely the varying and often incompatible meanings ascribed to the term, with the homophony of the name covering up the fundamental differences of the semantics. These depend on who is speaking, be it a philosopher, a sociologist, or a neurologist, with their disciplinary focus on issues of consciousness ranging from the social function of specific concepts of consciousness to physiological correlatives considered as indispensible for the functioning or even the emergence of consciousness. Thus 'consciousness' is conceptualised for example as a discursive construct, as the epiphenomenon of highly complex interrelations of material givens, cultural input, and individual predispositions, or as the phenomenal manifestation and format of neural processes in the brain. In Western cultures, consciousness is mostly understood as an intramundane – defining – feature of the individual human subject.[1] In various Indian philosophies, Consciousness – deliberately marked by a capital 'C'- denotes a transindividual, metaphysical ultimate reality, which defies any semiotic representation (cf. K.R. Rao 2005; also Kak in this volume). Accordingly, methods of studying consciousness vary profoundly.

Just as legitimately – and more in tune with science studies – one might formulate the same problem the other way round by saying: It is the regimes of specific disciplinary methodologies that bring to light radically different results, suggesting equally different concepts of the object of investigation – namely consciousness. More so, the specific methodologies – for example, of the natural sciences or the humanities – often preclude the acknowledgement of concepts that elude the normative rules of their respective strategies of investigation.[2] Karen Barad, eminent philosopher and physicist specialised in Theoretical Particle Physics and Quantum Field Theory – taking her start from the concept of complementarity in the 'physics-philosophy' of Niels Bohr – expounds: "To write matter and meaning into separate categories, to analyse them relative to separate disciplinary technologies, and to divide complex phenomena into one balkanized enclave or the other is to elide certain crucial aspects by design" (Barad 2007: 25). In this volume, Reinhard Blutner in his contribution tries to bridge such balkanised enclaves by spanning complementarity, quantum cognition, operational interpretation, and consciousness, and Mark Juergensmeyer in his essay points out instructive analogies between the function of knowledge production in religion and sociology.

Apart from disciplinary varieties of scientific and academic research methodology, there is a divide in the use of and the approach to the term 'consciousness' in East and West, and in the East there are significant variations between different religions and worldviews again – as the essays by Andrea Diem and Annette Wilke illustrate. Without wanting to cement monolithic cultural stereotypes and well aware of the impossibility of talking about 'the West' and 'the East' in any adequate way in just a few lines, I use these terms as marking points for some basic differences in culturally dominant semantics for the term 'consciousness'. To supply a conceptual frame for this juxtaposition, I will give the – in its pointedness certainly overstated – summary of foundational differences that K. Ramakrishna Rao outlines in historical and systematic detail in his book *Consciousness Studies. Cross-Cultural Perspectives* (2005). In general,

> the dominant trends that characterize Eastern and Western traditions [lie in] (a) the emphasis they place on the subjective and objective, i.e., the view from within and the view from outside, (b) on the presumed relations between the two, and (c) on the appropriateness of

---

[1] For a historical overview of various concepts of consciousness from the seventeenth to the twentieth century as conditioned by cultural context see Horatschek 2016.

[2] Methodologically highly self-reflexive essays concerning knowledge production in a range of academic disciplines are compiled in Horatschek 2020c.

> the methods to transform or translate subjective items of experience and knowledge into objective phenomena and vice versa. (Rao 2005: 5)

Concerning the methodology of investigation, Rao holds:

> In the East, the focus is on the person having the experience [...], the emphasis is on the psychological rather than on the physical, and the preferred method is first-person based introspection. [...] In the West [, t]he focus is on the object of experience, the emphasis is on the physical, and the preferred method is third-person based observation and measurement. (Rao 2005: 8)

Consequently, in the West

> [t]he approach is rational and intellectual [...] the preferred method is scientific and outward focused [with] outstanding achievements in understanding the physical reality. [...] the central focus in the Eastern tradition is on the inward. While reason is often used to explicate the inwardness, it is seen as secondary to the inner experience, which has its own content different from the outward material content of reason. (Rao 2005: 5)

And finally, the concomitant – or presumed – models of consciousness are delineated in the following way:

> In the Western scholarly tradition, we find (a) that consciousness is generally equated with the mind. If a distinction is made between them, consciousness is conceived as an aspect of the mind; (b) intentionality is regarded as its defining characteristic; (c) the goal is one of seeking a rational understanding of what consciousness/mind is. [...] In Eastern traditions, especially the Indian, we find [...] a different approach and emphasis. First, a basic distinction is made between consciousness and mind. Second, the existence of pure consciousness, which is believed to be nonintentional, is postulated whereas the mental phenomena are regarded as essentially intentional. Third, methods of accessing pure consciousness, as well as special disciplines to deal with higher states of the mind with tangible benefits, are suggested and developed. Whereas the Western perspective is limited to the phenomenal manifestations of consciousness, the Eastern tradition pays special attention to the transcendental aspects of consciousness through its concern with pure consciousness. (Rao 2005: 8)

While the still widely prevalent stereotype of the spiritual East and the secular West is continuously eroded by the spread of Western standards of scientificity in the East, the 'spiritualisation' of the West with yoga and meditation, and the scientification of religion in East and West, as Annette Wilke points out in her contribution to the present volume, there still remain the fundamental differences of consciousness models, and the concomitant strategies of knowledge production in the frame of the natural sciences, the humanities, and spiritual endeavours. And it is also a fact that

> the most salient features of the international knowledge system is its peculiar division of labor, in which key intellectual tasks, such as setting theoretical agendas and methodological standards, are the prerogative of [...] societies and institutions which are, almost without exception, located in the economically privileged regions of the world. (Weiler 2017: 3)

This division of intellectual labour culminates in an explicit hierarchy of validity ascriptions, namely that "higher status is conferred upon work that conforms to the evidentiary and analytical standards of Western [...] sciences [...]" (Weiler 2017: 7).[3]

[3]Amitav Ghosh in his novel *The Hungry Tide* dramatises the consequences of such a neglect of local knowledge traditions in favour of Western academic knowledge for environmental politics in India. See Horatschek 2017.

This state of affairs is the more surprising if one considers that the pronounced secularism ascribed to the West and its scientific methods of knowledge production is the offspring of "a globalised *elite* culture, [...] an international subculture composed of people with western-type higher education, which is the principal carrier of progressive enlightened beliefs and values but whose members are relatively thin on the ground" (Berger 1999: 10, quoted by Kirchhofer 2009:4). In a similar vein, the self-description of the West as secular has been analysed in the frame of Postcolonial Studies as a self-delusional identity marker, motivated by the desire to proclaim intellectual, mental, and emotional independence from a non-enlightened past steeped in 'self-imposed immaturity' (*selbstverschuldete Unmündigkeit*) in Kant's famous terms, and thus securing – or constructing – a positively connoted historical modernity compared with non-European cultures, in order to legitimate the ruthless exploitation under colonial rule by cultural superiority. However, as various researchers emphasise, "the secular vision is immersed in religious metaphors, diction, and perhaps even experiences" (Brown 2005: 748, quoted by Kirchhofer 2009: 4), and "the natural sciences adopted religious and metaphysical claims and integrated them in their framework of meaning, resulting in a special form of scientific religiosity that has gained much influence in the twentieth century" (von Stuckrad 2014: cover text),[4] with the result that "science is currently the predominating paradigm for creating meaning in our modern societies" (Schmidt and Walach 2014a, b: 4). In contrast, "the intellectual traditions once unbroken and alive in Sanskrit or Persian or Arabic are now only matters of historical research for most – perhaps all – modern social scientists in the region [that is Asia]" (Chakrabarty 2008: 6).

The persistent preferment of Western knowledge standards to Eastern knowledges in the face of these debunking diagnoses finds its explanation in the paramount role that political, economic, and social power structures and their interests play in the acknowledgement of what counts as legitimate knowledge in a specific time and place (Horatschek 2020b: 1–34). Especially Postcolonial Studies have exposed the political strategies to implement and affirm the Western "claim to superior, objective, and universal knowledge" (Nanda 2003, 153; see also Devy et al. 2014, Chakrabarty 2008). Concerning the spread of secular discourses in India's (intellectual) *elites*, Dipesh Chakrabarty holds that the colonial 'import' of Western concepts

> such as citizenship, the state, civil society, public sphere, human rights, equality before the law, the individual, distinctions between public and private, the idea of the subject, democ-

[4] See also Nate 2009, who shows in detail how the rhetorics endorsing the empirical methodology as a privileged way of knowledge production in the seventeenth century were steeped in religious metaphors and values. According to Latour (1993), the sixteenth-century Reformation by banishing God from the public spheres and privatising it in individualistic spirituality, and the seventeenth-century's "conjoined invention of scientific facts and citizens (Eisenstein, 1979)", succeeded to establish "an infinitely remote God who is simultaneously totally impotent and the sovereign judge" (Latour 1991: 33, 34).

racy, popular sovereignty, social justice, scientific rationality […] entail an unavoidable – and in a sense indispensable – universal and secular vision of the human (Chakrabarty 2008: 4).[5]

However, apart from these historical and political 'knowledge wars' (Weiler 2017) and the extensive research demonstrating the untenability of identity and alterity constructs between East and West along the lines of spirituality/religion and secularism/materialism, there remain systematic problems concerning the relation of religious/spiritual and scientific/academic knowledge production: While the former is bound to first-person perspective, individuality, and sensual/emotional experience, and comprises the possibility of trans-intellectual knowledge acquirement and production, metaphysical phenomena, and the freedom to communicate via media ranging from intellectual treatises through ritualised body regimes to poetry, music, and dance, the latter requires a third-person or bird's-eye perspective, replicability and generalisability, rational reasoning, a secular and material realm of applicability, and the rigid adherence to disciplinary rules of discourse, concepts, and strategies of communication. Given these normative conditions to produce scientifically valid knowledge, all sciences working on consciousness are confronted with the gap between consciousness as ephemeral and immaterial individual experience, and its conceptualisation as the emanation of material conditions like bodily senses and the brain. However, as Tim Parks observes, though "[…] neuroscience […] is directed at bridging this […] gap between brain cells and conscious experience, […] no one seems particularly worried that this isn't going to happen anytime soon, maybe because we're not talking about curing a killer disease, just understanding who we are" (Parks 2018: 132).

2 What Studies?

No matter what the underlying concept of consciousness is, all researchers with their various disciplinary methodologies, in East and West, are confronted with the same central questions theoretically and – in experimental research – also practically. In the following, I shall outline these methodological key questions in the humanities and in the natural sciences.

To make this wide field of interdisciplinary and intercultural approaches comprehensible, I shall concentrate on the following terms, which are vital for each exploration of consciousness today.

These are:

- experience
- experiment

[5] For a comprehensive stock-taking of the state of education at Indian schools and universities with these premises in mind cf. Devy 2017. About Indian Literature in English that critiques the European concept of 'history' as colonial legacy see Horatschek 2019.

- first-person reports
- representation
- interpretation

Experience and Experiment

Experience and experiment in English are closely related and were used interchangeably until the eighteenth century, as both stem from the same Latin root, namely "*experientia*, from *experiri* 'to travel, to try'" (OED 1998: 647) or 'to put to the test'. In Europe, inner experience became a topic from the eleventh century onward, when knowledge gained by experience was contrasted with knowledge gained by empirical experiments in astronomy and physics, especially in optics (cf. Walach 2014). The word *experimentum* "was used for the first time in the 13th century in connection with replicable scientific experience (McEvoy 1982; Crombie 1953) […] using defined methods of observations that can be repeated at will [and] looking at nature from outside, as an object, i.e. 'objectively'" (Walach 2014: 8). It was Roger Bacon (1220–1292), a Franciscan monk and philosopher in Oxford, England, who developed and established the empirical methodology in the thirteenth century. But he was also aware of the limits of gaining knowledge by this new methodology:

> Experience comes in two forms: One is through our outer senses, and this is how we experience what is in the heavens and below. This is human, scientific experience … but this experience does not suffice, because it does not give full evidence about material things on the basis of its difficulty, and about spiritual things it attains nothing. (Walach 2014: 10, transl. by Walach from Bacon's *Opus Majus*)

According to Uriah Kriegel, "[…] consciousness science differs markedly from other scientific disciplines [in that it deals with] phenomena [like emotions] with which we have pre-scientific acquaintance and for which we would like to provide scientific explanation" (Kriegel 2020: 3). This situation accounts for the fact that consciousness can be studied "from two different vantage points: from the 'inside' perspective of a conscious being, and from the 'outside' vantage point of any of the academic fields that study the mind" (Schneider and Velmans 2017: xix), namely through experience and through experiment. However,

> [w]hen we speak of scientific research into consciousness, we speak in truth of research which targets in the first instance salient *correlates* of consciousness. […] It is possible, of course, to hold that consciousness is in fact *nothing over and above* its neural correlates [, yet] nothing-over-and-above claims are paradigmatically philosophical claims and the considerations typically brought in their support tend to be adduced rather from the armchair. (Kriegel 2020: 5)

The foundational condition and simultaneously the biggest problem of CS that cannot be solved with armchair philosophy is the exclusive and indispensible manifestation of and access to consciousness, namely the individual experience.

Experience in Consciousness Studies

The only access to consciousness is through individual experience. This situation accounts for the central problem of CS, formulated by the Australian philosopher and cognitive scientist David Chalmers as 'the hard problem'. In his 1995 essay "Facing up to the problem of consciousness" he asks:

> Why is it that when our cognitive systems engage in visual and auditory information-processing, we have visual or auditory experience: the quality of deep blue, the sensation of middle C? How can we explain why there is something it is like to entertain a mental image, or to experience an emotion? It is widely agreed that experience arises from a physical basis, but we have no good explanation of why and how it so arises. [...] The really hard problem of consciousness is the problem of experience. When we think and perceive there is a whir of information processing, but there is also a subjective aspect. (Chalmers 2007: 226)

In other words: What we experience when seeing a colour is not the information processing going on in the brain, the firing of neurological synapses, the light waves meeting the retina of our eyes etc., but we know *what it is like* to feel the specific quality of blue, or of a sound, or of a smell, or of a touch. The term *qualia* describes this 'whatness' of the experience, and the status of *qualia* is a central topic in philosophical debates about consciousness, for example: Is this quality of experience we perceive when seeing something blue part of ourselves, which we project and thus add to the world out there, or is it part of the object we see? Ultimately the problem concerns the question about the transformation of a material world (perceived as) out there into the subjectively experienced interiority of thoughts, emotions, and perceptions, restricted to one human being: It thus concerns the relation between body and mind.

Another central question is: Who or what is doing the perceiving, and where and how does the change of lightwaves into a colour with the accompanying qualia like emotions, physical sensations etc. happen? Some philosophers like Rocco Gennaro in this volume espouse the Higher Order Theories or Thoughts (HOT) of consciousness, which assume that a hierarchically superior observing instance like 'a suitable higher-order thought is directed at that mental state' and is responsible for this awareness. But this seems to transfer the problem only onto a different level, because Gennaro has to posit that this instance simply is conscious – without having explained how this comes about.

The major problem of dealing with the individual experience of consciousness is that we have no access to any consciousness except our own, and no extensive description will help to know exactly the experience of another person. A keytext dealing with this problem is the 1974 paper "What is it like to be a bat?" by the philosopher Thomas Nagel. He asserts that "an organism has conscious mental states if and only if there is something that it is like to *be* that organism – something it is like *for* the organism" (Nagel 1974: 436). This assertion has achieved special status in CS as the standard 'what it's like' locution. Nagel comes to the conclusion that we cannot know what it is like to be a bat.

Experiment in Consciousness Studies

The experiment is the central methodology of the natural sciences to produce their knowledge, characterised by Karl Popper as "precise, unambiguous predictions about quantifiable, measurable phenomena" (Law 2007: 187). This definition includes the limits of this methodology, namely: The natural sciences can deal only with "quantifiable, measurable phenomena". Is consciousness such a phenomenon? In our experience, it certainly is not, and thus in CS, natural scientists like neurologists, biologists, and physicists have to adjust experience to the conditions that define a scientific fact. I shall come to this when I speak about first-person reports.

As it is, natural scientists like neurologists, biologists, physicists, or anaesthesiologists in CS do not address the 'hard problem' of how consciousness – understood as awareness of qualia of an experience – can come from matter. Since the 1990s, they have looked for the neural correlates of consciousness (NCC) by measuring brain activities to find out which neurobiological events occur simultaneously with specific experiences of subjective consciousness, like for example pain. In this way, they address the question of *which* neurobiological mechanisms are linked to consciousness, but the question of *why* these mechanisms should give rise to consciousness does not concern them. Metaphors like 'computing', 'emergence', or 'information' in many empirical theories of consciousness have to cover the ground between the physical brain and individual experience that experiments up to now have not been able to bridge in any convincing way.

Some scientists mistake the correlations they find for causation, claiming that the observed neural activity itself causes the experience of consciousness, without however being able to account for the transformation from the factual measurement of 'information processing' to subjectively experienced qualia of the phenomenon. Another rather widespread strategy to address the 'hard problem' formulated by Chalmers is to explain it away by declaring that the individual experience that we are conscious beings is only "folk theory" (Dennett 2017: 222) and "a user-illusion brilliantly designed by evolution to fit the needs of its users" (Dennett 2017: 222).[6] Dennett grounds this strong thesis on the claim that nowadays everybody accepts that goldness and silveriness are just differences in atoms, instead of any quality worthy of emotional reactions; similarly, Rovelli attributes our awe at a glorious sunset to "our bleary, customary eyesight" (Rovelli 2018: 13), which forgets that in reality the earth is turning. Both are wrong: We know about 'a difference in atoms' and the turning of the earth, but our experience of joy, admiration, and awe implies an entirely different subject–object relation and has a phenomenologically different quality from the knowledge of the natural sciences. "When you see the rich hues of a sunset or smell the aroma of your morning coffee, you are having conscious experiences. And it is this that makes it wonderful to be alive" (Schneider and Velmans 2017: xix).

Recent findings in the neurosciences support this differentiation between knowledge and experience: Hakwan Lau, a neuroscientist at the University of California, Los Angeles, with his colleagues has developed a treatment of post-traumatic stress disorder (PTSD) and phobias like the fear of snakes, by reprogramming the

[6] Similarly, when it comes to language, Dennett equates information with meaning, which leads to rather questionable theses concerning literature as a medium for 'meme replication', with memes being "informational structures that are normally valuable – they are worth copying – and copyright laws have been devised and refined to protect that value" (Dennett 2017: 228).

unconscious with the help of functional magnetic resonance imaging (fMRI) and electro-encephalography (EEG). But while he can thereby diminish the physical symptoms of fear – like sweaty hands – it does not affect how people feel about spiders and snakes. "If you ask the patients if they are actually afraid," Lau says, "they say yes." Joseph LeDoux, a neuroscientist at New York University in New York City, concludes from this observation that to really tackle fear might require targeting both unconscious and conscious pathways, which work in different ways in the brain. The unconscious pathway emerges from the amygdala, which is responsible for somatic reactions under stress, for example increase of heartbeat, etc. But those neural and physical reactions to threats, he suggests, should not be considered as fear at all. Instead, the conscious experience of fear involves cognitive awareness and the emotional interpretation of a situation (cf. Sohn 2019) – which brings us back to the experience of the individual.

First-Person Reports

In this situation, Fenwick claims that "[s]cientific materialism is appropriate only in areas that do not involve subjectivity. What is needed now is a new science of consciousness that does justice to subjective experience" (Fenwick 2004: 113). And the editors of *The Cambridge Handbook of Consciousness* suggest that "[…] it will not be possible to make serious headway in understanding consciousness without confronting the issue of how to acquire more precise descriptive first-person reports about subjective experience (Anthony and Roepstorff 2003, 2004)" (Zelazo et al. 2007: 2, quoting Anthony and Roepstorff 2003/2004). For first-person reports, test persons are asked to report about their experience of, say, pain or joy, and these reports are included in the interpretation of the data gained by neurological measurements. This inclusion of first-person reports in empirical investigations is prominently done by research conducted in Dayalbagh (Satsangi 2013), in the experiments of Price and Barrell (Price and Barrell 2012), in the Meditation or Contemplation Research of Schmidt and Walach (2014a, b), and in the frame of 'embodied mind' experiments (Varela et al. 2016). With the inclusion of first-person reports, individual experience has found a place in the methodology of CS – though a problematic one, because the phenomenon of individual experience as well as its rendering in the format of first- person reports disagree fundamentally with central tenets of scientific research in the Western sense, namely a third-person or bird's-eye perspective, replicability, and generalisability. This axiomatic absence of consistency with a scientific methodology becomes a problem in two central steps of any first-person report, namely the representation of the experience in language, and the transformation of individual experience into a scientific fact.

Representation and Interpretation

The individual experience as a first-person phenomenon is mostly narrated in verbal language. Already the transformation of experience into a narrative brings up foundational problems if we keep in mind that "[t]he genius of language is omission. It misses most things out, almost everything in fact", while "imposing some kind of shape and momentum on the precarious, barely describable business of actually being here, moment by moment, in the world" (Parks 2018: 16, 15). The shape imposed by language onto experience is deceptive and inadequate, because

> [e]xperience is more continuous and more happening at once […] than words, where only one thing can be said at a time. Words are linear. They sort things out, pull them out of the mix, conjure each object whole, though I don't see anything whole, ever. (Parks 2018: 9)

Additionally, we know that the 'shape and momentum' of each language, archived in the meaning and sound of words and the structure of sentences, implies a specific cultural worldview, which codes the perception and description of our experience, at the latest when we verbalise it. These cultural premises, internalised by language learning, form the bedrock of our values and concepts, in short: they supply our historically and culturally specific worldview that for the largest part is not even present to our waking consciousness. Instead, they form what philosopher Hans Georg Gadamer calls 'legitimate prejudices', namely "not a conscious bias but a sign that we think out of particular accretion of histories that are not always transparent to us" (Chakrabarty 2008: xiv). Accordingly, Kelso holds that "[a]n individual's knowledge is not a mere aggregate of perceptual data, but is divided and unified according to definite *prior* conceptual or theoretical points of views" (Kelso 1980: 21), and Indian philosopher J. N. Mohanty claims: "No bare experience tells its own tale. The experience […] has to play a certain role of fulfilling a prior conceptual intention, in order to be of cognitive value. […] experiences are made to support quite different theories" (Mohanty 1993, quoted in Rao 2005: 64). A case in point is the evaluation of visions during meditation in different meditation schools. Thus the author Tim Parks, an experienced meditator himself, upon asking his teacher about visions he had during extended *vipassana* meditations, was informed that visions are a sign of a sleepy and unfocused mind, in short, "an indication of my ineptitude as a meditator" (Parks 2018: 162). In contrast, many other religions, like for example Radhasoami Faith, attribute high significance to visionary experiences during meditation.

Historian Steven Theodore Katz, director of the Elie Wiesel Center for Judaic Studies at Boston University, is convinced that these different evaluations and verbalisations are not only a matter of interpretation, but that they shape the experiences of people growing up with these different appraisements:

> There are NO pure (i.e. unmediated) experiences. Neither mystical experience nor more ordinary forms of experience give […] any grounds for believing, that they are unmediated. […] This 'mediated' aspect of all our experience […] has to be properly acknowledged if our investigation of experience, including mystical experience, is to get very far. (Katz 1978: 26)

By way of illustration Katz points out that "[…] the Hindu experience of Brahman and the Christian experience of God are not the same. […] Thus we find, for example, in Hinduism monistic, pantheistic, and theistic trends, while Christianity knows both absorptive and non-absorptive forms of mysticism" (Katz 1978: 26).

In light of these fundamental problems concerning the relation of experience and the representation of experience, various philosophers and ethnologists are questioning the entire concept of experience as the rock-bottom of first-person knowledge acquirement, because "[w]hat counts as experience is neither self-evident nor straightforward; it is always contested, always therefore political. [Experience] is at once always already an interpretation and is in need of interpretation" (Scott 1992: 37).[7]

Recent research suggests that individual differences of experiencing comparable situations are not restricted to matters of verbal representation and feeling, but manifest in measurable data. Schmidt and Walach report from meditation experiments with

> two very experienced nuns from the Theravada tradition in our neurophysiological laboratory in Freiburg, Germany. [...] We spoke about techniques, meditation practices and different approaches and for us this all looked very similar in background, culturalization and practice of meditation. But the resulting EEGs of these two women were quite different. Thus, not only can we hardly compare inner experiences of meditators which [sic] each other, moreover, we even have to doubt that similar experiences result in similar third person data. (Schmidt and Walach 2014a, b: 3)

But what can 'similar' mean under these circumstances? Similar to what?

The philosophical as well as the empirical findings outlined above challenge models that suggest the ultimate convergence of esoteric, philosophical, and scientific discourses in one ultimate phenomenon, expressed by Prem Saran Satsangi in the following way:

> The object in view is to realize the highest degree of macrocosmic consciousness (spiritual potential) [...] i.e. God (the goal of religion) or Ultimate Reality (the goal of philosophy) or Truth (the goal of Science) which are but three names of the same Supreme (spiritual Essence) [...]. (Satsangi 2016: 18)

Or should we have to develop a model of C/consciousness that reaches beyond verbal representation and empirical data? Literature and philosophy have no problem dealing with such a concept in the realms of their disciplines. But the natural sciences – geared at the material and quantifiable world with their methodology – would have to redefine their norms of scientificity to include such a realm in the horizon of their research.

Individual Experience as Scientific Fact

James J. Barrell and his colleague Donald Price (Price and Barrell 2012) are aware of the entanglement of perception and interpretation in cultural concepts transported by language, and they try to overcome this in order to transform the first-person report about the individual experience of for example pain into a scientific object, by transforming the first-person report into the third-person perspective of an object of investigation for empirical methodology. In the words of Thomas Nagel, the problem "is that every subjective phenomenon is essentially connected with a single point of view, and it seems inevitable that an objective, physical theory has to abandon that point of view" (Nagel 1974: 437) in order to transform it into the normative third-person perspective of the natural sciences. Yet as the research of Hakwan Lau

[7] Corresponding discussions in Ethnology, Cultural Studies, and Narratology are gathered in Lehmann 2007.

showed, emotional states like fear or pain are largely determined by exactly those categories that format individual perception differently in different cultural surroundings, depending on family, education, religion, class, gender etc. Accordingly, Tim Parks emphasises "how absolutely crucial language is and how much of our perceptive experience is mediated and transformed [...] with the aid of language, so that any account of human consciousness is going to have to say something about language's role in it, even if almost nothing I have read so far has convinced me in this department" (Parks 2018: 135).

Additionally, there are no plausible models yet to accommodate or even take into account the awareness of the temporality of each experience.

> Is temporal information localized at all? At one extreme, one region of the brain may be the time-giver, [...] at the other extreme, time could be embedded in configurations of states, a distributed pattern of activation whose changes represent the evolving temporal landscape of this experiment. Phenomenology suggests that temporality is in every experience, suggesting a more distributed implementation. (Lloyd 2012: 14)

Narratology has dealt in depth with the problems of transforming the experience of time into a verbal account, yet to my knowledge these investigations have left no traces in the methodological self-reflection of first-person reports. The implications of such an elision of the individuality of experience, and the epistemological, social, and ethical consequences for the subsequent – or presumed – concept of consciousness are explored in the essays by Anna M. Horatschek in this anthology.

3 What to Do?

Despite a host of conceptual and methodological questions, the relevance of which for CS has only been touched upon in this introduction, there are promising attempts to bridge the gap between the material conditions and the immaterial aspects of consciousness in theory as well as practice. A first foundational step to 'merge' scientific and pre-scientific knowledge paradigms was the inclusion of first-person reports by Barrel and Price in their empirical studies of consciousness, considered by Max Velmans as "a paradigm shift in psychological research" (Velmans 2013: 213). Other ventures in the same direction are the Meditation or Contemplation Research conducted by Schmidt, Walach, Esch, and others (Schmidt and Walach 2014a, b), and the initiatory work of Francisco J. Varela, conjoining Cognitive Science, Biology, and Buddhist philosophy in his pursuit of *The Embodied Mind* (Varela et al. 2016). The institution most dedicated to research fusing material and metaphysical aspects of consciousness is probably the Dayalbagh Educational Institute (Deemed University) in Dayalbagh, India, founded as an educational institution of the Radhasoami Satsang Sabha in 1917, where spiritual practices and scientific investigation and conceptualisation have been a central field of research across various disciplines for decades. As Annette Wilke in her essay for this volume points out, this commitment to Western science was already noted by von

Glasenapp in 1928, and over the years assimilated modern Physics and Biology with conceptions and terminology like electricity, electrons, and kinetic energy into their unique sonic cosmology, espousing for example ideas of the material transmission of spiritual power.

> All these influences were melted together in an impressive sonic theology, the most consistent and comprehensive one in India that I am aware of. [...] The cosmic spheres are interpreted as different levels of consciousness and simultaneously as macrocosmic realities, both of which can be measured and ascertained scientifically. (Wilke)

Anand Srivastav's argument supporting the non-computability of consciousness in this volume focuses on the topological graph of Prem Saran Satsangi, the Eighth leader of Radhasoami Faith, Dayalbagh, India, which systematises and visualises the dynamic interrelatedness of microcosmic and macrocosmic levels in the frame of topological graph theory. From different perspectives, Juergensmeyer and Blutner also take their cue from these truly trans-disciplinary endeavours in CS: While Juergensmeyer explicates structural analogies between the function of religion and that of the sciences or the arts in evincing 'alternative realities', Blutner's article fruitfully functionalises concepts of quantum cognition for CS.

Bringing together what conceptually, pragmatically, and institutionally has been kept apart for centuries changes all parties involved, in this case the definition of 'science' as well as 'spirituality', 'metaphysics', and – as I pointed out before – the concomitant concepts of consciousness. By way of concluding the thematic introduction to the book, I shall outline some possibilities suggested by recent research in Consciousness and Science Studies.

In various theoretical and empirical research quarters, a change of the concept of 'science' – far beyond the often demanded and rarely practised interdisciplinarity – is deemed necessary, with far-reaching consequences for epistemology, ontology, social structures, and strategies of knowledge production: Thus Schmidt and Walach regarding Meditation Research expect not only that "research on meditation will change science [, but t]he scientific findings about the effects of meditation change our cultures and societies" (Schmidt and Walach 2014a, b: 5). Evan Thompson with his 'enactive approach' and the concept of 'the embodied mind' asks "for cognitive science and human experience to reshape each other in a transformative way beyond our scientific and habitual, experiential reifications of a separate self and an independent world" (Thompson 2016: xviii), and maintains that "applying the enactive ideas about cognition to science itself [makes it impossible to] hold on to the traditional realist conception of science as revealing the way things are in themselves apart from our interactions with them" (Thompson 2016: xxvii); and Karen Barad addresses foundational questions opened up by Quantum Theory "about the nature of nature, but also about the nature of scientific and other social practices" and asks for "a new ontology, epistemology, and ethics, including a new understanding of the nature of scientific practices" (Barad 2007: 25). Taking recourse to Niels Bohr, for her the central

> lesson of quantum physics [is]: *we are part of that nature that we seek to understand.* [...] our knowledge-making practices are social-material enactments that contribute to, and are a part of, the phenomena we describe. (Barad 2007: 26)

Replace 'nature' with 'consciousness' in this quote and we have a succinct description of one of the central problems of CS that hitherto has been addressed hardly at all. For a solution, Barad demands to eschew "anthropocentrist and representationalist assumptions" (Barad 2007: 27) as delusional discursive categories that lend themselves to impose an anthropocentric 'order of things' upon the phenomena of the world, by re-presenting 'things' in terms of the imaginary worldviews of historical subjects. Barad critiques this kind of representationalism, because it implies that a pre-existing thing is mirrored or rendered in the semiotic system – that "the representations and the objects […] they purport to represent are independent of one another" (Barad 2007: 28) – and opts for a strong performative approach. Such a move would change "the focus from questions of correspondence between descriptions and reality (e.g., do they mirror nature or culture?) to matters of practices or doings or actions" (Barad 2007: 28). This conceptual change means in practice that any semiotic enactment – be it drawing a graph, devising mathematical formulas, writing novels, performing theatre, or singing sacred hymns and *bhajans* – is not 'just' a re-presentation one step removed from the 'real reality' by the semiotic system, but the embodiment and production of situationally present responses, emotions etc. Barad terms the resulting event a 'phenomenon', and proposes a shift

> from a metaphysics of things to phenomena [that would] not merely mark the epistemological inseparability of observer and observed, or the results of measurements; rather, *phenomena* are the *ontological* inseparability of agentially intra-acting components. (Barad 2007: 33)

Interestingly, such a concept of embodied performativity that produces the presence of a specific event through 'agentially intra-acting components' is a central idea in Indian aesthetics, as Annette Wilke and Ralph Yarrow explicate in their essays with respect to 'sonic Hinduism' and theatrical improvisation.

Fusing a reality marked as 'metaphysical' in Western philosophy to the methodology of empiricism changes not only the concept of science and the object of investigation – in this case 'consciousness' – but relocates, perforates, and perhaps even erases the boundary between physics and metaphysics. As Barad observes with a view to the entirely new realm of research in 'experimental metaphysics':

> […] questions previously thought to be a matter solely for philosophical debate have been brought into the orbit of empirical inquiry. This is a striking development because it allows scientists to explore metaphysical issues in the laboratory (so much for the category 'metaphysical'). (Barad 2007: 35)

These radical shifts in the epistemology of scientific research challenge the ontology of subject/object relations and 'the nature of nature', the nature of things, and the nature of knowledge and ethics, and call for understanding "in an *integral* way the roles of human and nonhuman, material and discursive, and natural and cultural factors in scientific and other practices" (Barad 2007: 25). They shake the foundations of Western notions of realism, presentism, and representation, and seem to move them more or less perceptibly towards concepts of reality consisting not in things and subjects, but in interrelations, processes, and reciprocal interactions of fluctuating hierarchies and constellations, that remind of non-European world models, where everything is connected by and immersed in currents of energy, spirit, or consciousness. Some such models are represented in this volume by contributions that consider Radhasoami Faith (Srivastav, Wilke, Lane), Jain philosophy (Diem), or Panpsychism (Goff). In contrast to traditional Western world models, "*the notion of intra-action*" brings to light

> the world's radical aliveness [...] in an entirely nontraditional way that reworks the nature of both relationality and aliveness (vitality, dynamism, agency). This shift in ontology also entails a reconceptualization of other core philosophical concepts such as space, time, matter, dynamics, agency, structure, subjectivity, objectivity, knowing, intentionality, discursivity, performativity, entanglement, and ethical engagement. (Barad 2007: 33)

And how would knowledge production work in such a world, where interactively emergent 'phenomena' have replaced the conceptually segregated realms of subject and object? Barad suggests employing Donna Haraway's model of 'diffraction' as a new knowledge paradigm. The term refers on the one hand to the wave-particle duality paradox concerning the behaviour of light, and on the other hand takes up the function of 'light' as the central metaphor for knowledge and consciousness in Western and many other cultures. Yet while the notion of light and the semantically – and pragmatically – related experiences of seeing, perception, insight, reflection etc. as metaphors for knowledge cement the ontological separation and hierarchy of observer and observed, subject and object – most prominently explicated theoretically in Michel Foucault's treatment of Jeremy Bentham's Panopticum, a prison architecture of the eighteenth century that allowed to control the inmates by the regime of the unobserved observer (esp. Foucault 1977 and 2006)

> a diffractive methodology is respectful of the entanglement of ideas and other materials in ways that reflexive methodologies are not [, because it focuses on] the entanglement of the apparatuses of production [and] enables genealogical analyses of how boundaries are produced rather than pursuing sets of well-worn binaries in advance. (Barad 2007: 29–30)

Adopting a 'diffractive' perspective on knowledge production precludes any universally valid 'theory of everything' that fuses the knowledges produced in different disciplines, methodologies, and cultures. Indeed, the very concept of the universal comes under critical scrutiny as yet another colonial legacy of European traditions that functionalised the boundaries between the universal and the parochial, the individual and the general in their interests, and in the process turned 'the universal' into "a highly unstable figure, a necessary placeholder in our attempt to think through questions of modernity" (Chakrabarty 2008: xiii). Such a shift to metatheoretical issues would equally preclude the espousal of global validity claims of the natural sciences as the only reliable source of significant knowledge production, and instead perceive the knowledge of other cultures, gathered in Holy Scriptures and philosophical, religious, and spiritual treatises, as so many ways to transform the "phenomenal reality into knowledge formations attuned to a human scale and conditioned by epistemic and cultural conditions" (Sandkühler 2014: 59–60, transl. A.H.).[8]

[8] Sandkühler: "Die Problemstellung einer Epistemologie, die Wissenskulturen systematisch berücksichtigt, lautet nicht, wie unser Wissen eine substanziell verstandene 'Realität' nach dem Maß der Dinge abbildet, sondern wie phänomenale Wirklichkeit nach Menschenmaß in Wissensordnungen, epistemischen Konstellationen bzw. Wissenskulturen entsteht."

> [T]he crucial question of [such an] *epistemological pluralism* and *internal realism* [would not be] how our knowledge depicts reality according to the measure of things, but how the *phenomenal reality* emerges within knowledge, humanly speaking. (Sandkühler 2012: 174; see also Devy et al. 2014)

Recognising knowledge production not only in the frame of CS, but in general, as an endeavour that is shared by all "[h]uman beings [as] meaning making semiotic animals" (Walach 2014: 17) might instill a scientific ethos of humility to own up to the inconclusiveness of any scientific – and non-scientific – knowledge, to the entanglement of any kind of knowledge with conditions of language, culture, discipline, and historical place, and to the acknowledgement that the incompatibility of knowledges does not say anything about their relevance.[9] Because – in the words of Barbara Smith-Herrnstein:

> Incommensurability is, it appears, neither a logically scandalous relation between theories, nor an ontologically immutable relation between isolated systems of thought, nor a morally unhappy relation between sets of people, but a contingent experiential relation between historically and institutionally situated conceptual and discursive practices. Some radically divergent ideas never meet at all, at least not in the experience of mortal beings. (Herrnstein Smith 1997: 262)

References

Anthony, Jack and Andreas Roepstorff, eds. [2003] 2004. *Trusting the subject?* 2 Vol. Exeter: Imprint Academic.

Barad, Karen. 2007. *Meeting the universe halfway. Quantum physics and the entanglement of matter and meaning*. Durham and London: Duke University Press.

Chakrabarty, Dipesh. 2008. *Provincializing Europe. Postcolonial thought and historical difference.* With a new preface by the author. Princeton and Oxford: Princeton University Press.

Chalmers, David. 2007. The hard problem of consciousness. In *The Blackwell companion to consciousness*, ed. Susan Schneider and Max Velmans, 225–235. Oxford: Blackwell Publishing Ltd.

Dennett, Daniel C. 2017. *From bacteria to Bach and back. The evolution of minds.* London: Penguin Random House UK.

Devy, Ganesh N. 2017. *The crisis within.* Delhi: Aleph Book Company.

Devy, Ganesh N., Geoffrey Davies, and K.K. Chakravarty, eds. 2014. *Knowing differently. The cognitive challenge of the indigenous*. Delhi: Routledge.

Fenwick, Peter. 2004. Neurophysiology, consciousness and ultimate reality. In *Science, consciousness and ultimate reality*, ed. David Lorimer, 109–128. Imprint Academic: Exeter.

Foucault, Michel. [1975] 1977. *Discipline and punish. Birth of the prison.* Trans A. Sheridan. New York: Random House.

———. 2006. *Psychiatric Power.* Lectures at the College De France, 1973–1974. Ed. Jacques Lagrange. Trans Graham Burchell. New York: Picador.

[9] For a dramatic critique of the hierarchisation of knowledge traditions in India, and a successful attempt to interweave Western, classical Sanskrit, and oral Yakshagana folk knowledges see Girish Karnad's *Hayavadana* (1971) as analysed in Horatschek 2020a.

Herrnstein Smith, Barbara 1997. Microdynamics of incommensurability: Philosophy of science meets science studies. In *Mathematics, Science, and Post-Classical Theory*, ed. Barbara Herrnstein Smith and Arkady Plotnitsky, 243–266. Durham, NC: Duke UP.

Horatschek, Anna Margaretha. 2016. "'Inhabiting landscapes of the known and the unknown alike': Representation and consciousness in British literature." In *Consciousness: Integrating Eastern and Western perspectives*, ed. Prem Saran Satsangi, Stuart Hameroff, and Vishal Sahni, 155–187. Delhi: New Age Books.

———. 2017. 'Recasting the usual order of things': Competing knowledges in Amitav Ghosh's *The Hungry Tide. Literatur in Wissenschaft und Unterricht/ Literature in Research and Teaching (LWU)* 48.3 (2015, no annuity). (Theme issue: Literary knowledge production and the life sciences. Karin Hoepker and Heike Schaefer. Guest Eds.). 223–241.

———. 2019. Subalternity and Historicization in Amitav Ghosh's *The Hungry Tide* (2004) and Mahasweta Devi's 'Statue' (1979). In *Feminist Discourses in Indian Context. Assertion, Identity and Agency*, ed. Vandana Sharma, 68–95. New Delhi: Authors Press.

———. 2020a. Mixed attachments in Girish Karnad's Hayavadana. In *The transnational in literary studies. Potential and limitations of a concept*, ed. Kai Wiegandt, 21–43. Berlin and Boston: Walter de Gruyter.

———. 2020b. From knowledge to knowledges: An introduction. In *Competing knowledges. Wissen im Widerstreit*, ed. Anna Margaretha Horatschek, 1-34. Berlin/Boston: Walter de Gruyter.

———, ed. 2020c. *Competing knowledges. Wissen im Widerstreit*. Berlin/Boston: Walter de Gruyter.

Katz, Steven T. 1978. Language, epistemology, and mysticism. In *Mysticism and philosophical analysis*, ed. Steven T. Katz, 22–74. New York: Oxford University Press.

Kelso, James A. 1980. Science and the rhetoric of reality. *Central States Speech Journal* 31: 17–29.

Kirchhofer, Anton. 2009. The religious turn in literary and cultural studies: Introduction. In *Anglistentag 2008 Tübingen: Proceedings*, ed. Lars Eckstein and Christoph Reinfandt, 3–7. WVT: Trier.

Kriegel, Uriah. 2020. Introduction. What is the philosophy of consciousness? In *The Oxford handbook of the philosophy of consciousness*, ed. Uriah Kriegel, 1–13. Oxford: Oxford University Press.

Latour, Bruno. 1991. *We have never been modern.* Trans Catherine Porter. Cambridge: Harvard University Press.

Law, Stephen. 2007. *Philosophy*. London/New York: Cobaltid/DK Publishing.

Lehmann, Albrecht. 2007. *Reden über Erfahrung. Kulturwissenschaftliche Bewusstseinsanalyse des Erzählens*. Berlin: Dietrich Reimer Verlag GmbH.

Lloyd, Dan. 2012. Time after time. Temporality in the dynamic brain. In *Being in time: Dynamical models of phenomenal experience*, ed. Shimon Edelman, Tomer Fekete, and Neta Zach, 1–19. Amsterdam and Philadelphia: John Benjamins Publishing. https://books.google.de/books?id=u3y7hlwD6dwC&pg=PA1&lpg=PA1&dq=Time+after+time+Temporality+in+the+dynamic+brain+Dan+Lloyd&source=bl&ots=nhK1R4VX3d&sig=ACfU3U3PEgC_9H_tqzOPH6SGOG_kdFtEsQ&hl=de&sa=X&ved=2ahUKEwjZpK3FiI_2AhW9SvEDHSakAiIQ6AF6BAgQEAM#v=onepage&q=Time%20after%20time%20Temporality%20in%20the%20dynamic%20brain%20Dan%20Lloyd&f=false. Accessed 15 July 2022.

Mohanty, J.N. 1993. *Essays on Indian philosophy: Traditional and modern.* Oxford: Oxford University Press.

Nagel, Thomas. 1974. What is it like to be a bat? *The Philosophical Review* 83(4): 435–450. http://www.jstor.org/stable/2183914. Accessed 16 July 2022.

Nanda, Meera. 2003. *Prophets facing backward: Postmodern critiques of science and Hindu nationalism in India.* New Brunswick: Rutgers University Press.

Nate, Richard. 2009. *Wissenschaft, Rhetorik und Literatur: Historische Perspektiven.* Würzburg: Königshausen & Neumann GmbH.

Papineau, David. 2020. The problem of consciousness. In *The Oxford handbook of the philosophy of consciousness*, ed. Uriah Kriegel, 14–36. Oxford: Oxford University Press.

Parks, Tim. 2018. *Out of my head. On the trail of consciousness*. London: Penguin Random House UK/ Vintage.
Price, Donald D., and James J. Barrell. 2012. *Inner experience and neuroscience. Merging both perspectives*. Cambridge/London: MIT Press.
Rao, K. Ramakrishna. 2005. *Consciousness studies. Cross-cultural perspectives.* Jefferson/ London: McFarland Company, Inc.
Rovelli, Carlo. 2018. *The order of time*. New York: Penguin Random House USA.
Sandkühler, Hans J. 2012. Critique of representation: Cultures of knowledge – Humanly speaking. In *Rethinking epistemology*, ed. G. Abel and J. Conant, 173–193. Hamburg/Boston: de Gruyter. (Berlin Studies in Knowledge Research. Volume 1).
———. 2014. Wissenskulturen: Zum Status und zur Funktion eines epistemologischen Konzepts. In *Wissen: Wissenskulturen und die Kontextualität des Wissens*, ed. Hans J. Sandkühler, 59–72. Frankfurt/Main: Peter Lang.
Satsangi, Prem Saran. 2013. Vision Talk TSC 2013. https://www.dayalbagh.org.in/specialTalks/ index.htm.
———. 2016. Towards integrating Eastern arts and philosophy of consciousness and Western perspectives of science. (Rhythms of macrocosmic universe and human microcosm: Consciousness Systemology perspective). In *Consciousness. Integrating Eastern and Western perspectives*, ed. Prem Saran Satsangi and Stuart Hameroff, 3–43. New Delhi: New Age Books.
Schmidt, Stefan, and Harald Walach, eds. 2014a. *Meditation – Neuroscientific approaches and philosophical implications*. Heidelberg/New York: Springer International Publishing.
———. 2014b. Introduction: Laying out the field of meditation research. In *Meditation – Neuroscientific approaches and philosophical implications*, ed. Stefan Schmidt and Harald Walach, 1–6. Heidelberg/New York: Springer.
Schneider, Susan, and Max Velmans, eds. 2017. *The Blackwell companion to consciousness*. 2nd ed. Chichester: Wiley.
Scott, Joan W. 1992. Experience. In *Feminists theorize the political*, ed. Judith Butler and Joan Scott, 22–40. New York: Routledge.
Sohn, Emily. 2019, July 24. Decoding the neuroscience of consciousness. A growing understanding of consciousness could lead to fresh treatments for brain injuries and phobias. *OUTLOOK.* https://www.nature.com/articles/d41586-019-02207-1. Accessed 16 July 2022.
The New Oxford Dictionary of English. 1998. Judy Pearsall, ed. Oxford: Clarendon Press.
Thompson, Evan. 2016. Introduction to the revised edition. In *The embodied mind: Cognitive science and human experience*, ed. Francisco J. Varela, Evan Thompson, and Eleanor Rosch, xvii–xxxiv. Cambridge: The MIT Press.
Varela, Francisco J., Evan Thompson, and Eleanor Rosch. [1991] 2016. *The embodied mind. Cognitive science and human experience*, 2nd edn. New Foreword by Jon Kabat-Zinn. New Introductions by Evan Thompson and Eleanor Rosch. Cambridge: The MIT Press.
Velmans, Max. 2013. A review of Donald D. Price and James J. Barrell (2012) *Inner experience and neuroscience: Merging both perspectives*. *Journal of Consciousness Studies* 20: 208–214.
von Stuckrad, Kocku. 2014. *The scientification of religion. An historical study of discursive change, 1800–2000*. Berlin/Boston: de Gruyter.
Walach, Harald. 2014. Towards an epistemology of inner experience. In *Meditation – Neuroscientific approaches and philosophical implications*, ed. Stefan Schmidt and Harald Walach, 7–22. Heidelberg/New York: Springer International Publishing.
Weiler, Hans N. 2017, June 24. Whose knowledge matters? Development and the politics of knowledge. Homepage web.stanford.edu. n.d. Web. https://www.google.com/url?sa=t&rct=j&q=&esrc=s&source=web&cd=&cad=rja&uact=8&ved=2ahUKEwjClrzRl6uCAxXhXfEDHafmCPQQFnoECBMQAQ&url=https%3A%2F.%2Fweb.stanford.edu%2F~weiler%2FTexts09%2FWeiler_Molt_09.pdf&usg=AOvVaw2MRmBEMocMpi0hj4-GtURf&opi=89978449. Accessed 30 October 2023.
Zelazo, Philip David, Morris Moscovitch, and Evan Thompson. 2007. Consciousness: An introduction. In *The Cambridge handbook of consciousness*, ed. Philip David Zelazo, Morris Moscovitch, and Evan Thompson. Cambridge: Cambridge University Press.

The Book: Building Bridges – Exposing Gaps

Anna Margaretha Horatschek and Anand Srivastav

Abstract This chapter presents the structure of the entire book *Consciousness Studies in Sciences and Humanities: Eastern and Western Perspectives* as formal correlative to an implied model of consciousness that accommodates the research conducted in sciences and humanities, in East and West. It offers comprehensive summaries of the individual essays and their reciprocal thematic relatedness, and hints at a new model of knowledge production following reflections of philosopher of science Karen Barad.

Keywords Consciousness · Knowledge · Culture · Karen Barad

The plan to publish the contributions for the Dayalbagh Science of Consciousness (DSC) Conferences from 2019 to 2020 in a book germinated in 2021, and it took all of 2022 to complete this project. The book offers eighteen original essays from Philosophy through Comparative Theology, Medicine, Anthropology, History, Astrophysics, Neuroscience, Quantum Physics, Mathematics, Computer Science, and Literary Studies – to name a few. This variety of approaches bears the danger that the book is just an eclectic assembly of specific disciplinary perspectives. We therefore have devised a structure in which - after a comprehensive introduction to the vast field of Consciousness Studies in Part I - the essays are gathered according to a specific aspect of consciousness they focus on – irrespective of their methodology, though in some cases the essay would have fitted into another chapter as well. Accordingly, the essays are grouped under the following topics:

A. M. Horatschek (✉)
Department of English and American Literatures, Cultures, and Media, Kiel University, Kiel, Germany
e-mail: horatschek@anglistik.uni-kiel.de

A. Srivastav
Department of Mathematics, Kiel University, Kiel, Germany
e-mail: srivastav@math.uni-kiel.de

P. S. Satsangi et al. (eds.), *Consciousness Studies in Sciences and Humanities: Eastern and Western Perspectives*, Studies in Neuroscience, Consciousness and Spirituality 8, https://doi.org/10.1007/978-3-031-13920-8_2

Part I: **Introduction** outlines central questions and methodological problems of Consciousness Studies in the sciences and humanities in East and West, and supplies a summary of the individual essays.

Part II: **Models of Consciousness** offers comprehensive models of consciousness in holistic perspectives, often bridging historical and cultural traditions of philosophy in East and West.

Part III: **Consciousness and Artificial Intelligence** presents positive and sceptical views on Artificial Intelligence and its potential as a means of investigation in Consciousness Studies.

Part IV: **Consciousness and the Brain** gathers a study of EEG/MEG measurements on epilepsy children, and an essay on quantum biology in the frame of Consciousness Studies.

Part V: **Consciousness and the Senses** focuses on the function of sound in spiritual traditions of Hinduism to evoke states of consciousness, and on the effect of theatrical improvisation on consciousness.

Part VI: **Consciousness and the Intellect** focuses on the limits of intellectual understanding in the Negative Theology of medieval Western mysticism, on the potential of religion and the sciences to challenge the everyday perception of reality with alternative realities, on the impact of a cosmology founded on Einstein's theories of relativity for astrophysics and for the individual self-awareness, and on a case study of how knowledge about India changed the identitarian self-perception of the German Prince Frederick August of Augustenburg (1830–1881).

Part VII: **Consciousness and Pragmatics** addresses questions of the so-called operational interpretation of quantum theory and its applicability for quantum cognition, problems of first-person reports in light of reflections on strategies of verbal representation in English and Indian literature, and the role of religion concerning an ethics of 'conscious leadership'.

This very structure implies specific premises regarding the conceptualisation of consciousness, namely that consciousness cannot be reduced to brain functions, but comprises sensuality, ethics, emotions, and the awareness and experience of transcendence. In the words of Barad:

> Matter and meaning are not separate elements. They are inextricably fused together, and no event, no matter how energetic, can tear them asunder. [...] Mattering is simultaneously a matter of substance and significance, most evidently perhaps when it is the nature of matter that is in question, when the smallest parts of matter are found to be capable of exploding deeply entrenched ideas and large cities. Perhaps this is why contemporary physics makes the inescapable entanglement of matters of being, knowing, and doing, of ontology, epistemology, and ethics, of act and value, so tangible, so poignant. (Barad 2007: 4)

It also implies that the way we experience the world – or the way the world informs us – is real, and not just a form of self-delusion or an evolutionary hoax, as eminent researchers like Dennett claim. And finally, by gathering various cultural and disciplinary approaches with one focus in six main chapters, the anthology illustrates that what people mean when they use the term consciousness is largely determined by their cultural and disciplinary context. The form thus fuses matter and meaning, physics and philosophy, analysis and ethics – if not in one essay, then at least in one chapter.

In this way the book will hopefully highlight the often unacknowledged pre-theoretical assumptions and implied ethics of the respective disciplinary and cultural perspectives, as in this format they reciprocally comment on each other. Although most contributions in this book agree in firmly basing theories about consciousness in data gathered by scientific methodologies or first-person experiences, some considerations also include consciousness models based on substance dualism – found, for example, in many Indian philosophies and in most religions – that attribute a specific reality to non-material phenomena. The structure as well as various explicit cross-references among the essays of this volume underline the necessity and the theoretical and experimental productivity of interdisciplinary, transdisciplinary, and intercultural exchange and dialogue as mandatory conditions for a comprehensive approach to Consciousness Studies.

From a pragmatic perspective, we hope that the structure of the anthology with essays of various disciplines addressing one key aspect of consciousness gathered in specific main chapters will direct researchers in the humanities and the natural sciences interested in AI or sensuality and emotions outside CS proper to this book as well, as it offers commensurate essays with interdisciplinary perspectives already compiled in seven distinct Parts.

1 The Essays

The present anthology comprises an extensive introduction and eighteen essays, all of which – with one exception (Goff) – have been written for this anthology. The essays are based on lectures contributed to the DSC conferences from 2019 to 2021. The book gathers Eastern – namely Indian – and Western – namely German, British, American, and Canadian – perspectives on and approaches to consciousness from a range of disciplines in the natural sciences and humanities. The essays present ongoing research in the specific disciplines and thus offer a sound introduction to and overview of the broad field of CS.

The following synopsis will very briefly summarise the individual essays assembled in the seven Parts of the volume.

1.1 Part I: Introduction

The introductory essay by **Anna Margaretha Horatschek** surveys the often incompatible semantics attributed to the term 'consciousness' in different disciplines and in East and West, and outlines major research objectives of CS in the sciences and humanities. The focus lies on methodological challenges concerning the crucial phenomena of experiment, experience, first-person reports, representation, and interpretation, all of which are essential for the transformation of individual experience into a valid scientific fact. The second essay by **Anna Margaretha Horatschek and Anand Srivastav** outlines the epistemological significance of the structure of the book, and summarizes the individual essays.

1.2 Part II: Models of Consciousness

Different models of consciousness are created and discussed by Christine Mann, Rocco J. Gennaro, Andrea Diem-Lane, and Philip Goff, most of them bridging Indian and Western views. **Christine Mann** searches for a unique entity that unifies matter, energy, and what she denotes with the German word "das Geistige" (spirituality, mind, intellectuality, cognition). She draws on quantum theory to conceptualise a core entity in all of these phenomena, namely abstract quantum bits, because they form the constitutive elements for a comprehensive model of the world. She takes her lead from the work of C.F. von Weizsäcker and especially from the model developed by Thomas Görnitz – who studied with von Weizsäcker – and Brigitte Görnitz in their book *Von der Quantenphysik zum Bewusstsein. Kosmos, Geist und Materie* (Görnitz and Görnitz 2016), yet extends their model of consciousness by including issues of ecology, ethics, and spirituality. In Mann's model, "das Geistige" is a primary force that transforms into matter. Panpsychism (see the paper by Goff in this volume) would probably agree. Mann further concludes that though this entity transcends the purely material, it cannot exist in our world without a material basis, thus gesturing towards models of property dualism, yet clearly differentiated from many religions and Eastern philosophies with their substance dualism. **Rocco J. Gennaro** from a philosophical perspective highlights and discusses the central role of self-awareness for Higher-Order Thought (HOT) theory in Western science, compares this model with similar ideas concerning the inherently reflexive nature of consciousness and dual-aspect of cognition in some Indian philosophies, and works towards integrating both approaches. **Philip Goff** determines experimental data and a comprehensive and highly explicatory philosophy as two indispensible premises for any serious theory of consciousness. With these framing conditions in mind, Goff favors the philosophy of Panpsychism, which sees the whole universe permeated by consciousness, not only creatures with brains and nerve systems, as held by theories of materialism and assumed in the theses of strong artificial intelligence. Panpsychism is a radical proposition in the context of prevailing discourses of philosophy in the West, but if consciousness in Panpsychism would be conceptualised as a spiritual force, the model might be comparable with philosophies and cosmological models of Eastern provenance. However, many disputable aspects would still have to be discussed, for example which creatures are able to experience higher-order consciousness as discussed by Gennaro as well as in the cosmology of Radhasoami Faith, which has been presented in abstract mathematical equations and graph theoretical models. A very interesting perspective is opened up by **Andrea Diem-Lane** in her discussion of the principles of Anekāntavāda, Syādvāda, and Ahimsā in Jain philosophy of the sixth century BCE, and their function to model consciousness and self-awareness. Her in-depth presentation exposes that these Jain principles imply significant aspects that are constitutive for a scientific methodology, and very similar to those proclaimed by founding fathers of the empirical methodology like Francis Bacon and others, but developed far earlier than the emergence of Western science.

1.3 Part III: Consciousness and Artificial Intelligence

The section on consciousness and artificial intelligence offers views from very different disciplines, namely mathematics, computer science, virtual reality systems, and literary studies. The articles of Subhash Kak and Anand Srivastav both address the fundamental question whether consciousness is computable and thus can be acquired by artificial intelligence systems. Both authors argue for a non-computable nature of human consciousness. **Subhash Kak**'s considerations include a social perspective and issues of awareness, memory, the nature of reality, creativity, and theological questions concerning the origin of freedom in sentient beings. **Anand Srivastav** argues that the non-computability of consciousness might not be provable, because mathematical theorems and proofs are beyond the capability of artificial intelligence, an argument favored by Penrose in view of recent revolutionary developments in artificial intelligence (e.g. Ramanujan machine). However, third-person validations of first-person data, like the theoretical and experimental work on topological graph theory as an abstract mathematical method to describe macrocosm and microcosm in the work of Prem Saran Satsangi, offer new insights. It is demonstrated that analysis and modelling by mathematics is a descriptive and analytical key. **David Lane** perceives the human brain as the most sophisticated virtual reality, and with this premise discusses the role of (manufactured) virtual reality to understand human consciousness, by exploring the virtual reality accoutrements and the varied vistas they can create. At first sight, this perspective is a counterposition to a non-algorithmic nature of consciousness as emphasised in the articles of Kak and Srivastav, where human consciousness is something not creatable or understandable by AI machines. But all three authors certainly would agree that artificial intelligence and virtual realities may explain and approximate functions of the human brain. However, the ultimate understanding of the ontological nature of human consciousness remains an open question in CS. **Anna Margaretha Horatschek** supports the thesis of a non-algorithmic nature of human consciousness with an emphasis on issues of ethics and morals, as developed in Ian McEwan's novel *Machines Like Me And People Like You* (2019). In this text, the robot Adam, who is essentially a Turing machine, can adhere to the programmed rigourism of disembodied Kantian ethics, but he cannot act in a morally adequate way, because he lacks the embodied experience of an *ipse*-consciousness and the existential certainty of his personal mortality, which implies a clear repudiation of the theses of strong AI.

1.4 Part IV: Consciousness and the Brain

Apurva Narayan as well as Ulrich Stephani jointly with Ami Kumar discuss consciousness in the discourse of the life sciences. **Apurva Narayan** aims to connect quantum computing and the potential of graph-theoretical modelling that was introduced by Satsangi (2013) and developed by Srivastava et al. (2017) with quantum computing abilities of bio-molecules, e.g. cytostine. He points out that not only the storage of information in such systems, but the information transfer is key for an understanding. At the end of the essay, editor Prem Saran Satsangi comments on this approach and paves a way for further experimental and theoretical research in this direction by pointing out the importance of the bio-molecule Adrenaline in this context, and the necessity to connect neuro-scientific aspects with first-person reports in Consciousness Studies. **Ulrich Stephani** and **Ami Kumar** access consciousness from a neurobiological point of view. Their study of epileptic seizures in childhood absence epilepsy not only exhibits a new and non-invasive diagnostic methodology by way of electro- and magneto-encephalography (EEG and MEG), but opens experimental research paths for understanding consciousness phenomena that are detectable as activities in the neuronal network and in specific brain areas.

1.5 Part V: Consciousness and the Senses

The articles of Annette Wilke and Ralph Yarrow discuss the paramount importance of sensual perception in any work with and discussion of human consciousness. **Annette Wilke** describes the fundamental role and extraordinary cultural significance of sound and sonality, voice and hearing, the spoken and the sounding word in Hinduism, theorised and systematised to an exceptional degree in the Religions of Saints and Radhasoami Faith. She outlines the conceptual history and the pragmatic relevance of the Nāda-Brahman, literally "sound-Brahman" or "sonic Absolute", as one of the most powerful sonic conceptions of an ultimate being that has been appropriated, modified, and put into practice by various religious and spiritual communities through centuries. She closes her essay with a detailed account of Western appropriations of these auditory aesthetics and their religious resonances in music theory and meditation centers. **Ralph Yarrow** focuses on the function of sensual awareness for consciousness-changing practices of theatrical improvisation in India and the West. Writing as an experienced theatre practitioner in East and West, he explicates the philosophical, psychological, and aesthetic implications of theatrical improvisation, where sound, voice, and hearing, but also physical acting and moving are creating a unique sense of being in the present moment. Improvising and the improvisatory in this context is a state, a process of changing consciousness, and a mode of conscious being, or of being conscious. With reference to the work of Gary Peters, Adam Kahane, and Samuel Beckett, the theoretical discourse touches questions of 'group mind' and 'self- consciousness' as well as psycho-physiological and ecological parameters.

1.6 Part VI: Consciousness and the Intellect

This chapter focuses on the relation of intellect and consciousness from four disciplinary perspectives. **Andreas Müller's** article recounts the efforts of Christian mystics to communicate their experiences verbally. Taking the examples of two fathers of the so-called Negative Theology, namely Gregory of Nyssa in the late fourth century and Dionysius Pseudo-Areopagites in the fifth century, the essay expounds how these mystics on the one hand strove to transcend the boundaries of an intellectually graspable world in order to reach the deepest insights into the self and the world and eventually to be immersed in the divine, and how on the other hand they struggled with the problem of how to express such an experience in an adequate way. They ultimately acknowledged that representation is not possible in discursive language, but only in paradox images, specific sonic practices, and ritual. **Mark Juergensmeyer** is more optimistic about the possibility of opening up alternative realities via intellectual agency. He sees a similarity in religion and science in general and contemporary physics in particular, in that all of them find realties that significantly differ from everyday consciousness. Thus contemporary physics challenges the idea of a reality bounded by and embedded in the time-space world, and indicates that reality might be something completely different. In fact, modern string theories talk about multiple dimensions of the universe not comprehensible by the human brain, which in daily life functions as a three dimensional sensor with a "feeling" of time. He also observes that views of alternative realities have to be assessed critically considering their validity, for example pseudo-science "explaining" climate change. His analysis presents a hopeful view of a symbiotic relationship between progressive religion and critical science. **Wolfgang J. Duschl** starts his discourse on the platform of modern physics, namely Einstein's theories of relativity, and issues relating to the finite and universal value of the speed of light in astrophysics. He presents reflections on the importance of these laws in understanding the evolution of the universe, and the consequences for our self-awareness as parts of and participants in this universe. While the article by Duschl embeds the question of self-awareness in the time-space range of our universe, **Martin Krieger** addresses a similar topic from a historiographic perspective on intercultural communication and knowledge transfer between Europe and India. Taking the biography of the German Prince Frederick August of Augustenburg (1830–1881) as a case study, he illustrates how the prince re-shaped his self-perception through his travels and his intellectual pursuits in India that made him write the first Western biography of the Mughal emperor Akbar.

1.7 *Part VII: Consciousness and Pragmatics*

The section on consciousness and pragmatics presents an interdisciplinary and quite diverse spectrum of ideas oriented towards pragmatic aspects of CS. **Reinhard Blutner**'s article is rooted in quantum physics, but gives an independent motivation of the epistemic conception based on the so-called operational interpretation of quantum theory, applicable to quantum cognition. Examples illustrate the theoretical potency of complementarity in the domains of bounded rationality and psychological aspects of awareness. In this fashion he spans a topical circle comprising complementarity, quantum cognition, operational interpretation, and consciousness. **Anna Margaretha Horatschek** in her essay addresses the differences in concept and representation of consciousness in a materialistic and a spiritual worldview, as illustrated in Peter Watts' neuro-novel *Blindsight* (2006), and in poems by the Indian mystic Kabir (1440–1518). On the basis of these literary negotiations about the potential and limits of verbal communication, the essay explores the execution and scientific evaluation of first-person reports in CS. A very pragmatic aspect of consciousness lies at the center of the article by **Andrew Davies**, namely the concept of 'conscious leadership', and the question of what such leadership can gain from religion. In the present global crisis characterised by climate change and global warming, food shortage, vanishing biodiversity, armed conflicts, and refugee waves, such pragmatic considerations in a book about consciousness are absolutely necessary to put the theoretical deliberations into a practice that tackles the manifold global threats with which humanity is faced.

References

Barad, Karen. 2007. *Meeting the universe halfway. Quantum physics and the entanglement of matter and meaning*. Durham/London: Duke University Press.

Görnitz, Thomas and Brigitte Görnitz. 2016. *Von der Quantenphysik zum Bewusstsein*. Berlin, Heidelberg: Springer-Verlag.

Srivastava, Dayal Pyari, Vishal Sahni, and Prem Saran Satsangi. 2017. From n-qubit multi-particle quantum teleportation modelling to n-qudit contextuality based quantum teleportation and beyond. *International Journal of General Systems* 46(4): 414–435. https://doi.org/10.1080/03081079.2017.1308361.

Part II
Models of Consciousness

Is "das Geistige" the Basic of the World?

Christine Mann

Abstract First, I will explain the term "das Geistige" (spirituality), and why I use this German term in the title of my English text. I will illustrate with reference to the German term "Geisteswissenschaften" (Humanities) that I understand 'the spiritual' in a much broader sense than 'consciousness', 'mind', or 'wit', because ultimately with the emergence of life 'das Geistige' gained foundational relevance as a separate modality of the world alongside matter and energy, without which life would not exist; indeed, a preliminary form of the spiritual in this sense can already be discerned in the first single-celled organisms. I will then briefly discuss a new development in quantum physics, which shows that matter, energy, and "das Geistige" are based on the same foundational ground. Based on these premises, I will show how "das Geistige" has unfolded in the development of life, and how comprehensive this concept actually is.

Keywords Spirituality · Intelligence · Evolution · Quantum physics

This essay presents a new view on quantum theory, one that is not yet mainstream but offers a new understanding of the development of consciousness together with the development of mind and matter.

This would be much easier in the German language because we have one word, – das Geistige – that includes all the aspects of this phenomenon that in English means consciousness, spirit, mind, intellect, intelligence, cognition, cleverness for surviving, and information processing. One word would make it easier to understand that all these aspects developed out of one source together with matter and energy. Therefore, I used the German term "das Geistige" in the title of this essay.

C. Mann (✉)
Theologue, Psychologist, Munich, Germany

Independant Researcher, Munich, Germany
e-mail: mann.christine@gmx.de

P. S. Satsangi et al. (eds.), *Consciousness Studies in Sciences and Humanities: Eastern and Western Perspectives*, Studies in Neuroscience, Consciousness and Spirituality 8, https://doi.org/10.1007/978-3-031-13920-8_3

In this essay

- I will first show why I think that, from a scientific point of view, all the aspects of the phenomenon 'consciousness' that I mentioned above belong together.
- In the second part I will show how a new view on quantum physics provides a new perspective on consciousness.
- And, on that basis, I will show how information processing expanded alongside evolution of life to new concepts for surviving and more and more complex forms of cleverness. This opened new possibilities for life all the way up to the development of our mind and our consciousness.

1 What Is This Phenomenon, to Which Our Consciousness Belongs?

Our language and communication is not matter. It needs matter to exist. The brain in which we think a sentence, our mouth that forms the words and utters the sentence, the air that transports it to a listener, all that is matter. But the information that is given by the sentence isn't matter. It is something else. And it stays the same, irrespective of whether the sentence is spoken, is written down, or is broadcast on television by electromagnetic waves. And although it is combined with a tiny amount of energy, the vibrating air, for instance, it may have a huge impact. If in a crowd I were to point in one direction and shout: "Oh look, there is a fire", those who understand it were to turn around to verify what I was saying. We are made from matter. The content of the sentence, which is neither matter nor measurable energy, can yet have an impact on the matter of our body. The brain is a huge amount of nerve cells in our head, the mind is our understanding, our turning the biochemical activities of the brain into relevant meaning, as long as we understand the language. Our mind has to recognize the biochemical signals of the brain and combine them with stored information about the language, about reality, to understand them and therefore to be able to protect our life. If someone were in the crowd around me who did not understand what I said, he or she would turn around as well, only because of my frightened face and my finger pointing out to the distance. But that person might expect a person with a gun. That means we also use facial expressions and gestures for getting information. But language makes the information more precise.

But of course verbal information processing can also be much more simple and imprecise. If I say to a person: "Oh, you silly fool!" he, hopefully, would not believe himself to be a fool. But it is information processing nevertheless. He now knows what I think about what he did or said, or about his whole person. It is his intelligence that interprets my simple sentence. He will combine it with all the information of the situation, of my facial expression and my gestures, and he might react with a laugh, with a smile, or with just turning around and going away. Though there are not deep thoughts behind this sentence, though it is more emotionally spoken, it belongs to the same phenomenon as our consciousness. The vibrating air that transports these words from me to the listener is measurable energy. But this is not the energy that causes the reaction of the listener. If the listener did not speak English, this energy would not stimulate any reaction. It is the surplus that goes with the vibrating air, the sense of the phrase, that has to be understood and interpreted to incite the reaction. Mere information can have an impact on the biochemistry in our body.

This ability of communication is a privilege not just of humans. Dolphins, for instance, seem to have conversations over long distances. We don't know what goes on in the brains of the dolphins when they chat. But evidently, it is meaningful for them. And this decoding of the signals in their brains is an activity of their minds, just like when we listen to someone else talk.

Macaques, a species of monkeys, have three distinct warning calls: one for danger on the ground, another for danger from above, and one for danger at eye level. When a young macaque utters one of these signals, the group becomes alert. But only when an experienced macaque repeats the signal does the herd react with the necessary behaviour to protect itself. This shows that the young macaques have to learn the meaning of the signals and, maybe, also what constitutes a danger, like humans have to learn the language of their family. Though the language of the macaques consists only of three different signals for danger, it shows that this differentiation helps the macaques to survive – a cleverness for surviving.

It is presumed that blackbirds don't really chat with each other, but they give vocal information, like we do. With their melodious singing they inform rivals about their territory, as far as we know to date. When they see a cat or a fox, they utter an entirely different warning call, and the chicks in their nest will at once be silent. The siblings did not have to learn the right reaction; it is given to them through their genes. Is that the phenomenon we want to talk about in this essay? Well, if the nest is in a gutter above a terrace, the blackbird will warn the chicks when a cat creeps along the terrace, but when the family sits there having breakfast, the parents of the chicks will not warn their offspring. With this, the blackbirds give information to their siblings about what are enemies and what are not, and so help them to develop their intelligence for surviving. And intelligence is neither matter nor energy. It belongs to the category of mind.

There are other ways of informing each other of things. Bees, for instance, inform the other bees in their hive where to find food by dancing. That is also information processing. It is not matter, it is not pure energy, but the other bees observing the dance have to translate it into flying in the right direction for food. And, although it is most likely done by instinct, it is a cleverness for surviving that is even stored in their genes.

Actually, life is a very instable equilibrium, easily upset if the living being does not engage in many activities to keep it in balance. Even the single-celled organisms must constantly perform activities to keep up their metabolism, while they build up two nuclei and double their organelles to divide into two single-celled organisms in order to procreate. This means that, even for single-celled organisms, it is urgently necessary to receive information from outside and combine it with the information from inside to trigger the right activity for surviving. The single cell does not have a brain. Combining the information is a process that is not located in a special place in the cell but takes place in the whole cell. And as it concerns only this particular organism, no other being outside this organism can do this job for it. Every living being does it for itself. Therefore, every living being has a minimal self (Glasgow 2018).

These activities are neither pure matter nor pure energy. The minimal self uses the information aspect, the relationship aspect, and the openness of possibilities that are inherent in matter to stabilize their equilibrium and stay alive. These three aspects of matter, that have only become clear to us through quantum physics, derive some meaning for survival. I consider this capacity to combine the different signals and to adapt to the conditions from outside and inside for survival as the very beginning of cleverness, which, during evolution, will be developed into our brain and consciousness. It lies at the root of it.

2 So Now Let Us Look at the Quantum Physics That Made Us Recognize These Aspects of Matter During the Last Hundred Years

Around 1850, chemists believed that elements could not be divided infinitely, but only into atoms. Demokrit, in ancient Greece, had also already taught this view of matter. But when, at the end of the nineteenth century, physicists realized that atoms are not indivisible, but consist of smaller particles, they had to learn a new view of the world.

I will highlight three central aspects of this new view.

First of all: Each atom of the more than 90 chemical elements consists of the same kinds of particles. The cores of all atoms consist of protons and neutrons, and are surrounded by electrons, which seem to be arranged on virtual shells. And you must not imagine all these particles to be tiny balls of matter, like sand. No, they are neatly packed concentrations of energy. We must imagine matter as neatly packed bundles of energy in a certain structure.

There exist only a few different types of elementary particles. The different amount and the more or less complex structure of these few types of particles make the difference between, for instance, the gas oxygen, the metal gold, and the powder magnesium. Thus, structure is a very important aspect of matter. All the matter in our world is energy formed in a certain structure, bundles of energy in a certain relation to each other, in a certain order. Structure, relation, order, that is more than the mere energy. Matter is not pure energy, but energy plus an order, a structure that lets the energy form our matter. Matter is formed from structure plus energy, and thus structure is a surplus. And only with this surplus can our material world exist.

Nearly all the energy that keeps an atom together or that makes it combine with another atom to form a molecule is electromagnetic energy. Light is electromagnetic energy as well. The light we can see is only a small part of the huge range of electromagnetic waves. Radio waves, the waves that allow our mobile telephones to work, x-ray waves, all these are electromagnetic waves. The range of the electromagnetic waves reaches from the smallest wave with a length of the Planck constant up to more than hundred thousand miles. And about light, the electromagnetic waves we can see, we know that these electromagnetic waves give us information about our surroundings. In analogy, the electromagnetic waves that keep the atoms together can also be considered as a sort of information.

If you try to observe the movements of electrons in an atom, you realize that they don't move in a continuous way, but can jump discontinuously from one atomic shell to the next atomic shell. And, furthermore, from every point where you located them, there are infinite possibilities where they may be located next. Quantum physics can predict only the probability of an electron being found at this or that location at a certain moment. And that is the same with all elementary particles. You can never predict exactly where a certain particle will be next because, if you try to observe them, you can detect their location precisely, but then you don't know anything about their impulse. But when you observe their impulse exactly, then you cannot know their location. There is always a certain amount of indeterminacy. This does not mean that everything is possible. There is only a certain range of possibilities. There are certain probabilities of where the particle will occur next. And as in our everyday reality every object consists of a huge amount of elementary particles; the different probabilities balance each other out, so that the world of objects seems to be well determined. Only if you try to describe the subatomic reality extremely precisely will you recognize that there is the indeterminacy that is inherent in our reality and cannot be overcome.

Thus, quantum physics is physics of possibilities and probabilities. This means that the future is not entirely open, but there are always various possibilities.

The second interesting aspect is the following: The elementary particles have lots of properties that seem to have more in common with our thinking than with our experience of material reality. For instance, we can simultaneously think of different possibilities that are not compatible. We can think of going through one door or the other door. But as soon as we actually go through one door, we cannot go through the other door at the same time. The fact of choosing one door makes the other possibility impossible.

From our thinking we know: Open possibilities can have an effect on our present activities. If, for instance, the likelihood of contracting a life-threatening disease is high, one is likely to wear protective clothing and a mask. This means that possibilities and their probability can have a significant effect on our behaviour. In physics also the possibilities can have an effect. The possibility, for instance, to fill in missing electrons in an atomic shell can lead to the formation of molecules.

Another property of elementary particles is very interesting. Every elementary particle can be seen as a wave or a particle. It depends on how we look at it. And if we want to find a mathematical formula for it, we can start from the particle properties or from the wave function and will end up with the same formula. Light, for instance, we know as an electromagnetic wave with a certain wavelength. But at the same time it consists of particles, the photons. And the shorter the wavelength of the electromagnetic wave, the more energy is in the photon. In classical physics, such incompatible properties would not occur in any material object. However, in our mental and psychological reality, we find lots of complementary aspects. Justice and forgiveness are complementary reactions to an insignificant crime. In a dictatorship, honesty and cleverness may be such complementary properties. Nils Bohr, the physicist who formulated the principle of complementarity in physics, liked to find more and more such incompatible aspects in our everyday world.

If, in a collider, we let elementary particles be hit by other particles, they will not be broken into smaller parts by this particle. The elementary particles are indivisible indeed, but they can be transformed into other particles that are entirely new. They do not have the former particle as a part of them, but they are indivisible as well. An electron and a positron, for example, can be destroyed, and a pair of photons can be generated from their energy. Some of these generated particles will exist for only a short moment and transform into other particles again. The famous Higgs boson, for instance, which was already prognosticated by the physicist Higgs before 1970, in 2012 was finally ascertained once in the collider of CERN. But it existed for less than a billionth part of a second before it decomposed into other elementary particles. An analogy for such a fleeting existence can be found in our thinking. There are modes where a thought may come up for a very short time, vanish again, and give space for other, more stable thoughts.

In quantum mechanics, if you want to understand how two systems, like for instance two atoms, are combined to form a molecule, you cannot just add the different parameters of both systems, but you have to combine every parameter of the one system with every parameter of the other system to understand the new entity. This necessity made us recognize that the combination of systems forms a wholeness, which can be something entirely new. Water, for instance, can be split into its elements oxygen and hydrogen. But the quality of water is entirely new and does not seem to have anything to do with the gases oxygen and hydrogen.

Not only atoms can be combined to form a molecule, but elementary particles can also be combined with each other so that they form a new wholeness. We can view this as them being entangled. That wholeness can last even if the two poles of the entangled particles go in opposite directions over a distance of many miles.

So, quantum physics is physics of combinations that can form something entirely new.

The third aspect: Physicists try to find all elementary particles, even if these have much shorter wavelengths than the ones they have already found. And, in the end, they hope to find a simple formula underlying everything.

But this hope is in vain. The smaller the wavelength of a particle, the more energy is needed to generate it. And it will turn its energy into other well-known particles very quickly. The famous Higgs boson, for instance, needed a huge amount of energy to be generated. And there is no reason why "more energy" should also lead to a simpler formula for everything.

With this in mind, C.F. von Weizsäcker in the 1950s postulated that we should not look for the smallest particles but, instead, start at the other end and look for the simplest structure that might exist. And the simplest structure that can exist is the binary alternative of information yes/no, or in numbers: 1 or zero, existing or not existing.

This is a very abstract idea, which goes beyond all the ideas of quantum particles and waves I just described. But if structure is essential to form matter from energy, it seems a good idea to start with the simplest structure for finding a formula for the whole. But von Weizsäcker used the concept of information as understood in everyday life, and as I have used it so far, namely in the sense of “meaningful information”. But if we think of the examples of the dolphins and the bees, then it becomes clear that “meaning” is something subjective, depending on context, conditions, and point of view. The “subjective” however does not belong to physics, and hence “meaningful information” cannot be a basic concept for physics.

Therefore, it was necessary to become even more abstract than von Weizsäcker. His friend and co-worker, Thomas Görnitz, finally made this step from a “relative” to an absolute interpretation of quantum information, and that implies a concept of meaning-free quantum information (Görnitz and Görnitz 2008, 2016; Görnitz 2018).[1] “Absolute” here means that it is referring to the “whole”, to the cosmos. These very difficult steps of abstraction make it possible to connect these ideas about quantum information to the so far existing and established theories of physics. All scientific statements could be reduced to the quantum forms of binary alternatives.

*The simplest of the mathematically possible quantum structures in this model are therefore **A**bsolute Bits of **Q**uantum **I**nformation, AQIs. A single AQI is a cosmic quantum structure.*

The AQIs form a pre-structure, a pre-type – “Protyposis”, as Görnitz calls it.

In the course of cosmic evolution, the AQIs can combine to form material and energetic quanta. Later, in the course of biological evolution, they can also become meaningful information – this is what we usually mean by the term “information”.

Thomas Görnitz showed that this theory can be linked to a quantum cosmology and to today’s established physics.

The Absolute Quantum bits are equivalent to energy and mass. As the action of a single quantum bit corresponds to Planck’s quantum of action h, which is a fundamental natural constant, Görnitz showed that one photon is condensed from about 10^{30} AQIs, and that a proton needs about 10^{41} AQIs to be formed.

$$10^{30} = 1{,}000{,}000{,}000{,}000{,}000{,}000{,}000{,}000{,}000{,}000$$

If so many AQIs are needed to form one photon, or 10^{41} AQIs to form one proton, which together with neutrons constitute the core of an atom, we spontaneously imagine that one AQI must be extremely small. But this is not the case. Rather, if you visualise the process in an image, then an oscillation spread over the whole cosmic space would come close to its reality. When those oscillations superpose each other, they can form localized entities, like quantum particles without and with rest mass that form atoms and our matter.

[1] The following explications largely follow the model developed by Thomas and Brigitte Görnitz.

We humans and most animals have eyes. The light we can see is a small section of the huge range of electromagnetic waves. These electromagnetic waves as specific forms of energy give us all the information we need for orientation in our surroundings. This shows that energy in specific forms is information that can be used by living beings. This observation reinforces the idea of positing the binary alternative as the fundamental structure of our world, because yes/no is the simplest structure of information.

This expansion of quantum theory is not yet mainstream, but every radical change in our view of the world is not mainstream in the beginning. Yet this theory allows us to understand how our thinking and our consciousness can interact with matter. That means that we can think; we can plan an action and then do it. With this possibility everybody can influence the further development of the world a tiny bit.

3 Life and the Development of Consciousness

When we assume protyposis, a simple binary structure, is the fundamental entity of our world, and that that can model the formation of energy, matter, and meaningful information, it is easier to understand how, during the evolution of life, primitive cleverness for surviving developed into mind and consciousness in some animals and in human beings.

Living beings can be defined as unstable systems of matter and energy, which can stabilize themselves by internal quantum information processing.

Life means that the living being is capable of reproducing itself. Therefore, every living being must exchange energy and matter with its surroundings, and must be able to react to changing conditions in its surroundings and in itself. Thus, it must be able to respond to signals from inside and outside. This is the reason why unstable systems like living beings can be influenced by information alone. As energy and matter are ultimately information, the living systems can select the suitable particles for stabilizing their system. They only need to recognize what is helpful or harmful, and must trigger an energy supply that starts the appropriate activity. This ability to distinguish and react is the first primitive form of life, and it fills the information aspect of matter and energy with meaning for this special entity, the meaning being survival (Görnitz and Görnitz 2016).

Only the being that manages to use the information in the right way can survive and have offspring. This is how information becomes meaningful. The stabilizing evaluations will result in rules for further information processing, which will ultimately be stored in the genes. These rules are stored by means of material cell components. Thus, it became possible that during biological evolution, self-stabilizing information processing could develop to more complexity.

With this augmented complexity, new phenomena like coordination of movements, for instance, came into the world. Some single-celled organisms developed ciliates and learned to coordinate the movements of them. Coordination is neither pure energy nor matter, but a surplus, a cleverness, that enabled those one-celled beings to survive better and to develop into a new species. Other unicellular organisms managed to differentiate the signals from their environment and to react to them with different behaviour. The bell-shaped monocot, for example, developed the ability to detect sudden changes from light to dark, and to quickly contract its stem into a spiral to protect itself from predators. That was the beginning of the development of our eyes.

Darwin's description of evolution as the result of mutation and selection becomes much more convincing when we realize that matter and energy are information that can initiate the development of mutations. But the beginning of such a development is always some cleverness for surviving that helps the living being to adapt to changed conditions.

Biologists devised an experiment that illustrates this thesis very clearly: They took Drosophilas, tiny fruit flies, cut away their wings, and placed them on a metal plate which they heated slowly. There was a mechanism that made it possible for the Drosophilas to influence the temperature of the plate, which the flies did not know yet. When the warmth started to become uncomfortable for the flies, they could not just fly away as they usually would have done. Therefore, they began to try all the movement patterns they knew. They moved their legs in different ways, tried to preen their wings, and whatever other possibilities they had. The mechanism of controlling the temperature was that the plate cooled down when the fly turned to the right. As soon as the flies realized this, which only took them a few seconds, they could control the temperature and survived. That shows that living beings need cleverness and creativity for surviving. Only when the reaction turns out to be advantageous are mutations adapted to the new conditions possible. This is because only those beings that found a way of adapting to the alterations could survive and procreate.

In quantum physics, we learned that elementary particles can be entangled with each other to form a new wholeness. In the same way, we must assume that some AQIs of an atom can be entangled with other AQIs to form a wholeness of information that is something entirely new. Thus, the signals from the surroundings and the signals from inside the unicellular organisms could be entangled to form new signals, which trigger the necessary reactions. And with increasing complexity, this new entity of information can be used by the living being for better orientation and reactions to its surroundings.

If one proton is formed by 10^{41} AQIs, no physicist could tell whether 100 or 1000 AQIs of an atom, for example, are combined with other AQIs to form rough information about the surroundings of the living self that help to survive better.

The more complex the animals became, the more necessary it became that a central organ coordinated all the information, both internal and external to the animal. And, with more complexity, the information was combined to form an idea of the surroundings that made survival easier. We humans are so complex that we combine the information about ourselves and even about our brain and our thinking, and realize its existence. Thus, we are conscious about it and about our self.

If we compare our thinking with what we now know about the elementary particles, then it stands to reason to assume that both arise from the same primordial condition, the protyposis. Our thoughts are often very brief and jump discontinuously from one to the other. We can have very contradictory thoughts simultaneously. We can love and hate something, depending on how we look at it, resembling the complementarity of wave and particle. And, in thinking, we recognize various possibilities for the future, which cannot all be realized at the same time.

This does not mean that our thoughts are formed by elementary particles. But the AQIs, the most basic components of the world, form our thoughts as well as the elementary particles, which constitutes the basis for a fundamental similarity between them. And energy, with the surplus of more complex structure, forms matter, which is as stable as we experience it. It is interesting in this context to reflect on the fact that, at school, we try to give more structure to the thinking of our children, to help them to survive better.

But we should not ignore that we don't even realize most of the important information processing in ourselves. For instance, if you give a person a pair of spectacles that turns the picture of the surrounding world upside down (so that the sky is below and the floor above), after a few minutes, if the person moves and walks around, the person will see these surroundings as before wearing these spectacles. We cannot find any part in the brain where biochemical reactions have adjusted the image of the surrounding again. It was done solely by a mental activity, of which the person did not notice anything, except the effect.

Also, most of our internal organs usually work completely without our consciousness. Our intestines, for example, sort our food into those substances that we can utilize and convert into energy and those that are excreted. Only when something is not working correctly are signals sent to our consciousness, to which we react with appropriate activities.

In the Old Testament, the basis of the three Abrahamic religions, it is described that God breathed his spirit into man. This is a metaphor for the fact that people felt that spirit, thinking, and mind were given by God as the basis of life. With today's knowledge of the evolution of the world, we would say that God breathed his spirit into the first single-celled organism, thus enabling it to link information from outside and inside, and thereby initiate the right activities for survival. This rewriting of the myth takes into account that only this form of intellect, the linkage of information, made life on earth possible, and unfolded in the course of evolution up to our consciousness.

References

Glasgow, Rupert D.V. 2018. *Minimal selfhood and the origins of consciousness*. Würzburg University Press.

Görnitz, Thomas. 2018. *Protyposis – An introduction – Consciousness and matter from quantum information*. München: DAS NEUE DENKEN.

Görnitz, Thomas, and Brigitte Görnitz. 2008. *Der kreative Kosmos. Geist und Materie aus Quanteninformation*. Berlin/Heidelberg: Springer Spektrum.

———. 2016. *Von der Quantenphysik zum Bewusstsein*. Berlin/Heidelberg: Springer.

Consciousness and Implicit Self-Awareness: Eastern and Western Perspectives

Rocco J. Gennaro

Abstract Some contemporary Western theories of consciousness, such as the higher-order thought (HOT) theory and self-representationalism, hold that there is an implicit (pre-reflective) "self-awareness" that accompanies each conscious mental state. Important twentieth-century historical figures, such as Sartre and Brentano, have also embraced a similar position. This view, or something very close to it, can also be found centuries earlier in some Indian philosophy (such as in Dignāga) where conscious experience is thought of as "inherently reflexive" and cognition is understood to have a "dual-aspect." In this chapter, I lay out some of the similarities and differences among these views and explain why I prefer the way that HOT theory accounts for such self-awareness as compared to the others, including HOT theory's ability to avoid the much-discussed infinite regress threat and its more plausible explanation of introspection. I also explain how HOT theory, like many Buddhist accounts, is not logically committed to the existence of an enduring substantial "self."

Keywords Consciousness · Self-awareness · Higher-order thoughts · Self · Buddhism · Self-representation

Some contemporary western theories of consciousness, such as the higher-order thought (HOT) theory (Rosenthal 2005; Gennaro 2012) and self-representationalism (Kriegel 2009), hold that there is an implicit (pre-reflective) "self-awareness" that accompanies each conscious mental state. Important twentieth-century historical figures, such as Sartre and Brentano, have also embraced a similar position. This view, or something very close to it, can also be found centuries earlier in some Indian philosophy (such as in Dignāga) where conscious experience is thought of as

R. J. Gennaro (✉)
Political Science, Public Administration and Philosophy, University of Southern Indiana, Evansville, IN, USA
e-mail: rjgennaro@usi.edu

P. S. Satsangi et al. (eds.), *Consciousness Studies in Sciences and Humanities: Eastern and Western Perspectives*, Studies in Neuroscience, Consciousness and Spirituality 8, https://doi.org/10.1007/978-3-031-13920-8_4

"inherently reflexive" and cognition is understood to have a "dual-aspect." In this chapter, I lay out some of the similarities and differences among these views and explain why I prefer the way that HOT theory accounts for such self-awareness as compared to the others, including HOT theory's ability to avoid the much-discussed infinite regress threat and its more plausible explanation of introspection. I also explain how HOT theory, like many Buddhist accounts, is not logically committed to the existence of an enduring substantial "self."

Perhaps the most fundamental overall notion of 'conscious' among philosophers is captured by Thomas Nagel's famous "what it is like" sense (Nagel 1974). When I am in a conscious mental state, there is "something it is like" for me to be in that state from the first-person point of view. When I am, for example, smelling a rose or having a conscious visual experience, there is something it seems like from my perspective. This is the sense of conscious state that I have in mind in this chapter.

1 HOT Theory and Some Close Rivals

A central question that should be answered by any theory of consciousness is: What makes a mental state a *conscious* mental state? There is a long tradition that has attempted to understand consciousness in terms of some kind of higher-order awareness. For example, John Locke (1689/1975) famously said that consciousness is the perception of what passes in a man's own mind. However, this intuition has been revived by a number of contemporary philosophers in recent decades (Lycan 1996; Rosenthal 1997, 2005). The most general idea is that what makes a mental state conscious is that it is the object of some kind of higher-order *representation* (HOR). A "mental representation" is simply meant to describe when a mental state is "directed at" or "about" something. A mental state M becomes conscious when there is an HOR of M. An HOR is a "meta-psychological" or "metacognitive" state, that is, a mental state directed at another mental state. So, for example and generally, my desire to drink some water becomes conscious when I am (non-inferentially) "aware" of the desire. It seems that conscious states, as opposed to unconscious ones, are mental states that I am "aware of" being in some sense. This is sometimes referred to as the Transitivity Principle (TP):

TP: A conscious state is a state whose subject is, in some way, aware of being in it.

Conversely, the idea that I could be in a conscious state while totally *unaware* of being in that state seems odd or perhaps even contradictory. A mental state of which the subject is completely unaware is clearly an *un*conscious state. For example, I would not be aware of having a subliminal perception and thus it is an unconscious perception. Any theory which attempts to explain consciousness in terms of higher-order states is known as a higher-order representational theory of consciousness. It is best initially to use the neutral term "representation" because there are various higher-order theories depending on how one characterizes the HOR in question. Higher-order thought (HOT) theorists, such as David Rosenthal (2005), think it is better to understand the HOR as a thought containing concepts. HOTs are treated as cognitive states involving some kind of conceptual component. Higher-order theorists are united in the belief that their approach can explain the difference between unconscious and conscious mental states better than any purely "first-order" approach which attempts to explain consciousness in terms of first-order (FO) or purely "world-directed" states (such as Tye 2000).

My own view (Gennaro 1996, 2006, 2012) is that when one has a first-order conscious state, the HOT is better viewed as self-awareness that is *intrinsic* to the target state, so that we have a complex conscious state with parts. I take this to be a version of HOT theory that I call the "wide intrinsicality view" (WIV). Unlike Rosenthal and others, I think that it is best to individuate conscious states "widely", that is, to construe the HOT as an intrinsic part of an overall complex conscious state (Gennaro 2012: chapter 4). According to the WIV, there are two parts of a single first-order conscious state with one part directed at ("aware of") the other.

In a series of papers culminating in his 2009 book, Kriegel also argues that conscious mental states themselves are, in some important sense, reflexive or self-directed, which is itself a view that goes back a long way (e.g., to Brentano and perhaps even back to Aristotle). Thus, according to Kriegel, when one has a (focal) conscious desire for a cold glass of water, one is also (peripherally) aware that one is in that very state. The conscious desire both represents the glass of water and itself. It is this "self-representing" or "implicit self-awareness" which makes the state conscious.

In the end, there are at least four positions in the relatively recent literature regarding the structure of conscious states and self-awareness which invoke a representational relation between them. For the sake of clarity and frequent comparisons, I will use the following notation from this point on:

M = a world-directed (first-order) mental state
M* = the meta-psychological or higher-order state directed at M
M** = a third-order state directed at M*

I will use the acronym "**EHOT**" for Rosenthal's theory to emphasize that M* is entirely *extrinsic* to, or distinct from, M. We can call it "extrinsic HOT theory." So let us first distinguish three positions with respect to first-order world-directed conscious states:

(**EHOT**) A mental state M of a subject S is conscious if and only if S has a *distinct or extrinsic* (unconscious) mental state M* (= a HOT) that is an appropriate representation of M.

(**WIV**) A mental state M of a subject S is conscious if and only if S has a suitable (unconscious) meta-psychological thought, M* (= MET), directed at M, such that both M and M* are *proper parts of* a complex conscious mental state, CMS.

(**SRT**) A mental state M of a subject S is conscious if and only if S has a (peripherally) conscious M* directed at M, such that both M and M* are *proper parts of* a complex conscious mental state, CMS.

It is worth emphasizing that SRT ("self-representational theory") says that the M* in question is itself conscious but only peripherally as opposed to the focal attentive nature of M (Kriegel 2009).

A fourth view, which I have called "pure self-referentialism" or **PSR** (Gennaro 2006), maintains that M* = M, that is, M is also literally directed back at itself, which seems closest to Brentano ([874] 1973) and Kriegel's earlier view (2003). Thus, it is typically the M* which is thought to be a kind of "pre-reflective" self-awareness which accompanies conscious states.

Brentano, using an example of hearing a sound, explained that every mental act includes a consciousness of itself. Every conscious mental state has a double object, a primary and secondary object so, for example, one hears a sound *and* hears the hearing of the sound. A conscious state is thus directed both outward toward an object and back at the entire state itself. Mental states are essentially intentional. Note also, however, that on HOT theory and the WIV, M* is itself typically unconscious when one is having a conscious M.

I have critiqued SRT at length in previous work (e.g., Gennaro 2012: chapter 5). For example, it seems to me that our conscious attention is often so focused at the world and its objects that it is unlikely that we are continuously consciously self-aware. The clearest examples of inattentive (or peripheral) consciousness are *outer-directed*, for example, perhaps some of the awareness in one's peripheral visual field while watching a concert or working at one's computer. Indeed, Kriegel frequently uses these kinds of examples, by analogy, to support his view. But they obviously do not show that any such peripheral consciousness is *self-directed* at the *same time* when there is outer-directed attentional consciousness, let alone that such (conscious) peripheral consciousness *always* accompanies conscious states.

Further, and for what it's worth, I confess to not finding such inner-directed peripheral consciousness alongside my outer-directed attentive experience. Except when I am *introspecting*, conscious experience is so completely outer-directed that I deny we have such peripheral self-directed consciousness when in first-order conscious states. It does not seem to me that I am consciously aware (in any sense) of my own experience when I am, say, consciously attending to a play or the task of building a bookcase. We can of course often switch back and forth between outer-directed and more sophisticated introspective conscious states. This may contribute to the (false) sense that we are continuously consciously self-aware. On HOT theory, introspection occurs when M* itself becomes conscious, which requires the presence of M**.

2 Conscious States and Indian Philosophy

Something very close to SRT, or perhaps even PSR, is interestingly also found centuries earlier in Indian philosophy whereby consciousness (*citta, vijñāna*) is taken to be "inherently reflexive" and cognition is thought to have a "dual-aspect," such as in Dignāga and others (ca. 480–540 CE) (MacKenzie 2007; Arnold 2010; Kellner

2010; Coseru 2012, 2017; Chadha 2017). It is not clear that Indian philosophers had in mind "representationalism" at least in the sense that implies the mind-independent existence of an intentional object (Finnigan 2018). Still, it is at minimum a phenomenological claim that the content of conscious experiences consists (at least in part) in a phenomenal "object-appearance." Nonetheless, there is strong sympathy with the notion that all consciousness is "about" something or "intentional" as Brentano emphasized.

Coseru explains that "in general, philosophers of the Dignāga-Dharmakīrti school accept the reality of reflexive awareness [*svasaṃvedana or svasaṃvitti*] in the classical Yogācāra sense of self-luminosity (*svaprakāśa*) and support their arguments with the example of a lamp that illuminates itself while at the same time revealing other objects" (Coseru 2012: 34). Coseru also points out how issues of translation can arise. For example, different terms are translated as 'consciousness' such as *citta* and *vijñāna*, and, interestingly, 'introspective awareness' (*manovijñāna*) is used in some contexts. "There are also at least three terms for what is ordinarily designated as 'mind': *manas* ('mental faculty'), *vijñāna* ('consciousness faculty') and *citta* ('mind' or 'thought'). The term that most generally translates as 'mind' in the Abhidharma traditions is *citta*" (Coseru 2012: 14).

Coseru describes Dignāga's view as follows: "Each cognition arises having a double aspect: it appears as an apprehending subject and as an apprehended object. In terms of its appearance to itself, cognition manifests as self-awareness (*svasaṃvitti*)" (Coseru 2012: 35). Conscious states are thus both intentional and reflexive. Conscious experiences are essentially self-aware states. Thus, the similarities between this view and PSR (and SRT) become obvious. The subjective aspect of cognition (*grāhakākāra*) is the individual's self-awareness while the objective aspect (*grāhyākāra*) is the intentional character of cognition or its object-directedness. MacKenzie refers to this view as the "Reflexivity Thesis" (RXT), that is, "conscious states simultaneously disclose both the object of consciousness and (aspects of) the conscious state itself" (MacKenzie 2007: 40) in contrast to the "Reflection Thesis" which is more like standard HOT theory. MacKenzie (2008: 248) explains that "on Dharmakīrti's [ca. 600–660] view, conscious experience fundamentally involves self-awareness (*svasaṃvedana*)…" (MacKenzie 2008: 248). Finnigan tells us that Dignāga holds that conscious experiences have "two forms [or appearances] ... the cognition [or awareness] of the object and the cognition [or awareness] of that [i.e. the cognition itself]" (Finnigan 2018: 391).

This self-directed or reflexive aspect of conscious states is similar to Sartre's (1956) "pre-reflective self-consciousness" as many commentators have noted (Gennaro 2002; Zahavi 2007; Miguens et al. 2015) in contrast to "reflective self-consciousness" (see Gennaro 2002 for more in connection to HOT theory and the WIV). When it comes to pre-reflective self-consciousness, it does not really "posit" an object or at least not a *distinct* object. Thus, Sartre calls it "non-positional self-consciousness" (Sartre 1956: 26). It is also why he uses the word 'of' [de] in parentheses merely out of 'grammatical necessity' when speaking of such non-positional self-consciousness (of) self. The worry is that the mere use of the word 'of' implies a *distinct* mental state, which it clearly is according to SRT or PSR. Something like this grammatical worry can be found in the Buddhist literature in the context of the Sanskrit language (see Finnigan 2018: 396–397, for some discussion). Grammatical distinctness need not imply genuine distinctness of objects or events. Sartre explicitly says, for example, that "Consciousness (of) pleasure is *constitutive* of the pleasure as the very mode of its own existence…We understand now why the first consciousness of consciousness is not positional; it is because it is one with the consciousness of which it is consciousness" (Sartre 1956: 13–14, emphasis added).

Some have thus argued that it is best to construe the relationship between pre-reflective self-awareness (M*) and conscious states (M) as some sort of "acquaintance" relation as opposed to "representational" (see Hellie 2007; but also Gennaro 2015a). One problem here is that it still remains the case that 'acquaintance' is often described as a kind of *relation*, e.g. a relation between a subject and a conscious state or a relation between self-awareness and a conscious state. If it is not meant to be relational at all, then the onus is on the acquaintance theorist to explain it. I agree with Kriegel that this strategy is at best substituting one problem for an even deeper puzzle, namely, just how to understand the allegedly intimate and non-representational "awareness (of)" relation (Kriegel 2009: 106–113 and 205–208). I am not sure what to make of such a "sui generis" alternative. Indeed, "acquaintance" is often taken to be entirely unanalyzable and simple, in which case it is difficult to see how it could usefully explain anything, let alone the nature or structure of conscious states (Gennaro 2015a).

3 The Regress Threat

A common initial objection to HOT theory is that it leads to an infinite regress or is circular. It might seem that HOT theory results in circularity by defining consciousness in terms of HOTs (which is a kind of "higher-order awareness"), that is, we shouldn't explain or define a concept by using that very same concept. It also might seem that an infinite regress results because a conscious mental state must be accompanied by an HOT, which, in turn, must be accompanied by another HOT *ad infinitum*. However, the standard and widely accepted reply is that when a conscious mental state (M) is a first-order world-directed state, the higher-order thought (HOT), or M*, is *not itself conscious*; otherwise, circularity and an infinite regress would follow. When the HOT is itself conscious, there is a yet higher-order (or third-order), or M**, thought directed at the second-order state. In this case, we have *introspection* which involves a conscious HOT directed at a distinct inner mental state. When one introspects, one's attention is directed back into one's own mind. This is a crucial aspect of standard HOT theory. For example, what makes my desire to write a good paper a conscious first-order desire is that there is an unconscious HOT (M*) directed at the desire. In this case, my conscious focus is still directed at the paper and my computer screen, so I am not consciously aware of having the HOT from the first-person point of view. When I *introspect* that desire, however, I have a conscious HOT (accompanied by a yet higher, third-order, HOT, M**) directed at the desire itself.

Authors and critics alike immersed in the Buddhist and Eastern literature are fully aware of the threat of a vicious infinite regress of conscious states. Indeed, it is central to some of the main arguments in question (see MacKenzie 2007, 2008; Arnold 2010; Chadha 2017; Kellner 2010, 2011; Finnigan 2018). But I think that HOT theory (and the WIV for that matter) has an advantage over PSR, SRT, and "reflexivist" views because, according to the latter theories, the self-awareness in question (M*) is itself conscious in some sense. Interestingly, however, instead of viewing an infinite regress as a potential problem for the reflexive view, Dignāga tries to use it as an argument *for* this view. The idea seems to be that a regress can really only be avoided if one keeps the implicit self-awareness (M*) in question within the first-order state (M). If the self-awareness is truly distinct and higher-order, then it must also have a further self-awareness (M**), and so on *ad infinitum*. Thus, M* must be intrinsic to M. But, of course, as we have already seen, an infinite regress would only follow if mental states must be conscious, which is most certainly rejected by supporters of HOT theory (I'll leave aside the much-discussed "memory argument," but see also Ho 2007; Thompson 2011; Finnigan 2018; Kriegel 2019).

I also think that a real deficiency for the Indian reflexive approach to pre-reflective self-awareness is that it (somewhat like SRT and PSR) does not really attempt to answer the crucial question: What *makes* a mental state a conscious mental state? Both the WIV and HOT theory are, in part, trying to *explain* how an *unconscious* mental state becomes a conscious one. Of course, many are not satisfied by the explanation offered by HOT theorists (see, e.g., Gennaro 2012: chapter four, for one detailed attempt), but my point here is that PSR and SRT do not really offer an explanation at all.

Two reasons for this may be that defenders of PSR and SRT are inclined to reject outright all reductive explanations of consciousness and even reject the very existence of unconscious mental states (Brentano [1874] 1973). I won't argue here for the existence of unconscious mental states or for the view that reductive explanations are more desirable than non-reductive accounts (see Gennaro 2012: chapter 2), but to the extent that one agrees with one or both of these assumptions, reflexive views have a serious problem compared to their rivals. (Perhaps this is really where the rock bottom disagreement lies in the end.)

For example, if we ask a defender of PSR or SRT, "What *makes* M conscious?", the response cannot be that self-awareness (M*) is directed at M because M is supposed to be *identical to* M* (e.g., according to PSR). How can M* make M conscious or *explain* M's being conscious if M* = M or even if M* is one part or aspect of M? Circularity seems unavoidable.

Moreover, either M* is itself conscious or it is not, and then the familiar threat of regress rears its ugly head. If M* is itself conscious, then what makes *it* conscious, and so on? Alternatively, if M* is not conscious, then the defender of reflexive theory would first have to acknowledge the existence of unconscious mental states, but even worse, how could M be conscious and M* be unconscious if M = M*?

In addition, according to HOT theory, "reflection" or "introspection" is present only when the HOT is itself conscious. This is, we might say, it is a level up from pre-reflective self-awareness. The transition from (non-introspective) pre-reflective self-awareness to introspective self-awareness is clear. It occurs when an unconscious M* becomes conscious and is thus directed at M. However, the problem for PSR, SRT, and Dignāga is that it is difficult to see what introspection or reflection is. What would be the structure of an introspective state? Is it yet another self-awareness embedded within each conscious state or is it a level up and directed at a first-order conscious state? If the former, is there the threat of yet another regress now *internal* to that state? If the latter, is it a distinct conscious state that is also thus self-aware? How can this view be represented in terms of the abbreviations above, namely, M, M*, and M**? One answer along the lines of the latter might be that "Even when we reflect on some mental episode, we are pre-reflectively aware of the reflection itself. The same awareness is r*eflective* in relation to its object and *non-reflective* in relation to itself" (Ho 2007: 223). In many ways, an HOT theorist can agree, that is, M* requires the presence of M** in order for M* to be conscious and an instance of introspection. However, on the PSR or SRT, we do not have an account of how a specific mental state (e.g., M*) goes *from* being a pre-reflective self-awareness *to* a reflective state of introspection. That is, there isn't a self-aware state which is itself a stepping stone to an introspective state. For one thing, there wouldn't, and couldn't, be an *unconscious* mental state which becomes conscious; rather, there would presumably have to be a new and distinct conscious state directed at M (the "awareness is reflective in relation to its object"). One advantage of, say, HOT theory (and the WIV) is that the transition has a very straightforward and natural explanation, that is, the unconscious HOT becomes itself a conscious HOT. But, on Ho's account, it is unclear what the relationship is between the pre-reflective self-awareness and the reflective awareness. If no unconscious thought becomes conscious during such a transition, then does an entirely new state, M**, emerge as directed at M (and therefore also at M*)? Is M** itself also conscious (on pain of regress)? Would M** then also be "directed back at itself," which would then require also having an M*** directed at M**, and so on?

Having unconscious HOTs can thus also be understood, perhaps even from an evolutionary perspective, as a key stepping stone to the capacity for introspection. Unconscious HOTs can presumably become conscious more quickly, resulting in introspective conscious mental states. The ability of an organism to shift quickly between outer- and inner-directed conscious states is surely a crucial practical and adaptive factor in the evolution of species. For example, an animal that is able to shift back and forth between perceiving other animals (say, for potential food or danger) and introspecting its own mental states (say, a desire to eat or a fear for one's life) would be capable of a kind of practical intelligence that would be lacking otherwise.

4 Self and "I"

We should recall that HOTs are of the form "I am in mental state M," which obviously makes reference to "oneself" or "I" in some sense. Indeed, something like an "I" or "subject" of mental states is even embodied in TP (recall the Transitivity Principle): A conscious state is a state whose *subject* is, in some way, aware of being in it. But it is of course well known that a dominant theme of much of Buddhist and Indian philosophy is that there really is "no-self." This "not-self" (*anattā* or *anātman*) view is a rejection of a real and enduring or permanent self as the agent of mental activity and bodily behavior (MacKenzie 2008; Siderits et al. 2011). We might say, then, that the word "I" doesn't really *refer* to what many people believe it refers to (e.g., an enduring substance of some kind) given a more sophisticated metaphysical view about the nature of persons. However, this is not automatically to say that we can't have thoughts about our current mental states in a much less robust sense and so we can still attach some sort of "I-concept" from moment to moment ("self-awareness") to each of our mental states. It may seem paradoxical to say that there can be "self-awareness" without a "self" or even a sense of mental state ownership, but many authors have argued convincingly that this is not only a plausible view but the very one held by Dignāga and other Indian philosophers (e.g., MacKenzie 2008; see also Chadha 2018). The self-awareness in question is meant to apply only to individual conscious experiences or episodes. This need not imply that there is an enduring thing or substance within which they inhere. As we saw earlier, we should be cautious in drawing metaphysical conclusions solely on the basis of grammatical or linguistic facts. In this case, grammatical use of the term 'I' need not even imply a genuine referent.

Of course, even in Western philosophy, some have been very skeptical of the notion that there is a single enduring and unified self in the way that, say, Descartes and others have assumed there to be (such as Hume [1739] 1975, and, more recently, Dennett 1991 and Metzinger 2003). We could simply be wrong in supposing that there is a self to which all my individual mental states are "attached" in some sense, especially if this means some stable underlying substance or thing in addition to individual conscious states. Perhaps it is more important to explain the mere *appearance* of the same "I," that is, the subjective impression of a "subject unity." This line of argument is more consistent with, say, Dennett's view that the self is merely a "narrative" that we construct over time. Similarly, Hume's so-called "bundle theory," whereby a "person" is at best a bundle of sensations or perceptions, cannot be ruled out. All of this is consistent with the fact that we likely must use the word "I" as a practical matter of communication.

It seems to me that an HOT theorist can actually agree with the above line of thinking, including the Indian no-self view. For his own part, Rosenthal (2003) argues that the higher-order thought (HOT) theory of consciousness also need not settle this deeper metaphysical problem in order to explain conscious states. Still, he argues that HOT theory can help to account for the sense of subject unity, that is, why it at least *seems* to us that we are each a unified enduring self. HOT theory says that what makes a mental state conscious is that it is the object of some kind of a higher-order thought of the form "I am in mental state M." Indeed, unlike other theories of consciousness, self-reference through the self-concept "I" accompanies each and every conscious state. Rosenthal thus explains that "each HOT refers not only to such a [mental] state, but also to oneself as the individual that's in that state. This reference to oneself is unavoidable" (2003: 330). That is, when we have different conscious states, we are employing a self-concept ("I") in each case. This naturally leads us to think, or simply assume, that there is a single self or subject behind the appearances. Still, as Rosenthal points out, it doesn't automatically follow that there really is a unified subject or self. However, when we are engaged in *introspection*, the sense of "I" is increased because, according to HOT theory, one is now *consciously* thinking that "I am in a mental state" (see also Gennaro 2012: 294–302).

Chadha (2018) argues that the Abhidharma 'no-self' view can be plausibly interpreted as a no-ownership view according to which there is no locus or subject of experience and thus no owner of mental or bodily awareness. On the more phenomenological interpretation of this no-self view, Abhidharma Buddhist metaphysicians are committed to denying the ownership of experiences. That is, there is "nothing that it is like" to be an owner of experiences in the sense that there is no experiential phenomenology associated with the ownership of experience.

It is worth noting that some have also argued that various depersonalization disorders and psychopathologies of self-awareness, such as somatoparaphrenia, Cotard's syndrome, and thought insertion in schizophrenia, threaten the viability of the higher-order thought (HOT) theory of consciousness since it requires an HOT about *one's own* mental state to accompany every conscious state. Some of these objections and replies can be found in Gennaro (2015b, 2020) as well as in the references therein.

5 Conclusion

I agree that conscious states are always accompanied by an implicit or pre-reflective self-awareness. However, for the reasons above, I think it is far better to construe such self-awareness as unconscious in a manner similar to HOT theory, especially in order to avoid the infinite regress threat and to provide a clear account of introspective or reflective states. In addition, HOT theory, like many Buddhist accounts, is not necessarily logically committed to the existence of an enduring substantial "self."

References

Arnold, Dan. 2010. Self-awareness (svasaṃvitti) and related doctrines of Buddhists following Dignāga: Philosophical characterizations of some of the main issues. *Journal of Indian Philosophy* 38: 323–378.

Brentano, Franz. 1874/1973. *Psychology from an empirical standpoint*. New York: Humanities.

Chadha, Monima. 2017. Inner awareness is essential to consciousness: A Buddhist-Abhidharma perspective. *Review of Philosophy and Psychology* 8: 83–101.
———. 2018. No-self and the phenomenology of ownership. *Australasian Journal of Philosophy* 96: 14–27.
Coseru, Christian. 2012. *Perceiving reality: Consciousness, intentionality, and cognition in Buddhist philosophy*. New York: Oxford University Press.
———. 2017. Mind in Indian Buddhist philosophy. In *The Stanford Encyclopedia of philosophy*, ed. Edward N. Zalta. https://plato.stanford.edu/archives/spr2017/entries/mind-indian-buddhism/.
Dennett, Daniel Clement. 1991. *Consciousness explained*. New York: Little, Brown, and Co.
Finnigan, Bronwyn. 2018. Is consciousness reflexively self-aware?: A Buddhist analysis. *Ratio* 31: 389–401.
Gennaro, Rocco J. 1996. *Consciousness and self-consciousness: A defense of the higher-order thought theory of consciousness*. Amsterdam and Philadelphia: John Benjamins Publishers.
———. 2002. Jean-Paul Sartre and the HOT theory of consciousness. *Canadian Journal of Philosophy* 32: 293–330.
———. 2006. Between pure self-referentialism and the (extrinsic) HOT theory of consciousness. In *Self-representational approaches to consciousness*, ed. Uriah Kriegel and Kenneth Williford. Cambridge: MIT Press.
———. 2012. *The consciousness paradox: Consciousness, concepts, and higher-order thoughts*. Cambridge: MIT Press.
———. 2015a. The 'of' of intentionality and the 'of' of acquaintance. In *Pre-reflective consciousness: Sartre and contemporary philosophy of mind*, ed. Sofia Miguens, Gerhard Preyer, and Clara Bravo Morando, 317–341. New York: Routledge Publishers.
———. 2015b. Somatoparaphrenia, anosognosia, and higher-order thoughts. In *Disturbed consciousness: New essays on psychopathology and theories of consciousness*, ed. Rocco J. Gennaro. Cambridge: The MIT Press.
———. 2020. Cotard syndrome, self-awareness, and I-concepts. *Philosophy and the Mind Sciences* (Special Issue: *Radical Disruptions of Self-Consciousness*) 1: 1–20.
Hellie, Benj. 2007. Higher-order intentionalism and higher-order acquaintance. *Philosophical Studies* 134: 289–324.
Ho, Chien-hsing. 2007. Consciousness and self-awareness. *Asian Philosophy*. 17: 213–230.
Hume, David. [1739] 1975. *A treatise of human nature*. L. ed. A. Selby-Bigge and Peter Harold Nidditch. Oxford: Clarendon Press.
Kellner, Birgit. 2010. Self-awareness (svasaṃvedana) in Dignāga's Pramāṇasamuccaya and –vṛtti: A close reading. *Journal of Indian Philosophy* 38: 203–231.
———. 2011. Self-awareness (svasaṃvedana) and infinite regresses: A comparison of arguments by Dignāga and Dharmakīrti. *Journal of Indian Philosophy* 39: 411–426.
Kriegel, Uriah. 2003. Consciousness as intransitive self-consciousness: Two views and an argument. *Canadian Journal of Philosophy* 33: 103–132.
———. 2009. *Subjective consciousness. A self-representational theory*. New York: Oxford University Press.
———. 2019. Dignāga's argument for the awareness principle: An analytic refinement. *Philosophy East & West* 69: 143–155.
Locke, John. [1689] 1975. *An essay concerning human understanding*. Peter Harold Nidditch. Ed. Oxford: Clarendon.
Lycan, William G. 1996. *Consciousness and experience*. Cambridge: MIT Press.
MacKenzie, Matthew D. 2007. The illumination of consciousness: Approaches to self-awareness in the Indian and Western traditions. *Philosophy East and West* 57: 40–62.
———. 2008. Self-awareness without a self: Buddhism and the reflexivity of awareness. *Asian Philosophy* 18: 245–266.
Metzinger, Thomas. 2003. *Being no one: The self-model theory of subjectivity*. Cambridge: MIT Press.
Miguens, Sofia, Gerhard Preyer, and Clara Morando, eds. 2015. *Pre-reflective consciousness: Sartre and contemporary philosophy of mind*. New York: Routledge Publishers.

Nagel, Thomas. 1974. What is it like to be a bat? *The Philosophical Review* 83: 435–456.
Rosenthal, David M. 1997. A theory of consciousness. In *The nature of consciousness*, ed. Ned Block, Owen J. Flanagan, and Guven Guzeldere, 729–753. Cambridge: MIT Press.
———. 2003. Unity of consciousness and the self. *Proceedings of the Aristotelian Society* 103: 325–352.
———. 2005. *Consciousness and mind*. New York: Oxford University Press.
Sartre, Jean-Paul. 1956. *Being and nothingness*. New York: Philosophical Library.
Siderits, Mark, Evan Thompson, and Dan Zahavi, eds. 2011. *Self, no self?: Perspectives from analytical, phenomenological, and Indian traditions*. New York: Oxford University Press.
Thompson, Evan. 2011. Self-no-self? Memory and reflexive awareness. In *Self, no self?: Perspectives from analytical, phenomenological, and Indian traditions*, ed. Mark Siderits, Evan Thompson, and Dan Zahavi, 157–175. New York: Oxford University Press.
Tye, Michael. 2000. *Consciousness, color, and content*. Cambridge: MIT Press.
Zahavi, D. 2007. The Heidelberg school and the limits of reflection. In *Consciousness: From perception to reflection in the history of philosophy*, ed. Sara Heinämaa, Vili Lähteenmäki, and Pauliina Remes, 267–285. Dordrecht: Springer.

The Case for Panpsychism

Philip Goff

Abstract Panpsychism is the view that consciousness is a fundamental and ubiquitous feature of the physical world. This chapter outlines two major arguments for panpsychism, one in terms of its role in solving the hard problem of consciousness, and two the intrinsic nature argument. It also responds to the worry that panpsychism is too counterintuitive to be true.

Keywords Panpsychism · Consciousness · Materialism · Dualism · Hard problem

According to early twenty-first century Western common sense, the mental doesn't take up very much of the universe. Most folk assume that it exists only in the biological realm, specifically, in creatures with brains and nervous systems. Panpsychists deny this bit of common sense, believing that mentality is a fundamental and ubiquitous feature of the universe. Mind is everywhere (which is what 'panpsychism' translates as).

There have been panpsychists in Western philosophy since at least the pre-Socratics of the seventh century BC, and the view achieved a certain dominance in the nineteenth century. Panpsychism fared less well in the twentieth century, being almost universally dismissed by Western philosophers as absurd, if it was ever thought about at all.

However, this dismissal was arguably part and parcel of the anti-metaphysics scientism of the period: the attempt to show that any questions which cannot be answered by scientific investigation are either trivial or meaningless. This project failed, and metaphysics is back in a big way in academic philosophy. At the same time, there is a growing dissatisfaction with the physicalist approaches to

This essay is a slightly modified reprint of Goff (2017a).

P. Goff (✉)
Department of Philosophy, Durham University, Durham, UK
e-mail: philip.a.goff@durham.ac.uk

P. S. Satsangi et al. (eds.), *Consciousness Studies in Sciences and Humanities: Eastern and Western Perspectives*, Studies in Neuroscience, Consciousness and Spirituality 8, https://doi.org/10.1007/978-3-031-13920-8_5

consciousness which dominated the late twentieth century, and a sense that a radically new approach is called for. In this climate panpsychism is increasingly being taken up as a serious option, both for explaining consciousness and for providing a satisfactory account of the natural world.

1 The Essence of Panpsychism

Panpsychism is sometimes caricatured as the view that fundamental physical entities such as electrons have thoughts; that electrons are, say, driven by existential angst. However, panpsychism as defended in contemporary philosophy is the view that *consciousness* is fundamental and ubiquitous, where to be conscious is simply to have *subjective experience* of some kind. This doesn't necessarily imply anything as sophisticated as thoughts.

Of course in human beings consciousness is a sophisticated thing, involving subtle and complex emotions, thoughts, and sensory experiences. But there seems nothing incoherent with the idea that consciousness might exist in some extremely basic forms. We have good reason to think that the conscious experiences a horse has are much less complex than those of a human being, and the experiences a chicken has are much less complex than those of a horse. As organisms become simpler perhaps at some point the light of consciousness suddenly switches off, with simpler organisms having no subjective experience at all. But it is also possible that the light of consciousness never switches off entirely, but rather fades as organic complexity reduces, through flies, insects, plants, amoeba, and bacteria. For the panpsychist, this fading-whilst-never-turning-off continuum further extends into inorganic matter, with fundamental physical entities – perhaps electrons and quarks – possessing extremely rudimentary forms of consciousness, which reflects their extremely simple nature (Goff 2019).

2 Reasons to Believe Panpsychism I: Solving the Hard Problem of Consciousness

Panpsychism offers the hope of an extremely elegant and unified picture of the world. In contrast to substance dualism (the view that the universe consists of two kinds of substance, matter and mind), panpsychism does not involve minds popping into existence as certain forms of complex life emerge, or else a soul descending from an immaterial realm at the moment of conception. Rather, it claims that human beings are nothing more than complex arrangements of components that are already present in basic matter. The only way in which panpsychism differs from physicalism is that the basic components of the material world also involve very basic forms of consciousness, from which the more complex conscious experience of humans and other animals derives.

Physicalists believe that consciousness can be fully accounted for in terms of physical entities and processes. But many scientists and philosophers agree that at present we have not the faintest idea how to make sense of experience being generated from material activity such as the firings of neurons. This is the difficulty David Chalmers famously called 'the hard problem of consciousness'. Physical mechanisms are well-suited for the explanation of physical behaviour; but it's hard to make sense of a mechanistic explanation of *subjective experience*. No matter how complex the mechanism, it seems conceivable that it might have functioned in the absence of any experience at all, which seems to imply that mechanistic explanations shed no explanatory light on the existence of experience.

Of course there is much more to be said about whether or not physicalism is a viable project. But, given the deep difficulties associated with the attempt to fully account for consciousness in physical terms, and the deep philosophical doubts about whether this is even a coherent idea, it is perhaps a good idea to explore other options. And the panpsychist offers an alternative research programme. Rather than trying to account for consciousness entirely in terms of non-conscious elements, panpsychism tries to explain the complex consciousness of humans and other animals in terms of simpler forms of consciousness which are postulated to already exist in simpler forms of matter. This research project is still in its infancy. But a number of leading philosophers and neuroscientists are now finding that working within a panpsychist framework bears fruit. (To take one example, see Mørch's account of Integrated Information Theory, Mørch 2017.) The more fruit is borne by this alternative research programme, the more reason we have to accept panpsychism (Roelofs 2019).

Physicalists may object: 'Just because we haven't yet worked out how to give a mechanistic explanation of consciousness, it doesn't follow that such an explanation will be forever beyond our grasp. Scientists before Darwin had no explanation of the emergence of complex life, which led many to suppose that there must be something divine or miraculous in the existence of life. The genius of Darwin was to come up with the idea of natural selection, which removes the need for divine creation in the biological realm. We just need a "Darwin of consciousness" to come along and do something similar in the mental realm.' This kind of objection is often accompanied by a certain narrative of the history of science, according to which phenomenon after phenomenon was declared inexplicable by philosophers, only to be later explained by the relentless march of science.

However, to adopt panpsychism is *not* to give up on the attempt to explain consciousness scientifically. Rather, panpsychism is a scientific research programme in its own right. Panpsychists do not simply declare animal and human consciousness a sacred mystery which must have arrived by magic. Instead, they try to explain animal and human consciousness in terms of more basic forms of consciousness: the consciousness of basic materials entities, such as quarks and electrons. It is true that consciousness itself is not explained in terms of anything more fundamental: the basic consciousness of basic physical entities is a fundamental postulate of the theory. But there is no reason to think that science must always follow the most reductionist path. The scientific explanation of electromagnetism which eventually emerged in the nineteenth century involved the postulation of new fundamental properties and forces: electromagnetic ones.

Perhaps the scientific explanation of human consciousness, when it eventually arrives, will be similarly non-reductive in postulating fundamental kinds of consciousness.

3 Reasons to Believe Panpsychism II: The Intrinsic Nature Argument

In the public's mind, physics is on its way to giving us a complete account of the fundamental nature of the material world. It's taken to be almost tautological that 'physics' is developing the true theory of 'the physical', and hence that it is to physics that we should turn for a complete understanding of the nature of space, time, and matter. However, this commonplace opinion concerning the comprehensiveness of the explanatory reach of physical science comes under pressure when we reflect on the austere vocabulary in terms of which physical theories are framed.

A crucial moment in the scientific revolution was Galileo's declaration that "the book of the universe is written in the language of mathematics". From that point onwards mathematics has been the language of physics. The vocabulary of physics is arguably not *entirely* mathematical, since it involves causal notions (such as the notion of a *law of nature*); but the kind of *qualitative* concepts found in the Aristotelian characterisation of the universe before the scientific revolution – ideas, for example, of colours and tastes – are wholly absent from modern physics. Physical theories are nothing more than mathematical models of physical causation.

The problem is that it's not clear that such an austere vocabulary is capable of giving an adequate characterisation even of the nature of matter, let alone of the nature of experience, since mathematical models are mere tools for prediction. A mathematical model in economics, for example, abstracts away from the concrete reality of what is really going on – what is being bought or sold, and what actual jobs people are doing. It is simply an artificial representation that can be used for predicting certain outcomes. This is exactly what physics does to matter. Electrons are real, concrete entities. And yet physics abstracts from the concrete reality of the electron, presenting us with an abstract model that enables us to predict its behaviour. As Bertrand Russell put it, "Physics is mathematical not because we know so much about the physical world, but because we know so little" (Russell [1927a] 1951: 163; see also Goff 2017b).

This difficulty arising from the austerity of the physicists' way of speaking about the physical world might be evaded if we had a correspondingly austere conception of physical reality itself. *Causal structuralists* have such a conception. They believe that there is nothing more to the nature of a physical entity, such as an electron, than how it is disposed to behave: if you understand what an electron *does* you know everything there is to know about its nature. On this view things are not so much *beings* as *doings*. If you assume causal structuralism, it becomes plausible that the models of physics can completely characterise the nature of physical entities; a mathematical model can capture what an electron does, and in doing so will tell us what the electron is.

However, there are powerful arguments against causal structuralism. Most discussed is the worry that causal structuralist attempts to characterise the nature of matter lead either to a vicious regress or a vicious circle. According to causal structuralists, we understand the nature of a disposition only when we know the behaviour to which it gives rise when it is manifested. For example, the manifestation of flammability is burning; we only know what flammability is when we know that it's manifested through burning. However, assuming causal structuralism, the manifestation of any disposition will be another disposition, and the manifestation of that disposition will be another disposition, and so on *ad infinitum*. The buck is continually passed, and hence an adequate understanding of the nature of any property is impossible, even for an omniscient being. In other words, a causal structuralist world is unintelligible.

Let us try to make this clear with an example. According to general relativity, mass and spacetime stand in a relationship of mutual causal interaction: mass curves spacetime, and the curvature of spacetime in turn affects the behaviour of objects with mass (as matter tends, all things being equal, to follow geodesics through spacetime). What is mass? For a causal structuralist, we know what mass is when we know what it does, i.e. when we know the way in which it curves spacetime. But to really understand what this amounts to metaphysically, as opposed to being able merely to make accurate predictions, we need to know what spacetime curvature is. What is spacetime curvature? For a causal structuralist, we understand what spacetime curvature is only when we know what it does, which involves understanding how it affects objects with mass. But we understand this only when we know what mass is. And so we find ourselves in a classic Catch 22: we can understand the nature of mass only when we know what spacetime curvature is, but we can understand the nature of spacetime curvature only when we know what mass is. G.K. Chesterton said that "[w]e cannot all live by taking in each other's washing" (Chesterton [1920] 2020: n.p.), and Russell played on this idea in articulating his worry about circularity: "There are many possible ways of turning some things hitherto regarded as 'real' into mere laws concerning the other things. Obviously there must be a limit to this process, or else all the things in the world will merely be each other's washing" (Russell [1927b] 1992: 325).

This argument presses us to the conclusion that there must be more to physical entities than what they do: physical things must also have an 'intrinsic nature', as philosophers tend to put it. However, given that physics is restricted to telling us only about the *behaviour* of physical entities – electrons, quarks, and indeed spacetime itself – it leaves us completely in the dark about their intrinsic nature. Physics tells us what matter *does*, but not what it *is*.

What then *is* the intrinsic nature of matter? Panpsychism offers an answer: consciousness. Physics describes matter 'from the outside', that is to say, physics gives us rich information about the behaviour brought about by mass, spin, charge, etc. But there must be more to what something is than what it does; and according to panpsychism, mass, spin, charge, etc., are, in their intrinsic nature, forms of consciousness.

What reasons do we have to accept this proposal? Firstly, it's not clear that there's an alternative, as it's not clear that we have a positive conception of *any* intrinsic properties beyond those we know about in our own conscious experience (i.e. beyond the properties of the experiences themselves). So the available choice seems to be between the panpsychist view as to the intrinsic nature of matter, and the view that matter is, as John Locke put it, "we know not what" (Locke [1690] 2015: n.p.). So insofar as we seek a picture of reality without gaps, panpsychism may be our only option. The great physicist Arthur Eddington (the first scientist to confirm general relativity) thought this argument enough to embrace panpsychism, suggesting that given that we can know nothing from physics of the intrinsic nature of matter, it was rather 'silly' to suppose that its nature is incongruent with mentality, and then to wonder where mentality comes from!

Furthermore, panpsychism looks to be the most theoretically virtuous theory of matter consistent with the data. I call this the 'simplicity argument' for panpsychism. We know that some material entities – brains – have an intrinsically consciousness-involving nature (assuming that Descartes was wrong about the mind being separate from the brain). We have no clue as to the intrinsic nature of any other material entities. And so the most simple, elegant, parsimonious hypothesis is that the nature of the stuff outside of brains is continuous with that of brains, in also being consciousness-involving. Arguably, then, the reality of our consciousness supports the truth of panpsychism in much the same way that the Michelson-Morley discovery that the speed of light is measured to be the same in all frames of reference supports special relativity: in both cases the theory is the most elegant account of the data.

Of course, merely saying that the intrinsic nature of matter is 'consciousness' does not give us an understanding of the specific intrinsic nature of any given physical property. What kind of consciousness is mass, as opposed to the consciousness of negative charge? What is it like to be an electron? These are questions for the panpsychist research project to address over the long term. Panpsychism is a broad theoretical framework, and it will take time to fill in the details. Similarily, it took a couple of centuries for the Darwinian paradigm to get to DNA.

4 But Isn't It Crazy?

Panpsychism is increasingly being taken seriously in both philosophy and science, but it is still not unknown for panpsychists to receive the odd incredulous stare. The supposition that electrons have some form of consciousness, albeit extremely basic, is still thought by many to be just too crazy to take seriously.

This may be the result of a mixture of cultural factors. The rejection of idealism was one major motivation in the founding of analytic philosophy, and an intuitive distrust of related views such as panpsychism still hangs heavy. Another factor is the widespread public perception that physics is on its way to giving us a complete picture of the nature of everything. There is little understanding of the difficulties which arise when we reflect on the austere vocabulary of the physical sciences and of the dubious coherence of physicalist accounts of consciousness. In the mindset of thinking that physics is on its way to giving a complete story of the universe, a consciousness-filled universe seems extremely improbable, as this doesn't seem to be what physics is telling us. But if we accept that physics tell us nothing about the intrinsic nature of matter, and indeed that the only thing we really know about the intrinsic nature of matter is that some of it involves consciousness, panpsychism starts to look much more plausible.

It is certainly true that in popular culture views which sound a bit like panpsychism have been defended with rather unrigorous reasoning. But it shouldn't need to be pointed out that just because a view has been defended with all sorts of bad arguments, it doesn't follow that there are no good arguments for that same view. Serious philosophy requires us not to indulge in flights of fancy; but it also demands that we approach the arguments without prejudice.

At the end of the day, 'common sense intuition' should have little sway against a view which pulls its weight theoretically. The view that the world is (more-or-less) round; that we have a common ancestor with apes; that time slows down the faster you move – all of these ideas were or are wildly counter to common sense, but clearly that counts little, if at all, against their truth. If panpsychism can provide us with a plausible account of human consciousness and/or a coherent account of the intrinsic nature of matter, then we have good reason to take it very seriously indeed.

References

Chesterton, Gilbert Keith. [1920] 2020. *The superstition of divorce*. https://www.gutenberg.org/ebooks/62680. Accessed 1 Aug 2022.

Goff, Philip. 2017a. The case for Panpsychism. *Philosophy Now* 121: 6–8.

———. 2017b. *Consciousness and fundamental reality*. Oxford: Oxford University Press.

———. 2019. *Galileo's error: Foundations for a new science of consciousness*. New York: Pantheon Books.

Locke, John. [1690] 2015. *An essay concerning human understanding*. https://web.archive.org/web/20150309025534/http://oregonstate.edu/instruct/phl302/texts/locke/locke1/Essay_contents.html. Accessed 01 Aug 2022.

Mørch, Hedda Hassel. 2017. The integrated information theory of consciousness. *Philosophy Now* 121: 12–16.

Roelofs, Luke. 2019. *Combining minds*. Oxford: Oxford University Press.

Russel, Bertrand. [1927a] 1951. *An outline of philosophy*. London: Routledge.

———. [1927b] 1992. *The analysis of matter*. London: Routledge.

A Many-Sided Brain: The Jain Approach to Studying Consciousness

Andrea Diem-Lane

Abstract Science, by definition, is an open-ended approach to understand how the world works by observation, experimentation, and modeling. But underlining this human process is our ability to be wrong, to be corrected, and to change our views over time. As such, science invites contravening ideas that can compete with each other. This allows for its progressive nature, since we can then determine which hypothesis better describes a given phenomenon and its range of behavior. While Francis Bacon (1561–1626) is often regarded as the first Western thinker to properly codify what is commonly known today as the scientific method, it may come as a surprise to learn that Jain philosophy, which dates back to before the sixth century BCE, has three core beliefs that are contributive and elemental for any budding scientist: Anekāntavāda (अनेकान्तवाद), Syādvāda (स्याद्वाद), and Ahimsā (अहिंसा). These three core concepts of Jainism, I would suggest, are also – at least in some measure – part and parcel of a progressive scientific mind. In this chapter, we will be taking a Jainist approach to the current study of consciousness. Our goal is to survey the most viable theories on how and why self-awareness evolved and see what their respective strengths and weaknesses are.

Keywords Anekāntavāda · Syādvāda · Ahimsā · Jainism · Consciousness · Science

Science, by definition, is an open-ended approach to understand how the world works by observation, experimentation, and modeling. But underlining this human process is our ability to be wrong, to be corrected, and to change our views over time.

As such, science invites contravening ideas that can compete with each other. This allows for its progressive nature, since we can then determine which

A. Diem-Lane (✉)
Department of Philosophy, Mt. San Antonio College, Los Angeles, CA, USA
e-mail: adiem@mtsac.edu

P. S. Satsangi et al. (eds.), *Consciousness Studies in Sciences and Humanities: Eastern and Western Perspectives*, Studies in Neuroscience, Consciousness and Spirituality 8, https://doi.org/10.1007/978-3-031-13920-8_6

hypothesis better describes a given phenomenon and its range of behavior.[1] While Francis Bacon (1561–1626)[2] is often regarded as the first Western thinker to properly codify what is commonly known today as the scientific method, it may come as a surprise to learn that Jain philosophy, which dates back to before the sixth century BCE, has three core beliefs that are contributive and elemental for any budding scientist:

1 Anekāntavāda (अनेकान्तवाद), Syādvāda (स्याद्वाद), and Ahimsā (अहिंसा)

The first concept, Anekāntavāda, which means "many-sidedness," states that truth, like an ocean, has varying features and that our approaches to understand reality are inevitably partial or limited. Thus, whatever stance or position we take must be bracketed, keeping in mind that others may have perceived what we ourselves have not. It engenders an openness to listen to other points of view and seriously take them into consideration. The second concept, Syādvāda, is generally translated to mean that all final appraisements or judgments are tentative since what is theoretically proffered is understood to be potentially uncertain, as in it "may be" or "perhaps could be" or "let's wait and see." As in the famous parable of the blind men and an elephant, each believes that what they touch reveals the true nature of the animal. But each man, by his limited feeling, only grasps a small part of the totality of the animal. Likewise, humans in their quest for knowledge always come up short and thus should hold back on any final adjudication, lest they like the blind man confuse a trunk for an entire head, or a twig for a tree, or a wave for an ocean. The third concept, Ahimsā (अहिंसा), which literally translated means "not to injure or harm," is the most widely known Jain ideal and has far-reaching implications. While it is generally viewed as not hurting other sentient beings, Ahimsā also applies to how we treat the thoughts and ideas of others. (For a succinct introduction to Jainism and various studies of the religious philosophy for the general reader, see Diem-Lane (2015, 2016)).

Do we in sharing our ideas give wide berth to what others believe and do we engage with them in a respectful and considered fashion? These three core concepts of Jainism, I would suggest, are also – at least in some measure – part and parcel of

---

[1] Thomas Henry Huxley provides an apt definition of science which has wider implications than the typical narrow cliché of scientists in white coats doing experiments. He argues that everyone does science to some degree, whether baking a pie or testing out a new bicycle or navigating an open ocean in a boat – provided that they learn from their work and acknowledge their mistakes and try to find out how to do better (Huxley 2016).

[2] A famous quote from Bacon underlines his reasoning: "Primo enim paranda est Historia Naturalis et Experimentalis, suffidens et bona; quod fundamentum rei est: neque enim fingendum, aut excogitandum, sed inveniendum, quid natura faciat aut ferat." Translation A. D.-M.: "For first of all we must prepare a Natural and Experimental History, sufficient and good; and this is the foundation of all; for we are not to imagine or suppose, but to discover what nature does or may be made to do." See *The Philosophical Works of Francis Bacon* 1858: 127.

a progressive scientific mind. First, science is predicated to a large degree on tolerating and encouraging multiple points of view which can be tested over time (Anekāntavāda), Second, science is never absolute since any theory, even if well established for centuries (e.g., Newton's law of gravity), can be changed, corrected, and augmented by new discoveries and new information (e.g., Einstein's general theory of relativity). It is a systematic process which at its center is always tentative and potentially uncertain (Syādvāda). And, third, for any scientific endeavor to flourish it must treat alternative concepts and speculations without prejudice and give sufficient latitude so that new ideas can be properly tested and not dismissed prematurely. In other words, while many hypotheses may indeed be rejected for lack of convincing evidence, respect and toleration (Ahimsā) must be accorded lest science devolve into the cesspool of dogmatism. In this presentation, we will be taking a Jainist approach to the current study of consciousness, which far too often has been parochialized into various warring camps. Our goal is to understand how and why self-awareness evolved and see what respective strengths and weaknesses various theories hold. Furthermore, following the lead of Niels Bohr, the renowned Nobel Prize-winning physicist from Denmark, we want to see how the concept of complementarity (reveal one part only to conceal another and vice versa) can also be applied to different theories on consciousness and why the field is perhaps trapped in a Heisenberg-like uncertainty conundrum (Lane 2011).[3]

2 The Mysterianism Position | Ahimsā

We are currently at an impasse in the study of consciousness since no one theory has emerged as a clear front-runner. Rather, there are a number of viable contenders, each taking distinctively different approaches. Perhaps the key defining moment in the past 30 years has been the acceptance that the study of self-awareness is not only a worthy pursuit (which had too long been neglected in the hard sciences) but that it is a peculiarly "hard" problem (Chalmers 1997).

This is so because the interiority of consciousness, the subjective sense of what it is like to be something (Nagel 1974),[4] cannot easily be objectified or quantified. Indeed, some neuroscientists, such as Sam Harris, argue that we cannot find evi-

[3] David Christopher Lane and I have written extensively on consciousness and why any objective study of its innately subjective qualia appears paradoxical because of its Mobius-like nature. See Lane October 2011: "Perhaps the Mobius strip is a useful, albeit limited, metaphor here to invoke since its unusual properties on first sight boggle the mind: 'a surface with only one side and only one boundary component. It has the mathematical property of being non-orientable.' Analogously, the difficulty we have with studying consciousness is precisely that we cannot communicate what it is without losing the very quality that makes it such."

[4] As Nagel elaborates, "If physicalism is to be defended, the phenomenological features must themselves be given a physical account. But when we examine their subjective character it: seems that such a result is impossible. The reason is that every subjective phenomenon is essentially connected with a single point of view, and it seems inevitable that an objective, physical theory will abandon that point of view" (Nagel 1974: 437).

dence that consciousness as such exists in the world outside of our own unique experience of it. As Harris explains in his essay, *The Mystery of Consciousness* (Harris 2011):

> Physical events are simply mute as to whether it is 'like something' to be what they are. The only thing in this universe that attests to the existence of consciousness is consciousness itself; the only clue to subjectivity, as such, is subjectivity. Absolutely nothing about a brain, when surveyed as a physical system, suggests that it is a locus of experience. Were we not already brimming with consciousness ourselves, we would find no evidence of it in the physical universe – nor would we have any notion of the many experiential states that it gives rise to. The painfulness of pain, for instance, puts in an appearance only in consciousness. And no description of C-fibers or pain-avoiding behavior will bring the subjective reality into view.[5] (Harris 2011: n. p.)

Although Harris is not necessarily resistant to a purely physicalist explanation of self-awareness, he doesn't see how current scientific models can adequately do justice to the subject. David Chalmers, the distinguished Australian philosopher, argues that materialist science is insufficient to explain consciousness and that we are confronted with an "explanatory gap" between the objective and subjective worlds we witness.

Dovetailing, at least in part, with Harris and Chalmers, is philosopher Colin McGinn,[6] who argues that humans are incapable of properly explaining consciousness. This position, somewhat misleadingly (because it encompasses differing versions), has been called the "new mysterianism," since the mystery of self-awareness will remain such given our limited cranial apparatuses. What all three thinkers have in common, however, is that the study of consciousness cannot simply be reduced to something else (like water to hydrogen and oxygen molecules) without missing an elemental component. They sense that some philosophers (particularly Dennett 1991) are using verbal sleight of hand to explain away the phenomenal and ineffable aspects of awareness. More precisely, to accurately appraise self-awareness one must by necessity accept the phenomenon on its own terms and not "cheaply" (to borrow Daniel Dennett's pithy phrase, if slightly ironic in this context) reduce it piece by piece into something it is not. The Jainist concept of Ahimsā (non-violence and non-injury) seems appropriate here, since whenever we prematurely reduce a complex issue (thereby losing the unique feature intrinsic to it), we not only misunderstand and obfuscate the problem, but we also too easily damage future investigations by our rushed judgments. A good, historical example can be found in psychology, where an overreliance on Freudian theories and later the restrictive empiricism of behaviorism hindered research into the human mind and its potentials.[7]

[5] It is of interest to note that Sam Harris is a longtime meditator and has been highly influenced by Buddhist and Hindu insights on the nature of mind and its place in the cosmos. He is particularly keen on mindfulness and has written and talked about the subject extensively. Refer to Harris 2015.

[6] Colin McGinn and John Searle have written extensively on consciousness and have been wary of most attempts to solve the riddle of its emergence. Ironically, both have been discharged from their respective universities (Miami and Berkeley) because of allegations of sexual harassment.

[7] Freud's life and work are a double-edged sword. While many of his ideas have been severely criticized and much has been made about his "cult of psychoanalysis" (Crews 2017), Freud seems to

Science progresses best when it doesn't ad hoc explain away variegated structures with such non-informational starters as "hallucinations." John Lilly, famous for his work with dolphins and inventing water-based deprivation tanks, got in a heated debate with his friend, the distinguished Nobel Prize-winning physicist Richard Feynman of Cal Tech, over whether out of body experiences (OBEs) were apprehending a higher reality or were merely deceptions conjured up by a mind absent of incoming stimuli. Lilly believed that it was unscientific to dismiss such OBEs as hallucinations, whereas Feynman argued that they were precisely that:

> Feynman countered Lilly's assertion by pointing out that whatever out-of-body experiences he was having (and Feynman recalls becoming very good at) didn't correlate to the outside world, even when undergoing the dissociation, he thought they did. It was for this reason that he tried to convince Lilly that 'the imagination that things are real does not represent true reality. If you see golden globes, or something, several times, and they talk to you during your hallucination and tell you they are another intelligence, it doesn't mean they're another intelligence; it just means that you have had this particular hallucination.... I believe there's nothing in hallucinations that has anything to do with anything external to the internal psychological state of the person who's got the hallucination. I believe there's nothing in hallucinations that has anything to do with anything external to the internal psychological state of the person who's got the hallucination.'[8] (Lane 2020: 18–19)

3 Anekāntavāda and Multiple Pathways

In Jainist literature, the parables of *The Other Side of the Shield* and *The Blind Men and the Elephant* are deeply instructive ones and go to the very heart of how limited our knowledge is about any particular event, thing, or person. As Shri Jayatilal S. Sanghvi retells the two tales:

> In the outskirts of a village a statue was erected in honor of one of its hero. It had a sword in one hand and a shield in the other. One side of the shield was covered with gold while the other one was covered with silver. Two unknown persons came here each from the opposite direction and began expressing their views. One said that the statue was beautiful and more so because its shield was covered with gold. The other said that the shield was not covered with gold but was with silver. A quarrel ensued between them. A wise man came from the village by that time and said that the shield was covered with gold as well as silver. Let both of you just exchange your places and see the other side of the shield. Both realized their error and apologized to each other for fighting falsely.

always make a comeback in certain quarters, as we can see in Jordan Peterson's acceptance of some (but not all) of his elemental contributions and in the work of neuropsychologist Mark Solms of South Africa. Of course, Freudian concepts have so penetrated the psyche of western culture that it is impossible to ignore him, even if we wish to throw the baby out with some of the bathwater, to paraphrase a cliché.

[8] Feynman elaborates in the same text, "If you see golden globes, or something, several times, and they talk to you during your hallucination and tell you they are another intelligence, it doesn't mean they're another intelligence; it just means that you have had this particular hallucination.' Feynman's skepticism may cause disbelieving cackles among those who believe that what is experienced while having an OBE is something more than phantasmagoria" (Lane 2020: 21).

> Once a royal retinue was stopping at a village to spend their afternoon. The village folks came here and amongst them there were six blind men. All had heard a lot about elephants but none had ever been able to see one. They requested the care-taker to allow them to touch the elephant so that they may be able to make out what the elephant could be like. They were permitted to do so. The first who came across the ears stated that the stated that the elephant was like a tusk weeding tool (Supada). The other caught hold of the trunk and stated that the elephant was like a big wooden pestle. The third touched the tusks and said it was like a big windpipe. The fourth touched the legs and said it was like a big pillar. The fifth felt the stomach and said it looked like a water-bag. The sixth had a tail in his hand and said it appeared to him like a broom. Each thought that his version was right and others were wrong. The care-taker said that none of them had ever seen the elephant fully. Each one had merely seen one limb and from that data each one had given his surmises about the whole elephant. This was, therefore, the cause of their quarrel. He explained the whole position, and all the blind men became silent and departed.[9] (Neutralsurfer 2016)

Interestingly, the Jain idea of Anekāntavāda has some similarities to Niels Bohr's concept of complementarity, where in physics (particularly quantum mechanics) the revelation of one feature at a specific time and place conceals another aspect from view. A prime example of this, of course, can be found in Werner Heisenberg's uncertainty principle which posits that the more one ascertains the momentum of a subatomic particle/wave the less certain one becomes about its relative position and vice versa (Diem-Lane 2014).[9] It is as if the inner workings of the universe were a cosmic tease, forever revealing just as they are forever concealing. In this regard, Ken Wilber has been at the forefront of advocating a multi-sided approach to the study of consciousness. He has developed a four-quadrant model, "Upper Left (UL); subjective: individual, self, consciousness, experience – I Lower Left (LL); intersubjective: collective, community, culture, worldviews – WE. Upper Right (UR); objective: object, organism, thing, behavior – IT" (Wilber 2007).[10] As he is famous for saying, and which goes to the very heart of Anekāntavāda, "I have one major rule: Everybody is right. More specifically, everybody – including me – has some important pieces of truth, and all of those pieces need to be honored, cherished, and included in a more gracious, spacious, and compassionate embrace" (Wilber 1996).[11] One of the great difficulties in the purely academic study of consciousness has been a tendency toward recalcitrant parochialism, where each

[9] Cited from https://www.peacecorps.gov/educators/resources/blind-men-and-elephant/story-blind-men-and-elephant/#. Also see the film "ANEKANTAVADA: The Jain Version of Multiple World Views" which provides a nice illustration of the blind men and the elephant, https://youtu.be/bKfB3fepI30

[10] Ken Wilber is a prolific author. Although he has many insightful ideas about consciousness, his theories have been severely criticized because of his misunderstanding and, worse, misrepresentation of Darwinian evolution. See Lane 2014, and the numerous essays by Frank Visser on the Integral World website published in Netherlands (Visser 2022)

[11] Ken Wilber is a prolific author. Although he has many insightful ideas about consciousness, his theories have been severely criticized because of his misunderstanding and, worse, misrepresentation of Darwinian evolution. See Lane 2014, and the numerous essays by Frank Visser on the Integral World website published in Netherlands (Visser 2022).

philosopher tries to overly protect their own pet theory instead of genuinely understanding opposing purviews and trying to find a useful middle ground.

This is precisely why Anekāntavāda can serve as a constant reminder to adopt a many-sided perspective, even if it may contravene our own theories. Far too often, because we have invested so much time and energy in our own staked out positions, we end up creating defenses around our hypotheses instead of following Richard Feynman's wise dictum that we should be our own greatest critic when it comes to positing our theories, since given our myopic biases we don't always see from the vantage point of multiple angles. As Feynman elucidates, giving a modern twist to a long-held Jainist truism:

> Details that could throw doubt on your interpretation must be given, if you know them. You must do the best you can – if you know anything at all wrong, or possibly wrong – to explain it. If you make a theory, for example, and advertise it, or put it out, then you must also put down all the facts that disagree must also put down all the facts that disagree with it, as well as those that agree with it. There is also a subtler problem. When you have put a lot of ideas together to make an elaborate theory, you want to make sure, when explaining what it fits, that those things it fits are not just the things that gave you the idea for the theory; but that the finished theory makes something else come out right, in addition. In summary, the idea is to try to give all of the information to help others to judge the value of your contribution; not just the information that leads to judgement in one particular direction or another. (Feynman 1974: n.p.)

If we are to truly understand the multifarious nature of awareness, each and every discipline – from physics to sociology – will be useful, because like our proverbial blind men they capture one vital facet that others may not see. This is why Edward O. Wilson's book, *Consilience* (Wilson 1998), is a necessary reminder for us to take a wholistic approach to the subject even if we champion intertheoretic reductionisms whenever such is possible and viable.[12]

4 Syādvāda and the Uncertain Nature of Science

I find it remarkable that an ancient religion would establish the principle of uncertainty or tentativeness as one of its key tenets when positing claims about truth. It speaks volumes about Jainism's intellectual richness and profundity that its early seers understood a core principle of modern scientific thinking thousands of years before its more physical articulation. Syādvāda indicates that there is an almost infinite number of vantage points by which to appraise any event, experience, or specific thing (naya).

---

[12] When Wilson's book was first published in 1998 it generated a lot of heat controversy. Twenty-three year later it now generally accepted as a cornerstone to being well educated. The humanities cannot neglect the sciences and the sciences cannot neglect the humanities. C.P. Snow noticed this stark division back in 1959 when he gave his famous Rede lecture on the subject, which later became a highly influential text (Snow 1969).

As humans we have a limited cranial apparatus and thus we are bounded by a set of *neurally laced* components.[13] This simple fact is elemental and comes into sharper relief when we look at an animal, such as a squirrel or a beaver, and take a closer look at its particular anatomy. Do we believe that they are fully equipped to grasp Einstein's General Theory of Relativity or Gödel's Incompleteness Theorems? Likewise, we suffer from what may be rightly called the true original sin of humankind: the tendency to confuse neurology for ontology. Our brain states are in flux and every theory we postulate is but an incomplete model of what, by necessity, far exceeds its capability to encompass each and every unique particularity within it. Jainism understood this existential truth early in its history and made it a cornerstone of its spiritual outlook, which, when combined with its high ethic of non-injury and openness to multiple viewpoints, provides a necessary and advantageous beacon for anyone interested in the scientific study of consciousness. The history of science provides us with a rich array of examples of how important it is to be cautious in ever making claims about the future. One can only be reminded of Lord Kelvin's infamous reply about the future of aerial navigation: "Neither the balloon, nor the aeroplane, nor the gliding machine will be a practical success" (Lord Kelvin 1902: 4). By the turn of the nineteenth century, some physicists believed that all the fundamental laws had been discovered and that only further tinkering and refinements were needed. This turned out to be wrong in ways never imagined, especially when Max Planck threw physics upside down when he introduced the concept of quanta, which eventually gave birth to quantum mechanics and a complete revolution in understanding causality and probability (Diem-Lane 2014). One senses that the study of consciousness is also at a tipping point, ripe for a transformative breakthrough similar to what happened to molecular biology in 1953 when Francis Crick, James Watson, Maurice Wilkins, and (overlooked for the Nobel Prize because of her untimely death) Rosalind Franklin, discovered the structure and inner workings of DNA. Yet, even here, much work remains and no biologist worth his or her salt would postulate that any model was absolute. Syādvāda is a cerebral parenthetical, an admonitory "maybe." This intriguing and productive way of seeing the world is brilliantly illustrated by the oft-told story of a pot.

As explained by the renowned Indian philosopher, Bimal Krishna Matilal (Matilal 1981: 308):

1. From a certain point of view, or in a certain sense, the pot exists.
2. From a certain point of view, the pot does not exist.
3. From a certain point of view, the pot exists and does not exist.

[13] As the book by Lane elucidates: "A map by definition is less than the territory to which it points. If such a topographical depiction were exactly the same in detail, size, and description to the area it wished to cover, then it would be superfluous and redundant. Therefore, if we accept this definition of maps, we know that each of them (no matter how sophisticated) will have gaps. Which is another way of saying (perhaps less politely) that all models are in some sense wrong. They will have key pieces and vital information missing. So, like the recorded warning to oncoming passengers in the underground tube stations in London, England, we are told to 'mind the gaps'" (Lane 2019: 1).

4. From a certain point of view, the pot is inexpressible.
5. From a certain point of view, the pot both exists and is inexpressible.
6. From a certain point of view, the pot both does not exist and is inexpressible.
7. From a certain point of view, the pot exists, does not exist, and is also inexpressible.

Of course, the pot in question here is merely an analogous placeholder for any event or object. Yet, it is a helpful pathway to understand the vagaries in studying something as heterogeneous as consciousness.

On one end of the spectrum, we have philosophers who argue that consciousness is an illusion, and on the other end that it is the fundamental reality underlying the universe. In between, we have admixtures of both positions, ranging from an epiphenomenal spandrel to a panpsychic elemental. Ken Wilber, taking an Integral approach, would argue that each side is right and must be incorporated in any theory that wishes to properly appraise consciousness. The problem with Wilber's approach, however, is that he has shown a bias for dismissing ad hoc a purely Darwinian understanding of how self-awareness may have emerged from unconscious processes. Wilber's Integral theory is, ironically, less integral than he imagines.[14]

This is why I believe that Jainism's core ideals of Anekāntavāda (अनेकान्तवाद), Syādvāda (स्याद्वाद), and Ahimsā (अहिंसा) are so vital in any quest for knowledge, but most pointedly in our continued studies on the nature of consciousness and what it portends.

4.1 Postcript | Key Excerpts

The Uncertain Game of Science

> Science is founded on uncertainty. Each time we learn something new and surprising, the astonishment comes with the realization that we were wrong before. Lewis Thomas. (Thomas 1980)

> Science is a game – but a game with reality, a game with sharpened knives .... If a man cuts a picture carefully into 1000 pieces, you solve the puzzle when you reassemble the pieces into a picture; in the success or failure, both your intelligences compete. In the presentation of a scientific problem, the other player is the good Lord. He has not only set the problem but also has devised the rules of the game – but they are not completely known, half of them are left for you to discover or to deduce. The experiment is the tempered blade which you wield with success against the spirits of darkness – or which defeats you shamefully. The uncertainty is how many of the rules God himself has permanently ordained, and how many apparently are caused by your own mental inertia, while the solution generally becomes possible only through freedom from its limitations. (Erwin Schrödinger (in Moore 1994: 251))

---

[14] See the many critical articles by Frank Visser on the Integral World website which focus on Wilber's misunderstanding of Darwinian evolution and the role that chance/probability plays in the development and emergence of novel genotypes, http://www.integralworld.net/visser25.html

4.2 Jain Theory of Anekāntavāda

1. The Universe is the composite of groups consisting of adverse pairs like knowledge and ignorance, pleasure and sorrow, life and death and so on. Life depends on such adverse groups. All the groups have their own interests, which create clashes and conflicts in thinking among themselves. Religion is supposed to pacify these clashes through co-existence on socialistic pattern of society. The co-existence cannot be remained without relativity.
2. Jain philosophy is based on the nature of reality, which is considered through Non-absolutism or Many-fold Aspects (Anekāntavāda). According to this view, reality possesses infinite characteristics, which cannot be perceived or known at once by any ordinary man. Different people think about different aspects of the same reality and therefore their partial findings are contradictory to one another. Hence they indulge in debates claiming that each of them was completely true. The Jain philosophers thought over this conflict and tried to reveal the whole truth. They established the theory of Non-absolutistic standpoint (Anekāntavāda) with its two wings, Nayavada and Syādvāda. Proper understanding of the co-existence of mutually opposing groups through these principles rescues one from conflicts. Mutual co-operation is the Law of Nature.
3. Things are visible and invisible as well. We stand by visible objects and accept them as they surely are but do not recognize their invisible characteristics. Until and unless one does not recognize both these characters of an object, he cannot reach the truth and justice. None is absolutely similar or dissimilar, friend or enemy, good or bad. As a matter of fact, every entity hides in itself the innumerable possibilities. Coal can be converted into the state of the diamond or coal is the first stage of diamond. This is the conception of Anekāntavāda. It should be remembered here that total impossibility of becoming is very rare. Rational cannot be irrational and irrational cannot be rational. On the contrary, it can be converted into something else. One becomes desperate, as he does not understand the theory of relativity. He forgets that the modes are not imperishable. They are to be changed. Sorrow can be converted into pleasure. Absoluteness has no meaning in any field. Substance cannot be fully explained without the assistance of Anekāntavāda. Life itself cannot be properly understood without this philosophical notion. Pluralism, monotheism existence and nonexistence, eternality and non-eternality and so on go together. These characters of an entity can be comprehended with the help of real standpoint (Niscayanaya) and Practical standpoint (Vyavaharanaya). (Excerpts from Jainworld.com: n.p.)

4.3 Scientific Uncertainty

In the scientific process, absolute certainty is often difficult to achieve. Uncertainty is a factor of the process and does not mean that a result, hypothesis, or theory is wrong. Scientists have developed formal methods to address scientific uncertainty.

Outside the scientific community, these methods and associated terminology can be confusing. This may lead to incorrect conclusions about the validity of the science. Most people think of uncertainty as an absence of knowledge. However, in science, uncertainty is used as a measurement to tell us how well something is known.

Scientists almost always include the level of uncertainty in a discussion of scientific results. By quantifying how much uncertainty is associated with results, scientists are able to communicate their findings more precisely. Scientific uncertainty generally means that there is a range of possible values within which the true value of the measurement lies. Further research on a topic or theory may reduce the level of uncertainty or the range of possible values. At times, the public or the media interpret scientific uncertainties as synonymous with doubt. This can unfortunately discredit research findings, whereas reporting the level of uncertainty strengthens the research results and provides guidance for the focus of future research projects. Scientific uncertainty is a bedrock concept of science that makes the results stronger and more useful. Uncertainty in the scientific process includes several kinds of uncertainty: statistical uncertainty, natural variability, and true uncertainty (https://dosits.org/decision-makers/scientific-uncertainty/).

Sources for Further Reading

Chapple, Key, ed. 2002. *Jains and ecology*. Cambridge: Harvard University Press.
Diem-Lane, Andrea. 2015. *The Jain Path: An annotated guide*. Walnut: Mt. San Antonio College.
———. 2016. *Ahimsa: A brief guide to Jainism*. Walnut: Mt. San Antonio College.
———. 2016. *How to study the sacred: An introduction to religious studies*. Walnut: Mt. San Antonio College.
Jacobi, Hermann Georg, ed. 2008. *Jaina Sutras*. London: Forgotten Books.
Jaini, Padmanabh S. 2014. *The Jaina path of purification*. New Delhi: Motilal Banarsidass.
Long, Jeffrey D. 2009. *The Jains: An introduction*. London: I.B. Tauris.
Mardia, K.V. 2007. *The scientific foundations of Jainism*. New Delhi: Motilal Banarsidass.
Prime, Ranchor. 2006. *Mahavira: Prince of peace*. San Rafael: Mandala Publishing.
Tobias, Michael. 2000. *Life force: The world of Jainism*. Fremont: Jain publishing.

References

Bacon, Francis. 1858. Novum Organum. In *The Works of Francis Bacon*, ed. James Spedding, Robert Leslie Ellis, and Douglas Denon Heath, vol. 4. London: Longmans and Company.
Chalmers, David. 1997. *The conscious mind: In search of a fundamental theory*. Oxford: Oxford University Press.
Crews, Frederick. 2017. *The making of an illusion*. London: Profile Books.
Dennett, Daniel. 1991. *Consciousness explained*. Boston: Little Brown and Co.
Diem-Lane, Andrea. 2014. *Quantum weirdness: Einstein vs. Bohr*. Walnut: Mt. San Antonio College.
———. 2015. *The Jain path: An annotated guide*. Walnut: Mt. San Antonio College.
———. 2016. *Ahimsa: A brief guide to Jainism*. Walnut: Mt. San Antonio College.

Feynman, Richard Phillips. 1974. *Commencement speech*, Cal tech: Cargo cult science. *Engineering and Science* 37 (7): 10–13.
Harris, Sam. 2011. *The mystery of consciousness*. SAM HARRIS. https://www.samharris.org/blog/the-mystery-of-consciousness. Accessed 22 July 2022. n.p.
———. 2015. *Waking up: A guide to spirituality without religion*. New York: Simon and Schuster.
Huxley, Thomas Henry. 2016. *A scientific education: Essays from a naturalist*. Walnut: Mt. San Antonio College.
Kelvin, William T., and Lord. 1902. *Interview: Utter impracticability of aeronautics & favorable opinion on wireless*, 4. The Newark Advocate.
Lane, David Christopher. 2011. *Inside-outside*. Integral World. https://www.integralworld.net. Accessed July 23, 2022. n.p.
———. 2014. *Cosmic Creationism: Ken Wilber's theory of evolution, A Skeptic's critique*. Walnut: Mt. San Antonio College.
———. 2019. *On being unknowing: The enlightenment of limits*. Walnut: Mt. San Antonio College.
———. 2020. *The astral plane debates: Richard Feynman vs. John Lilly*. Walnut: Mt. San Antonio College.
Matilal, Bimal Krishna. 1981. *The central philosophy of Jainism (Anekāntavāda)*, L.D. Series 79. Ahmedabad.
Moore, Walter J. 1994. *A life of Erwin Schrödinger*. Cambridge: Cambridge University Press.
Nagel, Thomas. 1974. *What is it like to be bat? The Psychological Review*. Vol. 83, 430–450. Durham: Duke University Press.
Neutralsurfer. 2016. *ANEKANTAVADA: The Jain version of multiple world views*. Educational Video, 5:57. https://youtu.be/bKfB3fepI30. Accessed 23 July 2022.
Snow, Charles P. 1969. *The two cultures and the scientific revolution*. Cambridge: Cambridge University Press.
Theory of Anekantavada. n.d.. Jainworld.com. Https://jainworld.com/library/jain-books/books-on-line/jainworld-books-in-indian-languages/jainism-a-religion-of-asceticism-new/theory-of-anekantavada/. Accessed 23 July 2022.
Thomas, Lewis. 1980. *Qtd. In Famous Scientists. The Art of Genius*. http://www.famousscientists.org/what-is-science-quotes/. Accessed 23 July 2022. n.p.
Visser, Frank. 2022. http://www.integralworld.net/visser25.html. Accessed 23 July 2022. n.p.
Wilber, Ken. 2007, 1996. *A brief history of everything*. Boulder: Shambhala Publishers.
Wilson, Edward O. 1998. *Consilience: The unity of knowledge*. New York: Knopf.

Part III
Consciousness and Artificial Intelligence

On the Non-Computability of Consciousness

Subhash Kak

Abstract The chapter will consider the problem of the computability of consciousness and provide evidence from different fields that supports the position that it is non-computable. If consciousness were computable then the present, which includes sentient agents, is completely determined by the past, and so one can emulate it and, therefore, conscious or sentient machines will be built. The arguments in favor of the non-computability of consciousness are: first-person accounts of creativity, the fact that mathematical logic is associated with incompleteness, and that consciousness appears to have the capacity to halt any computation in the brain. The non-computability position is consistent with the Orthodox Copenhagen Interpretation of quantum theory, in which the subject and the object are forever separated by the Heisenberg Cut, which, in turn, implies that materiality and consciousness are like two sides of a coin. The question of how the two can interact, as in controlling the quantum state through observation, will be discussed. Larger philosophical questions related to the problem of consciousness will be considered and assumptions behind the building blocks of reality as in space, time, and matter will be examined together with recent results from logic that are relevant to this discussion.

Keywords Computability of consciousness · Sentient machines · Creativity · Subject-object interaction

1 Introduction

Present-day computers are not conscious, whereas the brain is. If consciousness is a consequence of computation, then there are two possible reasons behind the discrepancy: first, that the manner in which the brain performs computations is

S. Kak (✉)
School of Electrical and Computer Engineering, Oklahoma State University, Stillwater, OK, USA

P. S. Satsangi et al. (eds.), *Consciousness Studies in Sciences and Humanities: Eastern and Western Perspectives*, Studies in Neuroscience, Consciousness and Spirituality 8, https://doi.org/10.1007/978-3-031-13920-8_7

different from the way digital computers work; second, computers are deficient because their hardware is not powerful enough, but at some point, as they become more powerful, they will eventually become conscious. Many computer scientists accept the latter position. Thus, Baars and Franklin (2009) assert that the conscious as well as the non-conscious aspects of human thinking, planning, and perception are produced by adaptive, biological algorithms, and that machine consciousness may be produced by similar adaptive algorithms running on the machine.

The idea of the ecological view of biological computation is found in Gibson (1979) where perception is understood in terms of perceptual systems rather than channels of sensation. This may be understood as computation that is not according to the Turing machine model, which is at the basis of digital computers. Although there have been some attempts to cast the model in a Turing machine framework, these attempts have not been satisfactory. A more serious objection to this view is that an ecological view does not in itself provide a foundational basis for consciousness. Thus, consider the ant colony in which the colony cannot be viewed as a sentient entity although the individual ants may be seen that way.

To consider whether the "ghost of the self" in the body-machine can be fully explained in terms of computation, one must distinguish between computation and awareness of the computation within the agent. There are cognitive tasks during which the agent does not have a sense of subjective awareness, and such tasks are done by the computer, and it does so faster and with more reliability. The question of consciousness in machines is whether such machines can be "aware" of the computation or, in other words, whether they have ego, or a sense of self.

Some argue that brains, unlike classical computers, perform quantum operations deep in the neural circuitry and that's what leads to consciousness (Penrose 1989; 1994; Stapp 2007), but there seems to be no plausible explanation for the rise of the phenomenon of awareness in this approach as well (Kak 2019b). Furthermore, it is known now that the quantum computation paradigm also fits into the Turing machine model (Nielsen and Chuang 2000).

Consciousness is experienced differently by different observers, and this difference is created by the fact that the individual minds processing the experience use different lenses. The lens depends on the individual's storehouse of memories, the degree of attention, and the expectations related to the experience. Its apparent plurality is a consequence of its projection in different minds.

An entirely different way to look at the problem is to assert that consciousness is the ground on which memories and experiences form and this ground is not part of our physical bodies. The same awareness within the individual suffering from dissociative identity disorder is bound to different disjoint memories but each of the alters (alternate selves) remains consistent and whole. Perhaps even more dramatically, the same individual might be the most loving family man at home and a heartless individual in a different social environment. We think of ourselves as being outside of the physical world, and even our conceptions of the universe are as if we are not a part of it.

If this sense of being outside of the physical world is true, it would be impossible to emulate it by hardware and processing by any kind of computing elements. It also follows that it will not be a computational property of the modules that comprise the system.

We can also look for the non-computability of consciousness from its parallel to the unsolvability of the halting problem of computer science, according to which you cannot tell whether a given program will run endlessly or halt. Let us define "consciousness" as some privileged state of the mind that makes its processes halt (we don't bother to specify it beyond this description) and its contents registered (which is what we imply by awareness). Humans can get into the state of "awareness" at any time, which means that the earlier computation has halted, and this is irrespective of the initial state of the immediately preceding process. The exceptions to this are if a person is sleeping or unconscious as in coma. But since halting to arbitrary input is impossible to predict from a computability point of view, it follows that consciousness is not computable (Kak 2016, 2019a, 2022a).

Mathematically, consciousness, C, as a property, P, emerging from the structure and the organization of a physical system, S, may be written as $C = P(S)$. But as a dependent variable, it cannot have any sway on the evolution of the system, which goes against our intuition about consciousness. While the workings of personal consciousness are correlated with activity in the brain, it doesn't follow that it arises in the brain circuitry (Zeki 2003). A parallel example, although not quite equivalent, is that while the image on the TV screen requires the receiver circuitry to work correctly, it doesn't follow that the primary source of the image is the electronics of the receiver. Thus, the argument that consciousness is based on the brain does not rule out the possibility that its nature transcends brain processes.

Thus, we can claim that consciousness itself is not a computation, although the conscious mind oversees computations and has access to computations performed by the brain. This essay will present further evidence in support of this hypothesis. But before doing so some issues related to the relationship between computation and consciousness are sketched from the social perspective, awareness and memory, the nature of reality, and creativity. Implications of this to the theological question of the origin of the freedom of sentient beings will also be summarized.

2 Computers and Consciousness, a Social View

Were computers to become conscious and given the fact that biological evolution is slower than technological evolution, it is certain that biological life would be unable to compete with sentient machines, and become extinct rather quickly. But if we model the earth ecosystem as a quantum system, then such a sudden extinction is contrary to a linear Schrodinger-equation-driven evolution of the system.

Some futurists and sci-fi writers are imagining a marriage of sorts between AI and brains so that, someday, technology will make it possible for humans to become "posthuman," transcending the limits of the human condition (Nayar 2018). There are others who believe that the only way to make sense of all the scientific facts is to take reality as a simulation. Another scenario is to imagine that once humans learn how to completely characterize brains, they will be able to copy themselves into computers, creating their emulations, or *ems*, in the process. Such *ems* could be replicated easily and thus could quickly outnumber real people. But *ems* are states and their evolution must be governed by the Schrodinger equation (Kak 2009) and therefore, this scenario may also be dismissed.

3 Awareness and Memories

Let us explore the matter of awareness a bit further from an operational perspective. If awareness is some kind of a measurement, it should have a reference. If we postulate a single transcendental consciousness, the individual's empirical consciousness is a projection and the referent for it must be the universal. In Vedanta, the analogy of the same sun getting reflected in a million different pots of water as little suns is provided to explain the empirical consciousness of the individual (Kak 2016). In this view, there is a single consciousness that gets manifested in the minds of sentient beings (Schrödinger 1967). The mind is an instrument consisting of the complex of ego, intelligence, and memory. Without the inner lamp of consciousness, this instrument cannot have awareness, and the different awareness states depend on the clarity of the surface of the mind. The inner experience is always dual: a sequence of impressions and detached awareness of this sequence.

In modern Western tradition, William James echoed the Vedic conception of two kinds of selves: the self as knower (the "I"), and the self as known (the "me") (James 1950). He believed that as knower, the self is comprised of different mental states. Thought has no constant elements and every perception is relative and contextualized. States of mind are never repeated, and whereas objects might be constant and discrete, thought is constantly changing and mental states arise out of choices that are made by the mind. James believed that thought flows, and thus he could speak of a stream of consciousness.

If one were to find the boundaries between the "me" and the "I" of consciousness, it becomes essential to find a "minimal" sense of self. It is easy to speak of the intuition that there is a basic or primitive something that is the true self, and much harder to provide evidence for such belief. Conceptually, there must be something permanent – a bedrock – underlying the stream of consciousness. The permanent "minimal" sense of self cannot be the result of a computation, because then we can add it to the "me" associated with the individual.

Each person's self is partly subjective ("I") and partly objective ("me"). The objective self itself may be described in its three aspects: the material self, the social self, and the spiritual self. Narrative self-reference is in contrast to the immediate knowing "I" that supports the notion of momentary experience as an expression of selfhood.

To deal with the empirical pre-conscious or conscious awareness, one can postulate a hierarchical model of consciousness with independent and distributed neural structures at the lowest level (Kak et al. 2016). The speed of the binding of the attributes would depend on the complexity of the communications and the relationships between the modules (Kak 2022b). Interference between various levels and the tangled nature of the information flow can help explain many illusions of perception.

Memory is one element that leads to the nature of the corresponding consciousness state. The mind must select from the pool of memories, and this selection may not be made consciously and is determined by the stream of previous consciousness states and the emotional state of the subject. It is to be expected that the executive control processes play an important role in the selection.

Furthermore, repeated selection of certain memories at the expense of others affects the recall process, causing unwanted memories to be pushed back into the unconscious. One may imagine the existence of mechanisms that prevent unwanted declarative memories from entering awareness, and this cognitive act has enduring consequences for the rejected memories.

4 The Nature of Reality

Is the world a machine described by its parts and their interconnections (the ontic view), or is it fundamentally knowledge (the epistemic view)? In philosophy, these are the positions of two different schools, one believing that reality is *being*, and the other that it is *becoming*. The conception of the world as *being* is associated with materialism, while that of *becoming* assigns a more significant role to the observers.

In the standard neuroscience view, mind emerges from the interoperation of the various modules in the brain, and its behavior must be completely described by the corresponding brain function (Freeman 1999; Koch 2004; Pribram 2004). These descriptions are associated with paradoxes when an attempt is made to create an overarching model (Eccles 1990; Kafatos and Nadeau 2000). One must note that logical systems suffer from paradoxes, which is why we are forced to compartmentalize our knowledge of reality into different fields or subjects that cannot be fully reconciled (Kak 2016).

No specific neural correlate of consciousness has been found (Zeki 2003). Some argue that counterintuitive characteristics of the mind may be ascribable to underlying quantum processes (Penrose 1989). But although quantum mechanics might indeed play a role in brain processes (Bohr 1958; Penrose 1994; Stapp 2007) there is no reason to assume that it throws any light on the phenomenon of consciousness.

In our intuition, consciousness is a category that is dual to physical reality. We apprehend reality in our mind and not in terms of space, time, and matter, and this experience varies based on brain states. It is significant that in our conscious experience we are always outside of the physical world and witness ourselves as apart from our bodies. Even in scientific theory, as, for example, in classical mechanics, the observer is apart from the system, and there is no explanation of the observer within the theory.

Our models of the universe come with many unstated assumptions that may not be valid. For example, we take space to be three-dimensional, but it may not quite be so, as suggested by a new theory (Kak 2020, 2021). If physical space is not integer-dimensional, it may have significant implications for how brain processing is modeled. For example, the notion that the brain is a machine like a computer may be incorrect.

An important viewpoint on consciousness comes from quantum theory, which is the deepest theory of physics. According to the orthodox Copenhagen Interpretation (CI), which was popular with the pioneers of quantum theory such as Niels Bohr, Werner Heisenberg, and Erwin Schrödinger, consciousness and the physical world are complementary aspects of the same reality (Bohr 1958; Schrödinger 1967; Moore 1994). CI assumes complementarity at different levels and this includes the duality of particle and wave or object and subject.

If consciousness and the material world complement each other, the influence of one on the other may be explained by the Quantum Zeno effect (Misra and Sudarshan 1977). It is significant that the pioneers of quantum mechanics thought their theory was consistent with Vedanta (Moore 1994), in which the experienced dualism of reality is only apparent for there exists a unitary ground that is consciousness.

This traditional Vedic model of the mind and consciousness, which is described at length in the Upaniṣads and other texts, includes elements such as ego (the sense of the autobiographical self) and awareness (Kak 2009, 2016). Mind in this model is an inner processor that carries out operations associated with different cognitive capacities in a systematic manner. Such a model cannot be implemented since we don't know how to consider ego and awareness in a formal system, and there is no reason to assume that either could ever be put within a formal framework.

5 Will Computers Be Conscious?

Nearly 5 years ago, I was part of a series of weeklong workshops organized by SRI International, Menlo Park, in different locations in the United States and in Cambridge, UK, to consider this question of whether machines at some future time will become conscious. The 30-odd participants in these workshops included computer scientists, physicists, neuroscientists, and philosophers (Rushby and Sanchez 2018).

We took stock of the many difficulties with the conception of selfhood in a machine paradigm. Standard neuroscience accepts the doctrine of the identity of brain and mind, taking the latter to emerge from the complexity of the interconnections. In this view, behavior must be completely described by the corresponding brain function, leaving no room for agency of the individual.

The selfhood of humans leads to paradoxes related to autonomy and freedom. Humans reject the idea that they are mere machines, yet they often equate their "self" to the machinery of the body. On the other hand, the human's self-image is that of the body, together with transient thoughts, which is overseen by an observing "I" within.

We make a distinction between the "autobiographical self" related to one's memories and relationships, and the "core self," which is rooted in the momentary present. The "autobiographical self" is partly the result of one's imagination since it is an interpretation of the past and it includes hopes for the future. The "core self" is elusive; it is the light that shines on things around and associates with them in time and space.

The experiments of Benjamin Libet showed how decisions made by a subject arise first on a subconscious level and only afterward are translated into the conscious decision (Libet 1985). Upon a retrospective view of the event, the subject arrives at the belief that the decision occurred at the behest of his will. In Libet's experiment the subject was to choose a random moment to flick the wrist while the associated activity in the motor cortex was measured. Libet found that the unconscious brain activity leading up to the conscious decision by the subject began approximately half a second before the subject consciously felt that he had taken his decision. But this is not to be taken as an example of retrocausation; rather, this represents a lag in the operation of the conscious mind in which this construction of reality by the mind occurs.

The participants at the workshops agreed that AI machines of the future will be able to emulate all cognitive tasks and by implication able to replace humans at nearly all kinds of jobs. But they were split on whether machines will be conscious like humans. The split turned out to be based on the idea that the phenomenon of consciousness could come in two different varieties that I call Little-C and Big-C (Kak 2022a).

Little-C is the emergent phenomenon of consciousness that is completely based on brain processes, whereas Big-C is consciousness that transcends the brain although its experience by the subject is dependent on the brain processes. If all there is to consciousness is Little-C then machines will be conscious. But if human consciousness is actually Big-C, then machines will fall short.

6 Big-C and the Vedic View

The Big-C position assumes consciousness is something that is apart from the physical reality. Philosophically, this is the position of Vedanta, in which the mental and the physical phenomena are two aspects of the same reality (Kak 2016, 2022c).

The Vedic texts claim to be *ātmavidyā*, "science of self" or "consciousness science." The most ancient of these is the cryptic Ṛgveda. But prose commentaries, called the Brāhmaṇas and the Upaniṣads that appeared in the centuries following the Vedas, provide a framework to decode its narrative, establishing its central concern with consciousness. In the Vedic view, reality is unitary at the deepest level and it is called Brahman (neuter gender). Brahman engenders and, paradoxically, transcends the mind/matter split. It is identical to consciousness at the cosmic scale and it informs individual minds. Turning focus to the very nature of the mind provides insight about consciousness (Kak 2022c).

Since language is linear, whereas the unfolding of the universe takes place in a multitude of dimensions, language is limited in its ability to describe reality. Because of this limitation, reality can only be experienced and never described fully. All descriptions of the universe lead to logical paradox, and Brahman is the category transcending all oppositions. It follows that it is impossible to develop a language-based "theory of consciousness."

Knowledge is classified in two ways: the higher or unified and the lower or dual, which mirrors the dichotomy of consciousness and mind. The "higher knowledge" concerns the perceiving subject (consciousness), whereas the "lower knowledge" concerns relationships of objects and concepts. The higher knowledge is arrived at through intuition whereas the lower knowledge is analytical and represents standard sciences with its many branches. There is a complementarity between the higher and the lower, each being necessary to define the other. The complementarity of these two kinds of knowledge is reminiscent of the complementarity of subject-object in quantum mechanics, and this fact has been noted by historians of science (Moore 1994).

The Vedic texts assert that physicality and consciousness are interwoven (Sanskrit *ota-prota*) with each other (e.g., in Bṛhadāraṇyaka Upaniṣad 3.8.11). To clarify the meaning of this assertion, if we use the other metaphor of consciousness and materiality being two sides of the same coin, then this holds at all scales. In other words, one could say that they are entangled. But whereas entanglement in quantum theory is viewed as a relationship between objects that in themselves are well-defined, *ota-prota* is entanglement that is true at all scales. If *ota-prota* were true, then isolating objects is well-nigh impossible.

One can see consciousness at work in the material world in awareness that goes beyond contact through fields associated with the objects. One might speculate this is behind the random fluctuations associated with the workings of the uncertainty principle in microsystems. One may further speculate that living organisms are unique because their memory provides them the capacity to interact with organisms and the physical world in an autonomous way, and in doing so they are *channeling* the agency to act freely that is made possible by their access to Big-C consciousness.

7 Creativity and Consciousness

The limits to the machine paradigm become clear when we consider the question of creativity. From what we know it appears that the creative moment is not at the end of a deliberate computation (Kak 2022a). There are many anecdotal accounts of dreams or visions that preceded specific acts of creativity. Two famous examples of this are Elias Howe's 1845 dream on the design of the modern sewing machine, and August Kekulé's similar discovery of the structure of benzene in 1862.

The life of Srinivasa Ramanujan, who died in 1920 at the age of 32, is evidence in favor of Big-C consciousness. His long-forgotten notebook, which was published in 1988, contains several thousand formulas that were well ahead of their time, without explanation of how he had arrived at them. When he was alive, he claimed that formulas were revealed to him in his sleep (Kanigel 1991; Berndt and Rankin 1995).

Even if one were to dismiss accounts of creativity as nothing but coincidence, the ontic understanding of reality becomes problematic when one brings information into the mix, as is done extensively in modern physics. This is because information implies the existence of a mind, which category lies outside of the realm of physics.

Information or entropy cannot be reduced to local operations by any reductionist program. It requires the use of signs derived from global properties and the capacity to make choices, which, in turn, implies agency. If consciousness is not material, then the question of how it interacts with a physical system becomes relevant. While consciousness cannot intervene in physical law, it can change the probabilities in the evolution of quantum processes, as in the Quantum Zeno effect.

The current crisis of physics in which 96 percent of the universe is supposed to be dark matter and dark energy, for either of which no evidence exists, may be a consequence of pushing the ontic program too far. It is possible that a shift in perspective that provides a greater role to the conscious agent will be able to resolve the crisis, just as a similar shift solved the problem of the speed of light being constant in all frames. This has been recently shown for the problem of Hubble tension, which is different values at which the universe is expanding by considering the early and late universe models (Kak 2020).

8 Discussion

The arguments in favor of the non-computability of consciousness help us look at an important theological problem that has bedeviled the world for centuries. "Does God exist?" has been debated by philosophers, theologians, and even prominent scientists and mathematicians, e.g., Kurt Gödel (see Wang (1997)). These debates are unsatisfactory because the term "God" is not well defined, and there is no agreement on what kind of evidence should be used in the debate. Some theologians make the case that the judgment on the existence of God need not be based on evidence, and that one can use arguments based on morality or necessity.

By "God" I don't mean someone who sits in paradise and watches over the world, meting out punishment or reward on the Day of Judgment, for such suppositions are not subject to refutation, and are outside scientific and rational argumentation. By "God" I mean its literal meaning, traced by some linguists to the Sanskrit original "*svatavas*" meaning "*self-powered*" or "*self-free*," which is attested in the literature as in the Ṛgveda 3.1.1. Just as the Sanskrit *svar* ("sun") of the Ṛgveda becomes *xwarə* of Avestan and *khar* of Persian, the word *sva-tava* (self-powered) became *xwatāw* in Avestan, and *xudā* and *khudā* in Farsi, and via Iranians it became German *Gott* and English *God* (Kak 2022c). In other words, I take "God" to mean the Self within each individual that represents the freedom within each sentient being.

The proof that God exists is to show that consciousness cannot be derived from materiality. If it were a property of matter, one could measure it. As property of matter, it would be subject to diminution or expansion and determined by the past, whereas in reality consciousness is associated with freedom.

We can conclude by saying that there exists a core self within each sentient being that has freedom. Materiality and consciousness are then two aspects of reality, but its experience by the mind is contingent on the proper working of the brain circuitry. Assuming there exists another aspect to reality beyond materiality that is expressed in the workings of mind and consciousness is consistent both with the Vedic view and the ideas of the founders of quantum theory. In this view, the ordinary experiences of matter and mind are like the two wings of a bird in flight.

References

Baars, Bernard J., and Stan Franklin. 2009. Consciousness is computational. *International Journal of Machine Consciousness* 1: 23–32.

Berndt, Bruce C., and Robert A. Rankin. 1995. *Ramanujan: Letters and commentary*. Providence: American Mathematical Society.

Bohr, Niels. 1958. *Atomic physics and human knowledge*. New York: Wiley Interscience.

Eccles, John C. 1990. A unitary hypothesis of mind-brain interaction in the cerebral cortex. *Proceedings of the Royal Society of London* B240: 433–451.

Freeman, Walter J. 1999. *How brains make up their minds*. New York: Columbia University Press.

Gibson, James Jerome. 1979. *The ecological approach to visual perception*. Boston: Houghton Mifflin.

James, William. [1890] 1950. *The principles of psychology*. New York: Dover Publications.

Kafatos, Menas, and Robert Nadeau. 2000. *The conscious universe: Parts and wholes in physical reality*. New York: Springer.

Kak, Subhash. 2009. The universe, quantum physics, and consciousness. *Journal of Cosmology* 3: 500–510.

———. 2016. *The nature of physical reality*. Mississauga: Mt. Meru Publishing.

———. 2019a. Is consciousness computable? *NeuroQuantology* 17: 71–75.

———. 2019b. Are there limits to artificial intelligence? *Current Science* 116: 1951–1952.

———. 2020. Information theory and dimensionality of space. *Scientific Reports* 10: 20733.

———. 2021. Asymptotic freedom in noninteger spaces. *Scientific Reports* 11: 3406.

———. 2022a. The limits to machine consciousness. *Journal of Artificial Intelligence and Consciousness* 9: 59–72.

———. 2022b. Number of autonomous cognitive agents in a neural network. *Journal of Artificial Intelligence and Consciousness* 9: 227–240. https://doi.org/10.1142/S2705078522500023

———. 2022c. *The vedic tradition: Cosmos, connections, and consciousness*. Bengaluru: SVYASA University.

Kak, Arushi, Abhinav Gautam, and Subhash Kak. 2016. A three-layered model for consciousness states. *NeuroQuantology* 14: 166–174.

Kanigel, Robert. 1991. *The man who knew infinity*. New York: Scribner's.

Koch, Christof. 2004. *The quest for consciousness*. Englewood: Roberts & Company Publishers.

Libet, Benjamin. 1985. Unconscious cerebral initiative and the role of conscious will in voluntary action. *The Behavioral and Brain Sciences*. 8 (4): 529–566.

Misra, B., and E.C.G. Sudarshan. 1977. The Zeno's paradox in quantum theory. *Journal of Mathematical Physics* 18: 756–763.

Moore, Walter J. 1994. *Schrödinger: Life and thought*. Cambridge: Cambridge University Press.

Nayar, Pramod K. 2018. *Posthumanism*. Hoboken: Wiley.

Nielsen, Michael A., and Isaac L. Chuang. 2000. *Quantum computation and quantum information*. Cambridge: Cambridge University Press.

Penrose, Roger. 1989. *The Emperor's new mind*. London: Penguin Books.

———. 1994. *Shadows of the mind*. Oxford: Oxford University Press.

Pribram, Karl H. 2004. Consciousness reassessed. *Mind and Matter* 2: 7–35.

Rushby, John, and Daniel Sanchez. 2018. Technology and consciousness. *Technology and consciousness workshops report*. Menlo Park: SRI International. http://www.csl.sri.com/~rushby/papers/techconscwks2017.pdf. Accessed 8 Aug 2022.

Schrödinger, Erwin. 1967. *What is life? Mind and matter*. Cambridge: Cambridge University Press.

Stapp, Henry P. 2007. *Mindful universe*. New York: Springer.

Wang, Hao. 1997. *A logical journey: From Gödel to philosophy*. Boston: MIT Press.

Zeki, Semir. 2003. The disunity of consciousness. *Trends Cognitive Science* 7: 214–218.

Consciousness and Mathematical Sciences

Anand Srivastav

Abstract In this chapter we discuss the connection of consciousness studies with mathematics. We address third-person validation of first-person experiences in meditational practices of *Surat Shabdh Yoga* and topological graph-theoretical descriptions of the cosmology of Radhasoami Faith, using experiments of measuring topological brain waves with modern technologies like MRI, initiated by Prem Saran Satsangi, who coined mathematical modeling and rigorous analysis for third-person validation.

Keywords Topological graph theory · Spectral methods · First-person experience · Artificial intelligence and consciousness

1 Visiting Mathematics: Is the Human Mind Non-Algorithmic?

Questions of the nature of the human mind, its relation to consciousness, its roots in the human brain, but also the role of soul, spirituality, and higher levels of consciousness are among the fundamental attempts of humanity to gain knowledge about existence itself.

With the advent of modern physics, biology, mathematics and computer science, and the ability to build machines that may mimic human capabilities (called artificial intelligence (AI)) the question emerged whether machines can have a mind, or even some kind of consciousness. Proponents of the latter strong AI thesis are convinced that this in fact is or will be the case (e.g., Hofstadter 1981; Schmidhuber 2015).

Supplementary Information The online version contains supplementary material available at https://doi.org/10.1007/978-3-031-13920-8_8

A. Srivastav (✉)
Department of Mathematics, Kiel University, Kiel, Germany
e-mail: srivastav@math.uni-kiel.de

P. S. Satsangi et al. (eds.), *Consciousness Studies in Sciences and Humanities: Eastern and Western Perspectives*, Studies in Neuroscience, Consciousness and Spirituality 8, https://doi.org/10.1007/978-3-031-13920-8_8

In his book *The Emperor's New Mind*, Sir Roger Penrose counters the thesis of strong AI in a very principal and fundamental way by arguing with the nature of mathematical insight. Let us briefly describe the key arguments of Penrose, before in the next section reviewing and challenging this argumentation in view of the present revolution in AI. Finding proofs of mathematical theorems is not only a sometimes hard and painful journey, but seems to be a mystery even for mathematicians. One of the greatest and exceptional mathematicians of the twentieth century, Paul Erdös (see Ciscery 1991) described this mystery in a quite Platonic way. According to him, the most beautiful proofs of mathematics are written in THE BOOK, and that is why he calls them book-proofs (see Aigner and Ziegler 2010). A mathematician before his birth gets a glimpse of a small portion of THE BOOK, and throughout his life he reproduces by remembering what he has seen as theorems and proofs. This narrative implies that mathematical truth is there a priori, in an existential form, and accessed in an unexplainable, mysterious way. As a consequence of this model, since machines are not taking birth, they would not be able to produce book-proofs. But let us look more closely at the nature of computing machines leaving the more or less categorical Platonic view.

Computers and computing machines are nothing other than Turing machines (invented by Alan Turing, see Turing 1937). Further, any algorithm is a Turing machine. We may ask whether there are other concepts to design computing machines, a legitimate and natural question. The Church-Turing thesis says: NO, in terms of computability all are equivalent to the Turing machine. In fact, all alternative computing concepts known today, for example the Lambda calculus, have been proven to satisfy the Church-Turing thesis. If we accept the Church-Turing thesis for a moment, we may ask how powerful a Turing machine is or can be. The attempt to answer this question leads to a paradoxical dilemma, as we will shortly see. The key is the ingenious idea of Georg Cantor to prove the uncountability of the set of real numbers, namely that there are not enough natural numbers like 1,2,3,... to count the real numbers. This idea is known as Cantor's diagonal method, which is taught in the first weeks of any lecture on calculus to undergraduate students in mathematics and computer science. So, infinity is not infinity; there is an infinite tower of infinities starting with the infinity of the natural numbers.

The famous Halting Problem asks whether or not a Turing machine once getting an input and starting the computation will ever halt. Is there a way, or let's say an algorithm, to decide this problem with an answer YES or NO? Turing observed that this is in fact a special case of David Hilbert's Entscheidungsproblem. The striking result is that the Entscheidungsproblem has no solution. The proof uses Cantor's diagonal argument, and a coding of the problem by enumerating all Turing machines. The paradoxical situation is obvious: Turing machines, though considered as the universal concept of computing, cannot decide whether they will ever halt, in general. Thus their ability is restricted by principal means. In the same spirit is Kurt Gödel's undecidability theorem on certain mathematical systems. We may ask whether an AI system based on Turing machines would have found Cantor's diagonal method and proved its own incapability.

Penrose concludes: "If we can see that the role of consciousness is non-algorithmic when forming mathematical judgements, where calculation and rigorous proof constitute such an important factor, then surely we may be persuaded that such a non-algorithmic ingredient could be critical also for the role of consciousness in more general (non-mathematical) circumstances" (Penrose 1999: 538).

Since the year 1999, two decades have passed and we may re-open the debate on the non-algorithmic nature of consciousness in the emerging era of surprisingly powerful new AI systems.

2 The Advent of AlphaGo and Friends

Many scientists have been quite skeptical about any success of AI systems. In fact, since W. Grey Walter's 'tortoise' – made in the early 1950s – until the end of the 1990s, capabilities of AI systems were less impressive. The scenario suddenly changed with AlphaGo, a self-learning neuronal network (or machine learning system) playing the Japanese board game GO (Silver et al. 2016). The remarkable observation is not only its win of 4 out of 5 games against the world champion Lee Sedol in 2016, or the unbeatable successor AlphaGo Master introduced in 2017, but the surprising 'creativity' of moves, for example move 78 in the fourth game against Sedol, qualified by professionals as a 'divine move' beyond human intuition. The innovation of AlphaGo compared to chess programs is the ability to learn, maybe at an exponential speed, surpassing classical brute-force computation. Today, the mathematical engine of any machine learning AI system (and its neural network) consists of highly efficient stochastic gradient methods for minimization of non-linear functions. Intensive research for better gradient methods, like the program Katyusha X of MIT graduate Allen-Zhu (Allen-Zhu 2018) may further advance the potential of machine learning. AlphaGo can be considered as a quantum leap in AI.

It might be too early to speculate on machine learning and neural networks embedded in quantum computing systems. However, in the field of quantum computing after years of incremental success, the Sycamore quantum processor of Google consisting of 53 qubits claimed quantum supremacy in 2019, solving an artificial mathematical problem in 200 seconds, for which a supercomputer would need 10,000 years (Arute et al. 2019). The potential seems to be immense!

In view of Penrose we may say that the success of AlphaGo is impressive but Go is only a rule-based game with a small number of rules, and human capabilities are far more diverse, navigating us through unknown domains. Well, this is true. But what about the nature of mathematical proof finding, which according to Penrose is non-algorithmic?

Recent advances of AI for finding mathematical proofs are challenging the non-algorithmic nature of mathematical insight, for example the AI project called the Ramanujan machine. The authors write: "The Ramanujan Machine is a novel way to do mathematics by harnessing your computer power to make new discoveries. The Ramanujan Machine already discovered dozens of new conjectures" (see http://www.ramanujanmachine.com/). So clearly, the domain of intuition and mathematical creativity is reached by this AI. A. Davies et al. in a 2021 *Nature* paper showed that AI helped to find and prove new theorems in knot theory, a very abstract branch of pure mathematics, for the first time in the history of mathematics (Davies et al. 2021; Raayoni et al. 2021). If this is the beginning, where is this development leading us? It is perhaps not too speculative to predict that in due time mathematics may change its face, and this means Turing machines are doing mathematics as well as and perhaps better than humans. If this becomes reality, the thesis of the non-algorithmic nature of general consciousness must be discussed without the assumption of the non-algorithmic nature of mathematical insight.

In this chapter, we adopt a different point of view. Religions have claimed to have access to higher levels of consciousness and realities beyond this world, and they teach methods to experience such realities. Most prominent is meditational practice found in all major religions, Hinduism, Buddhism, Christianity, Islam, and in practices of the aborigines in Australia. The perhaps most methodological system in this regard is Radhasoami Faith and the Religion of Saints, where the meditational practice is called *Surat Shabdh Yoga*. It is embedded in a theoretical description of the macrocosm as well as the microcosm in any human, from the gross world to ultimate reality, perceived as the region of pure spirituality. Both macrocosm and microcosm are not only linked to each other, but the microcosm mirrors the macrocosm in humans. Humans through *Surat Shabh Yoga* are in principle able to experience (or inwardly see) the highest region ascending from lower regions toward the highest loci.

The experience itself is a first-person experience. If many persons independently of each other describe the same form of experience, this is already more than an indication. David Chalmers in a pioneering work stated: "As I see it, the science of consciousness is all about relating third-person data about brain processes, behavior, environmental interaction, and the like to first-person data about conscious experience. I take it for granted that there are first-person data. It's a manifest fact about our minds that there is something it is like to be us – that we have subjective experiences – and that these subjective experiences are quite different at different times. Our direct knowledge of subjective experiences stems from our first-person access to them. And subjective experiences are arguably the central data that we want a science of consciousness to explain" (Chalmers 1999: Abstract).

In the next sections we discuss such third-person validations of first-person experiences in Radhasoami Faith and its cosmology, pursued by Prem Saran Satsangi. We will see that mathematics is a descriptional key.

3 Mathematics Revisited: Consciousness and Topological Graphs, Qudits Modeling, and More

Prem Saran Satsangi (2012, 2014) presented a system model for spiritual domains as a topological graph, which means as a drawing of a graph in the plane (see Figure A from Satsangi 2012) based on first-person phenomenology. It shows (we quote from Satsangi 2012): "Apart from the physical universe (the Region of Universal Matter), in the macrocosm, it envisages the Region of Universal Mind and the Region of Universal Spirit."

From this graph a directed subgraph, the system linear graph model, showing the different domains of the macrocosm and spiritual potential sources, is extracted (see Figure B from Satsangi 2012).

We further quote from Satsangi 2012: "Eventually, as in any quantum system, any observations turn out to be classical ones, that is determinate ones. Therefore,

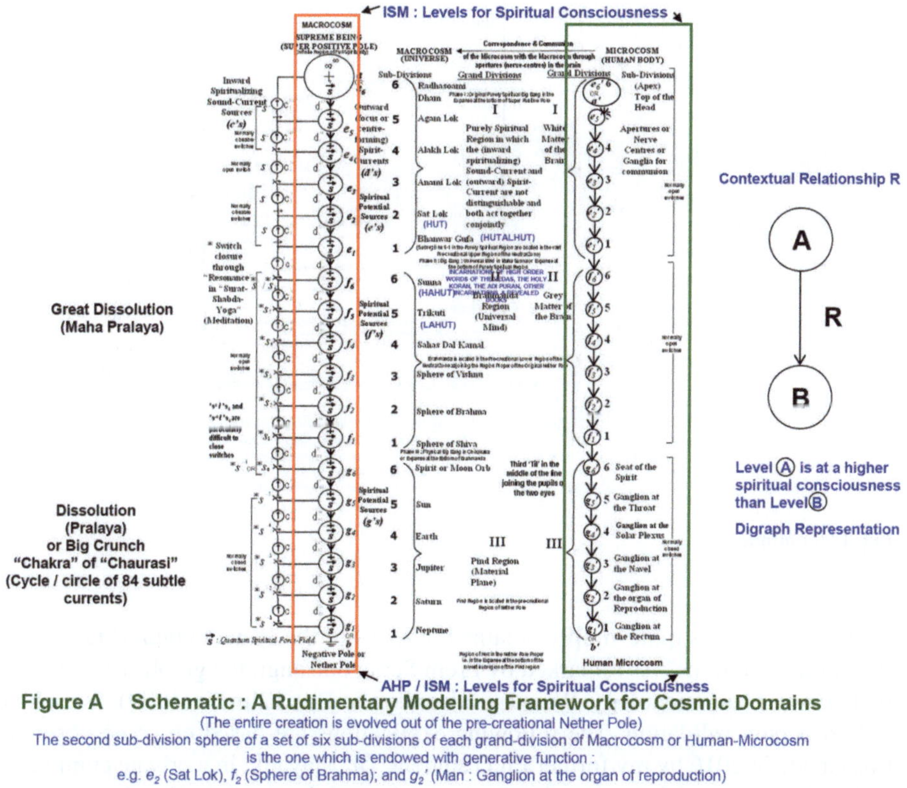

Fig. 1 Spiritual domains as a topological graph (Figure A from Satsangi 2012). Figure A is the original graphic. A high-resolution new drawing with some changes is added as well (see Figure A1)

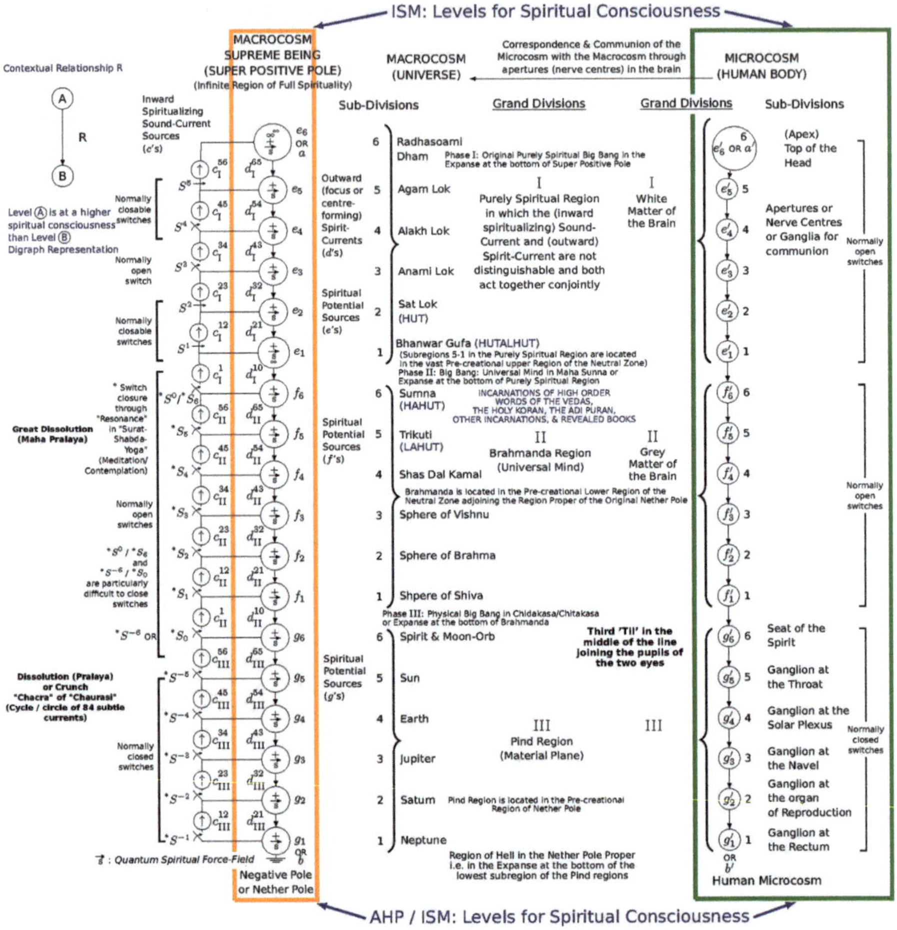

Figure A1 Schematic: A Rudimentary Modelling Framework for Cosmic Domains
(The entire creation is evolved out of the pre-creational Nether Pole)
The second sub-division sphere of a set of six sub-divisions of each grand-division of Macrocosm or Human-Microcosm is the one which is endowed with generative function:
e.g. e_2 (Sat Lok), f_2 (Sphere of Brahma), and g_2' (Man: Ganglion at the organ of reproduction)

Fig. 1 (continued)

the final solution is quite simply obtained in terms of the system graph (Fig. 2) and its implied solutions."As remarked by Prem Saran Satsangi, the graphs came out of first-person phenomenology. We may, according to D. Chalmers (1999), ask about a third-person validation. The following explications are based on some observations made in 2016 by my former Ph.D student, Dr. Mayank, in a private communication. They approach frequencies (and energies) of the graph in Fig. 2 – as stated by Prem Saran Satsangi – by a mathematical method, namely by calculating the

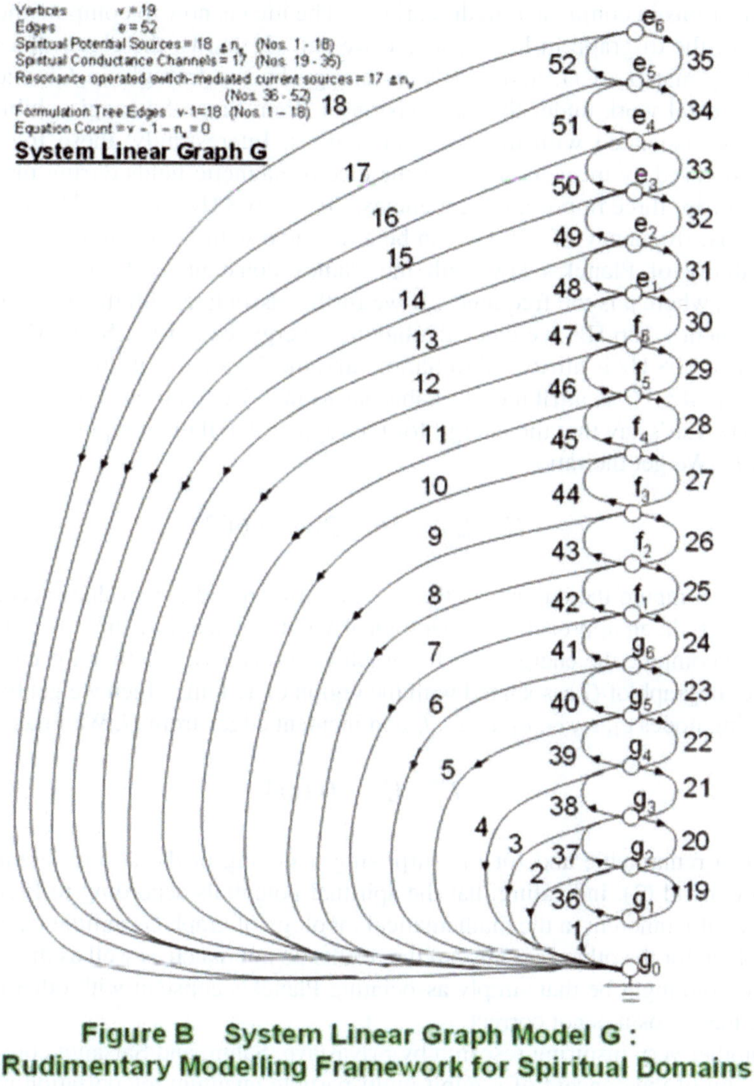

Fig. 2 System graph. (Figure B from Satsangi 2012)

energy via algebraic graph theory and (this is essential) using only the di-graph in Fig. 2.

The energy of a graph has been extensively studied in the literature (Gutman and Trinajstić 1973, Graovac et al. 1977, Li et al. 2012), in particular the correspondence between the graph eigenvalues and the molecular orbital energy levels of π-electrons in conjugated hydrocarbons. The idea is now to compute the eigenvalues of the di-graph in Fig. 2, which we call G, that is, the eigenvalues of the adjacency matrix of G, and finally its energy by the formula in Gutman et al. Experimental work under the supervision of Prem Saran Satsangi exhibited frequencies associated with the different regions. Interestingly, these frequencies were observed by measurements of the electromagnetic fields during meditation and there are three important frequencies: 76 Hz, 108 Hz, and 126 Hz. It is established that frequency of 126 Hz can be associated with the mentor. If we assume the validity of Planck's law with the Planck constant h, the energy satisfies $E = h \cdot \nu$, where ν is the frequency. If we further associate with the whole graph G the frequency 126 Hz, we can calculate the energy, called E_G. Similarly, the frequency of 108 Hz is attributed to Satlok, and the frequency of 76 Hz is attributed to Trikuti, 4.5 Hz is attributed to Sahasdal Kamal. Let us focus on the frequency of 76 Hz. Let's say that the energy for the region of Trikuti computed with $\nu = 76$ HZ is E_t. We get the ratio

$$E_t / E_G = 76/126 = 0.603 \tag{1}$$

Now we compute the energy by the energy formula and eigenvalues according to Gutman et al., an approach methodologically independent of using the ratio in (1). First we compute the energy E_G^* for graph G in this way, and then the energy E_t^* for the subgraph of G associated with the region of Trikuti, which we get by simply removing nodes e_1, e_2, e_3, e_4, e_5, e_6, f_6 and incident edges from G. We finally get

$$E_t^* / E_G^* = 0.604 \tag{2}$$

This is a remarkable and for me surprising matching of the two different energy ratios (1) and (2), indicating that the spiritual potentials according to Prem Saran Satsangi are inherent in the mathematical topology of graph G. Similar calculations were done for the other regions, but the results do not match as well as they do here. The reason might be that simply associating Planck's constant with other domains of the macrocosm is not correct.

Another very inspiring research by Srivastava, Sahni, and Satsangi concerns the extension of the concept of n-qubit multi-particle quantum teleportation modeling to n-qudit contextuality-based quantum teleportation (Srivastava, Sahni, and Satsangi 2011, 2016, 2017). I quote from the abstract of this paper: "We are heading towards point-sized loops or fine-grained particles of nature, which have been rejected out of hand by string theorists. It is not physical reality when measured from the sense of Planck's length of 10^{-35} meters, but finer-grained particles than Planck's length 10^{-35} meters may exist, although, they are not matter anymore. We argue that why stop at degree of freedom of three, but instead pursue quantum odd-prime based units with higher degree n such as 5, 7, 11, 13, 17 and so on till nth degree of freedom even tending to infinity."

The authors aim to open a path for modeling particles smaller than 10^{-35} meters, and distances greater than 10^{10} light years, though both limits are out of reach in state-of-the-art physics. The known quantum teleportation circuit for qubits is extended and analyzed to three-dimensional qudits ($d = 3$) with a two-player (Alice and Bob) communication. Already in this case the table of measurement outcomes of Alice and action required by Bob contains nine entries. For higher-dimensional qudits (dimension d) the complexity may grow exponentially with d. It is a great challenge to model and understand such communication protocols. The authors link the mathematical modeling of finite dimensional qudits and by taking a limit of the dimension d to infinity express their hope to approach "the threshold of being face to face mathematically in abstract terms with the ultimate source of consciousness" (Srivastava et al. 2017: 428).

The vision of a certain limit behavior of d tending to infinity is – as far as I see – at the moment out of reach. If we consider qudits as finite-dimensional functions, when d is finite, the formal transition to functions in infinite dimensional spaces, say certain Hilbert spaces, is natural. The crucial point here is not this kind of mathematical abstraction, but an understanding of the evolving physics in the limit process. There is an immense potential for modeling with qudits. One question would be whether one can gain advantage in quantum computing, for example for factorization of a natural number in primes in the pioneering work of the Nevanlinna Prize (1998) and Gödel Prize (1999) winner Peter Shor (1997). Furthermore, qudit-based quantum modeling of communication protocols could also be interesting to advance two-person games on graphs and may exhibit new phenomena not existent in classical non-quantum communication. One example of such a game is the so-called Maker-Breaker game on graphs, perhaps the most interesting positional game (Glazik and Srivastav 2022).

4 Experimental Evidences: Scanning the Brain

Prem Saran Satsangi (Satsangi 2022) gave a diagram of the human brain with reference points for locations in the brain ("brain apertures") corresponding to the domain in the macrocosm and reached in the meditational practice of Surat Shabdh Yoga (see Fig. 3). The finding of a power law of meditational consciousness by measuring low-frequency brain waves connects the level of meditational consciousness with the brain apertures (Satsangi 2022). This is an extremely interesting result, and a third-person validation per se.

I wonder if there is a simple and cost-efficient way to relate the brain activities during meditation to respective regions of the brain. It might be worth exploring the applicability of EEG measurements for source location. The results and findings in

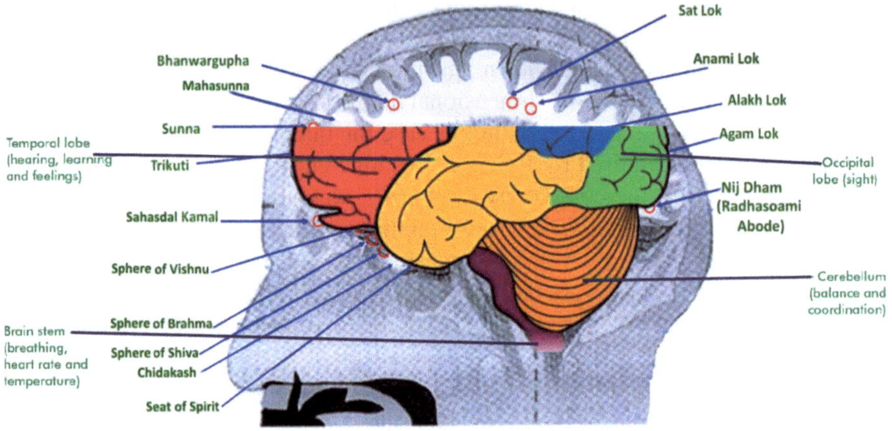

Fig. 3 From Satsangi (2022)

the doctoral dissertation of Dr. Ami Kumar (Columbia Univ.) (Kumar 2020; Kumar et al. 2021, 2023), supervised by myself and Prof. Ulrich Stephani at Kiel University in 2020, could be helpful. Let us briefly look at the work and innovative aspects thereof.

The thesis introduced new methods for the diagnosis of childhood absence epilepsy based on EEG measurement data, applying theories and methods from medical science, statistics, and numerical mathematics. The results may impact and facilitate the clinical diagnosis, whether or not absence of consciousness observed with children, like "day dreaming," or staring for some seconds, which appear in epilepsy, but also in non-epileptic children, is really caused by epilepsy or not, in a quantitative way.

A central question of the thesis work was the reconstruction of neuronal activity during, before, and after absences in the deep brain.

The experimental set-up is an EEG with 31 electrodes at the scalp and 5003 voxel positions with the primary source. The goal is to detect the regions of activity in the deep brain by measuring the electric potential at the scalp.

At present, the only way to achieve this goal is the mathematical modeling of electromagnetic fields in the brain, and the numerical solution of the inverse problem derived from suitable discretization. The starting points are the Maxwell equations in the human brain, assuming that the magnetic permeability is constant in the brain. This leads to a Poisson problem with boundary conditions, the so-called forward problem, and discretization transforms it to a matrix equation $HJ = \Phi$,where Φ is the electrode potential and J is the unknown current density at the voxels. The inverse problem now is to compute J. Unfortunately, wrong best approximation solutions for this problem have been circulated in literature for 20 years. We were

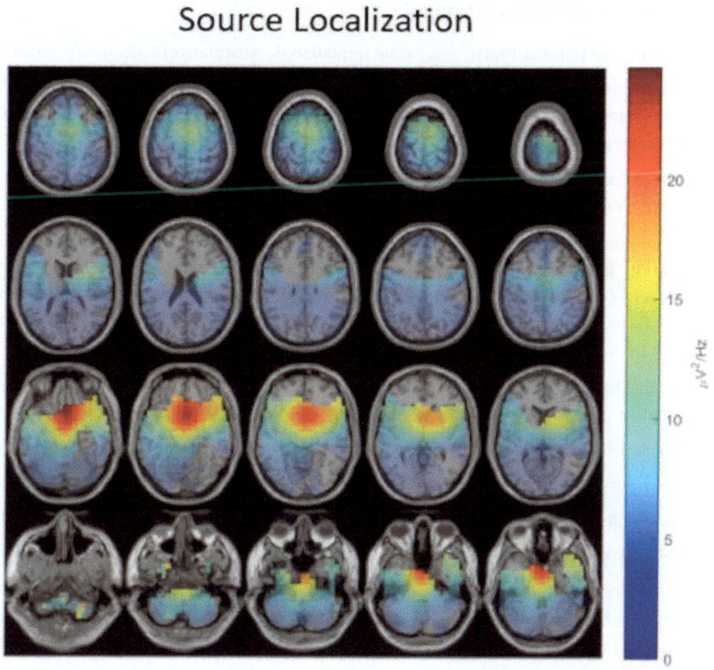

Fig. 4 From Kumar (2020)

able to give the correct best approximation formulation, and the proof of the correctness of the iterative eLORETTA algorithm finding J using the theory of the Moore-Penrose inverse in linear algebra. Figure 4 shows a typical image of source localization for the EEG measurement of childhood absence epilepsy (Kumar 2020; Kumar et al. 2021, 2022).

It would be interesting if the EEG source localization approach could be applied to detect and localize brain activities during meditation, and in this way, to validate first-person experiences by third-person observations.

A very interesting neuro-scientific research line has been initiated by Prof. Dr. Anirban Bandyopadhyay[1] and colleagues (see Agrawal et al. 2018). It exhibits geometric-topological visualization of nested rhythms of every single part of the brain, where sensory signals are converted in terms of a few geometric shapes.

[1] Prof. A. Bandyopadhyay has largely contributed to the different conferences and workshops in consciousness studies organized at the DEI. This time he kindly informed me in June 2022 that he was not able to contribute to this book due to several other commitments.

Comment by Rev. Professor Prem Saran Satsangi :

Ultra-transcendental Meditational Practices of Contemplation & Articulative/Articulatory Words of Five Basic/Fundamental Constituents Ra-Dha-Sva-Aah-Mi/ रा-धा-स्व-आ-मी (Syntax) as distinct from Descriptive Words like Benaras/Banaras/Kashi/Varanasi, Dayalbagh, Konkani, Murar, Orissa & Eastern States of India, Soami Bagh, Tamil, Telegu, etc. (Semantics)

Dayalbagh Science of Consciousness (DSC- 2022) refers to Super-arching Role of Conscientiousness including Spiritual aspects e.g. Meta Mindfulness.

Chairman/Vice-Chairman "ACE" (Advisory Committee on Education), a Non-statutory Body for Consensus-Building for utilising Idea Engineering: ISM- Interpretive Structural Modelling, TQM- Total Quality Management, Benefit-Cost Analysis (in particular AHP- Analytical Hierarchy Process, leading to CGPA- Cumulative Grade Point Average: - Peaking at 4.00/4.00 (FPS System) → 10.000/10.000 (Metric System : vide Wikipedia.org)

[Scientific American October 2022 (Accessed September 15, 2022) Gas Lighting {generalising to Green-Lighting to SPECTRUM (V I B G Y O R) - Lighting} to ULTIMATE REALITY/TRUTH OF SCIENCE {Characterised by Ra-Dha-Sva-Aah-Mi/रा-धा-स्व-आ-मी Faith, originating from Panni Gali, Agra (INDIA) as being progressively Supervised by Infinite Conglomerate of Saints (संत-समूह) :- "Anami Purush" (अनामी पुरुष) of Pure Spirituality, leading further advancement to Regions of "Alakh Purush" (अलख पुरुष), "Agam Purush" (अगम पुरुष) and "Nij Dham" of The Apex-Entity/"Supreme Being" Ra-Dha-Sva-Aah-Mi/रा-धा-स्व-आ-मी as formulated by Radhasoami Satsang Sabha Hqs. Dayalbagh, Agra- 282005, Uttar Pradesh (India), Registered as Religious & Charitable Society with DEI (Deemed to be University) as its Sponsored Endeavour (Vide GOI Notification Dated 16.05.1981 issued by Shri M.P. Kolhatkar, the then Joint Secretary of the Govt. of India)}

Vide Discourses on Radhasoami Faith by the Third Revered Leader Param Guru Maharaj Sahab with Revised Second Edition issued by Shri Prem Kumar*, President RS Sabha, Dayalbagh with Supplement Authored by the Eighth "Acclaimed" Waqt/Present Sant Satguru i.e. Leader of Radhasoami Faith, Dayalbagh Hqs., Titled as "Rudimentary Modelling Framework for Spiritual Domains"]

Latest Innovation for cost-effective & cost-efficient implementation for the aforesaid strategy is the Concept of coupling Consciousness with Conscientiousness (Derived from CONSCIENCE, which is a popular and well understood Phrase) by All & Sundry as discussed with Visiting National & International Delegates to DSC- 2022 on the very first day of their visit on 25.09.2022 during their scheduled meeting with Chairman ACE in the Evening Session at Agroecology-cum-Precision Farming Site of DEI Bio Diversity Park, Located in front of DEI International Seminar Hall Complex, Dayalbagh.

Brain apertures in the context of meditational practice of
(Sahaj) Surat Shabda Yog

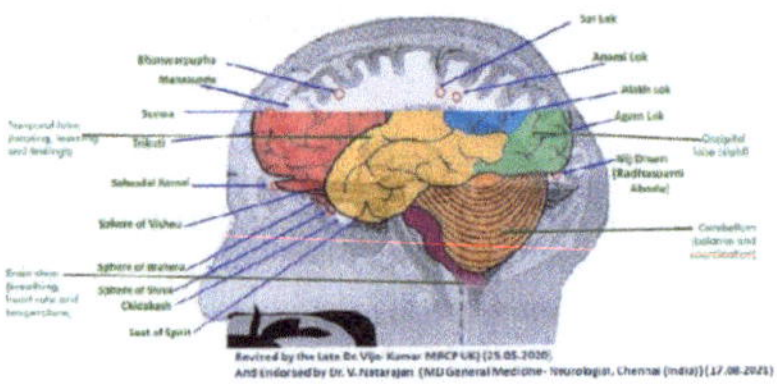

*India's Permanent Representative at GATT (General Agreement on Tariffs and Trade) at Geneva, Switzerland and Chairman Tea Board in Australia, etc.

Friday, 23-09-2022, 05:26 AM

Acknowledgement I sincerely thank Revered Professor Prem Saran Satsangi for giving me the opportunity to co-organize the Dayalbagh Science of Consciousness conferences since 2020, and for the benevolent support of the cooperation between Kiel University and the Dayalbagh Educational Institute (DEI), Deemed to be University, Agra, India. I am grateful for the continued and generous support of Kiel University to advance the cooperation with DEI. I thank my colleague Dr. Volkmar Sauerland and the secretary of my chair, Mrs. Annette Gottwald-Müller, for all their help in editing this book. Last not least I thank my colleague Prof. Dr. Anna Margaretha Horatschek for her helpful comments on the manuscript.

References

Agrawal, Lokesh, Rutuja Chhajed, Subrata Ghosh, Batu Ghosh, Kanad Ray, Satyajit Sahu, Daisuke Fujitaand, and Anirban Bandyopadhyay. 2018. Fractal Information Theory (FIT)-Derived Geometric Musical Language (GML) for brain-inspired hypercomputing. In *Soft computing: Theories and applications, advances in intelligent systems and computing*, ed. Millie Pant et al., 584. Singapore: Springer. https://doi.org/10.1007/978-981-10-5699-4_33.

Aigner, Martin, and Günter M. Ziegler. 2010. *Proofs from THE BOOK*. 4th ed. Berlin/Heidelberg: Springer.

Allen-Zhu, Zeyuan. 2018. Katyusha X: Practical momentum method for stochastic sum-of-nonconvex optimization. In: Jennifer G. Dy and Andreas Krause. Eds. *Proceedings of the 35th International Conference on Machine Learning (ICML 2018), Stockholmsässan, Stockholm, Sweden, July 10–15, 2018*. Proceedings of Machine Learning Research (PMLR) 80, 179–185.

Arute, Frank, et al. 2019. Quantum supremacy using a programmable superconducting processor. *Nature* 574 (7779): 505–510. https://doi.org/10.1038/s41586-019-1666-5.

Chalmers, David J. 1999. First-person methods in the science of consciousness. *Consciousness Bulletin*.

Csicsery, George Paul. 1991. Paul Erdös: N is a number. A documentary film.

Davies, Alex, et al. 2021. Advancing mathematics by guiding human intuition with AI. *Nature* 600 (7887): 70–74. https://doi.org/10.1038/s41586-021-04086-x.

Glazik, Christian, and Anand Srivastav. 2022. A new bound for the Maker–Breaker triangle game. *European Journal of Combinatorics* 104: 14. https://doi.org/10.1016/j.ejc.2022.103536.

Graovac, Ante, Ivan Gutman, and Nenad Trinajstić. 1977. *Topological approach to the chemistry of conjugated molecules*. Lecture Notes in Chemistry (LNC) 4. Berlin, Heidelberg: Springer-Verlag.

Gutman, Ivan, and Nenad Trinajstić. 1973. Graph theory and molecular orbitals, the loop rule. *Chemical Physics Letters* 20 (3): 257–260. https://doi.org/10.1016/0009-2614(73)85170-X.

Hofstadter, Douglas R. 1981. A conversation with Einsteins's brain. In *The Mind's I*, ed. Douglas R. Hofstadter and Daniel C. Denett, 430–460. New York: Basic Books.

Kumar, Ami. 2020. *Characterization of ictal/non-ictal EEG patterns and neuronal networks in childhood absence epilepsy*. Christian-Albrechts-Universität Kiel. https://macau.uni-kiel.de/receive/macau_mods_00000794. Accessed 27 June 2022.

Kumar, Ami, Ekaterina Lyzhko, Laith Hamid, Anand Srivastav, Ulrich Stephani, and Natia Japaridze. 2021. Differentiating ictal/subclinical spikes and waves in childhood absence epilepsy by spectral and network analyses: A pilot study. *Clinical Neurophysiology* 132 (9): 2222–2231. https://doi.org/10.1016/j.clinph.2021.06.011.

———. 2023. Neuronal networks underlying ictal and subclinical discharges in childhood absence epilepsy. *Journal of Neurology* 270 (3): 1402–1415. Published online November 12, 2022. https://doi.org/10.1007/s00415-022-11462-8.

Li, Xueliang, Yongtang Shi, and Ivan Gutman. 2012. *Graph energy*. New York: Springer.

Penrose, Roger. 1999. *The emperors new mind.* Oxford: Oxford University Press.

Raayoni, Gal, Shahar Gottlieb, Yahel Manor, et al. 2021. Generating conjectures on fundamental constants with the Ramanujan Machine. *Nature* 590 (7844): 67–73. https://doi.org/10.1038/s41586-021-03229-4.

Satsangi, Prem Saran. 2012. Vision Talk. Neuro-environmental cognitive spiritual phenomenology. *Special Forum on 'Quantum Theory and Science of Consciousness' at the International School on Quantum and Nano Computing Systems and Applications (QANSAS 2012).* Organized at Quantum-Nano Systems Centre, Dayalbagh Educational Institute, Deemed to be University, Agra, India, November 28, 2012.

———. 2014. Vision Talk. Satyam-shivam-sundaram revisited in the perspective of grand macro/micro cosmology. *Special Forum on Consciousness : Integrating Eastern and Western Perspectives at International School on Quantum and Nano Computing Systems and Applications.* Organized at Quantum-Nano Systems Centre, Dayalbagh Educational Institute, Deemed to be University, Agra, India, November 26, 2014.

———. 2022. Vision Talk. (Toward Evolutionary) Science of Consciousnes. *Winter Session of DSC 2022.* Dayalbagh Educational Institute, Deemed to be University, Agra, India, January 1, 2022.

Schmidhuber, Jürgen. 2015. Deep learning. *Scholarpedia* 10 (11): 1527–1554. https://doi.org/10.4249/scholarpedia.32832.

Shor, Peter. 1997. Polynomial-time algorithms for prime factorization and discrete logarithms on a quantum computer. *SIAM Journal of Computing* 26 (5): 1484–1509. https://doi.org/10.1137/S0097539795293172.

Silver, David, et al. 2016. Mastering the game of GO with deep neural networks and tree search. *Nature* 529 (7587): 484–489. https://doi.org/10.1038/nature16961.

Srivastava, Dayal Pyari, Vishal Sahni, and Prem Saran Satsangi. 2011. Graph-theoretic quantum system modelling for information/computation processing circuits. *International Journal of General Systems* 40 (8): 777–804. https://doi.org/10.1080/03081079.2011.602016.

———. 2016. Modelling microtubules in the brain as n-qudit quantum Hopfield network and beyond. *International Journal of General Systems* 45 (1): 41–54. https://doi.org/10.1080/03081079.2015.1076405.

———. 2017. From n-qubit multi-particle quantum teleportation modelling to n-qudit contextuality based quantum teleportation and beyond. *International Journal of General Systems* 46 (4): 414–435. https://doi.org/10.1080/03081079.2017.1308361.

Turing, Alan. 1937. On computable numbers, with an application to the Entscheidungsproblem. *Proceedings of the London Mathematical Society* 42 (1): 230–265. https://doi.org/10.1112/plms/s2-42.1.230.

The Rendered Universe: Why Virtual Reality Unlocks the Secret of Consciousness

David Christopher Lane

Abstract Today with the advent of ever-increasing scientific advances, we are on the threshold of better understanding why consciousness evolved and how it works. This became apparent when the computational semblance of the mind faltered when neuroscientists and philosophers realized that awareness was not merely digital. However, with virtual reality (VR), augmented reality, and mixed reality (MR), our models of how consciousness functions are becoming much clearer to grasp since we now have the ability to simulate in four dimensions an all-encompassing environment artificially. Consciousness is a forging mechanism, which by its attentional posturing enables it to fully immerse in a world of its own making, even as it remains mostly unaware of how such a magical performance occurs. Using VR and MR as touchstones allows us to better appreciate how our brains construct the world around us. By exploring manufactured VR accouterments and the varied vistas they can create, we have—perhaps for the first time in our history—the necessary tools to synthetically reconstruct how the brain perceives and interacts with reality. This essay is an examination of how the advent of virtual reality changes our understanding of human consciousness and how future researches will benefit by employing its many iterations.

Keywords Virtual reality · VR · Consciousness · Simulation theory · Artificial intelligence

> Most of us have spent some time wondering how our brain works. Brain scientists spend their entire lives pondering it, looking for a way to begin asking the question, How does the brain generate mind? The brain, after all, is so complex an organ and can be approached from so many different directions using so many different techniques and experimental animals that studying it is a little like entering a blizzard, the Casbah, a dense forest. It's easy enough to find a way in – an interesting phenomenon to study – but also very easy to get lost. (*Susan Allport*)

D. C. Lane (✉)
Departement of Philosophy, Mt. San Antonio College, Walnut, CA, USA
e-mail: dlane@mtsac.edu

P. S. Satsangi et al. (eds.), *Consciousness Studies in Sciences and Humanities: Eastern and Western Perspectives*, Studies in Neuroscience, Consciousness and Spirituality 8, https://doi.org/10.1007/978-3-031-13920-8_9

Although Plato didn't know about virtual reality headsets or what such a new-fangled technology could evoke, he did understand better than most that humans do not see reality as it is, but rather how we filter it moment to moment. His famous *Allegory of the Cave*, which is arguably the single greatest thought experiment in the history of philosophy, explains (via the mouthpiece of Socrates) very simply that we are prisoners who confuse shadows on the wall, which are artificially manufactured by an unseen burning fire, as if they were real "men and other living things" (Plato, section 514a–520a).[1] It is an illusion, and as such the imprisoned men are duped. Today with the advent of ever-increasing scientific advances, we are on the threshold of better understanding why consciousness evolved and how it works.

Professor Donald Hoffman, the controversial cognitive scientist at the University of California, Irvine, has taken *Plato's Allegory* and given it a computational update, by using the metaphor of a desktop computer, where all we see is the user interface, not the underlying software programming or the hardware circuitry of electron exchanges. Accordingly, what we see/hear/smell/touch around us is a filtering mechanism developed by evolution to ensure that we focus on what will allow us to live long enough to pass on our genetic code (Hoffman 2000, 2019a, b).[2] In the past century we have come up with various models to understand consciousness – from Dennett's *multiple drafts* to Leary's *eight circuits* to Dehaene and Changeux *global neuronal workspace,* etc. – yet each has been hindered by their use of limited metaphors.

For instance, Roger Penrose, the Nobel prize-winning mathematician-physicist (2020, for his work on black holes and the theory of general relativity), has long argued that consciousness is non-algorithmic and thus cannot be explained by digital (or purely computational) models. In his two contentious books on the subject (Penrose 1989, 1996), Penrose employs Kurt Gödel's famous *Incompleteness Theorem* (circa 1930) to demonstrate that any purely computational explanation of the mind will fall short. Instead, he argues that quantum theory must be taken into any account of consciousness and the quantum world is probabilistic and non-digital, given Heisenberg's Uncertainty principle. Although Penrose has found his supporters (particularly the American anesthesiologist Stuart Hameroff with whom has co-authored and co-presented a number of pregnant essays and talks), his views have been severely critiqued on different fronts from both professional philosophers, e.g., Churchland and Grush (1998), and mathematical physicists, particularly Max Tegmark (1999). Stuart Hameroff, however, has come out strongly in defense of Penrose's insistence on a quantum basis for understanding consciousness citing counter-factual studies that upend both Churchland and Tegmark, especially in his rebuttal to Churchland (Hameroff 1996), and more recently in his co-authored response to Max Tegmark.

[1] Also refer to the short film elaborating on how the cave analogy can be likened to the brain and its limitations, https://youtu.be/lTHLptra6WA

[2] See the filmed lecture I presented to the philosophy club at the University of California, Irvine, which elaborates on his desktop metaphor in contradistinction to a more updated, VR analogy, https://youtu.be/0DJuOXRs38M

However, with the advent of virtual reality (VR), augmented reality (AR), and mixed reality (MR), our models of how consciousness functions are becoming much clearer to grasp, since we now have the ability to simulate in four dimensions an all-encompassing environment artificially. Using VR and MR as touchstones allows us for the first time to better appreciate how our brains construct the world around us via incoming data streams. Consciousness is a forging mechanism, which by its attentional posturing enables it to fully immerse in a world of its own making, even as it remains mostly unaware of how such a magical performance occurs.

Attentional awareness is a cornerstone to understanding how our awareness operates moment to moment. While I am typing away here on the computer, my attention is focused on the screen to see how the letters are conjoined with occasional glances at my keyboard. But what is behind me at this moment (the numerous books, the art work, and the closed door) remains unnoticed. The human field of view (FOV) is slightly about 210 degrees, and thus our brains need only render within that parameter and then only when we focus in any specific direction. Knowing this limitation has been very helpful in developing VR headsets, since the smaller the frame of reference the less CPU (central processing unit) is necessary. Currently, the Oculus 2 FOV is just under 100 degrees, which is less than optimal in terms of simulating our normal range of vision. But what is key here is eye tracking since that directly correlates to our attentional span. These technical details are important elements for any future understanding of consciousness, and this why knowing the mechanics behind VR is so helpful in providing us with a rich and useful parallel modeling system.

Simply put, we live not in an objective cosmos distinct from our interactions with it, but rather in a rendered universe, where our participation and our observations are fundamental to our interpretations of it. This becomes transparently obvious when we realize that the most sophisticated virtual reality headset known to exist is our own brain. However, by exploring manufactured VR accouterments and the varied vistas they can create, we have (perhaps for the first time in our history) the necessary tools to synthetically reconstruct how the brain perceives and interacts with reality. This essay is an examination of how the advent of virtual reality changes our understanding of human consciousness and how future researches will benefit by employing its many iterations.

1 The Virtual Brain

Brian Greene, the physicist polymath at Columbia University, recently made a parous observation about why human beings cannot visualize the quantum world or extraordinarily large time scales. Our bodies evolved to a middle-range environment, where what was vital for our evolutionary survival was dependent on resources within a certain physical parameter. Because of this Darwinian dictum of eat or be eaten, our brains didn't develop the capacity to truly envision what ten dimensions may be like or what occurs at the level of a photon or time durations or what was before the Big Bang, etc. Yes, we can imagine and we can speculate, but even here we have limited models with which to work (Greene 2020).

Yet, Greene went on to speculate that with the advent of virtual and augmented reality, intertwined with Artificial Intelligence (AI), it may be possible in the future for humans to stretch their cognitive abilities in ways unimaginable before the emergence of such technologies. Elon Musk's championing of neural lace/link is but a stepping stone for a wholly different way of thinking and being. It may well be that the very reason we have yet to see a major breakthrough in the study of human consciousness is precisely due to our present cranial limitations. Just as the field of astronomy achieved greater success with building better and more refined telescopes, and molecular biology blossomed with more powerful microscopes, the study of the brain and awareness necessitates radically new tools.

Our future devices and instrumentations need to go far beyond electroencephalography (EEG), positron emission tomography (PET), magnetic resonance imaging (MRI), and functional magnetic resonance imaging (fMRI). The immense complexity of the brain with its 86 billion or so neurons curled up in a compacted mound of gray matter weighing just three pounds cannot be approached, much less comprehensively understood, with crude instrumentation. Given that our technologies must evolve to properly inspect such an intricate compound, so too must our own mental maps, lest we be too inadequate to grapple with nature's greatest secret. This is why I believe that virtual and augmented reality is a dramatic stride in the right direction.

Lest there be any confusion, I am not arguing that VR is a perfect transparency for understanding consciousness. Rather, I am pointing out all the pertinent details about virtual reality hardware and software that help us improve our models of how consciousness arises in the way it does. VR is a progressive tool in our arsenal, and as such it can and will lead us to better grasp the underpinnings of self-awareness. This includes studying why our critical fusion frequency (CFF, where perception of light moves from sporadic to continuous) is within the range of "7 Hz to ~200 Hz" and not smaller or greater. The very nuts and bolts of VR are predicated upon a functional reconstruction of these neural limits, and because of this, it acts as both a subjective and objective pathway to simulate a key feature of our own awareness.

For example, in 2019 I attended the Oculus Connect 6 Conference held in San Jose, which highlights the latest developments in virtual reality hardware and software. One of the keynote speakers was the video gaming pioneer, John Carmack, who pointed out two of the most important and fundamental "hard" problems in getting virtual reality to work seamlessly: spot-on rendering and time corrected latency. First, when you don a VR headset (such as the standalone *Oculus Quest 2* or HTC Vive's *Cosmos Play*), wherever you look the scene becomes clearer by a process technically known as foveated rendering, "which uses an eye tracker integrated" with whatever headset you may be using that is specially designed "to reduce the rendering workload" systematically compressing the visual field "in the peripheral vision (outside of the zone gazed by the fovea)." Some headsets use a "less sophisticated variant called fixed foveated rendering [which] doesn't utilize eye tracking and instead assumes a fixed focal point." John Carmack and Michael

Abrash, chief scientist for Oculus at Facebook (now known as Meta), who also spoke at the Oculus 6 Conference, argued that perfecting rendering is the key to making virtual reality almost indistinguishable from our regular everyday sense of reality. What is most intriguing about this in relation to our own particular form of attention is how our brain is literally a rendering machine, which provides us with only a small smattering of information via our senses whenever we pay attention to a particular occurrence or scene.[3]

We don't ever see objects or persons or events in their totality, but merely the tiniest of data slivers, which are evolutionarily filtered for our survival and not necessarily for us to better understand the cosmos at large. Yet, we remain oblivious of how such neural rendering works and how it delimits the world we inhabit. In a very real sense we are born wearing the universe's most sophisticated VR headset and live as if we aren't aware of this requisite fact. Understanding the mechanics behind VR rendering and its various permutations is undoubtedly helpful in better appreciating how our own brains render reality via our five senses. It may seem odd to realize that the universe we behold is "versional" and not a clear transparency of what is. This becomes readily apparent whenever we suffer a case of vertigo, drink too much alcohol, or suffer severe depression. Perhaps the most telling case is when we are sleep deprived. Stay up too long and the so-called "real" world turns magically hallucinogenic.

Other species have different brain-induced versions, and we cannot accurately gauge what they experience if we only have limited knowledge about their respective rendering apparatuses. Attention and rendering are correlative, and this can have startling implications. As James B. Glattfelder explains:

> Most people who have journeyed to the DMT realm argue that what they have experienced is just as real as experiences made during the sober waking state – if not much more so. As mentioned, it is tempting to disregard such experiences as hallucinations – nothing more, nothing less. However, this raises philosophical questions. Recall that neuroscientists are quite clear: Our perception of reality is a hallucination tethered by a bit of sensory input. What I experience through sober waking consciousness is an elaborate virtual reality rendering in my brain. The nature of this hallucination can be modulated by the chemical composition of the brain. Crucially, the DMT produced by my own body has the potential to intrinsically and naturally modify the contents of my consciousness – by "teleporting" my mind to the DMT realm. Indeed, simply by shutting down the left hemisphere of the human brain appears to induce travels of the mind, allowing consciousness to access a "place" of peace and euphoria. How can I then be certain that the fidelity and truth content of the one hallucination is superior to the other? What epistemic guarantee do I have that would allow me to negate the reality of the DMT realm? How can I exclude the possibility that reality is indeed "queerer than we can suppose" and everything I have ever known about is metaphorically restricted to a tiny isolated island in a vast archipelago of transcendental existence? (Glattfelder 2019: 568–69)

---

[3] In the short film, *The Radiant Form: How Mixed Reality Will Transform Us and the World We Live In*, I elaborate on how computer scientists are demonstrating how augmenting our day-to-day world (via glasses or some other device) will transform how we experience the world. See https://youtu.be/2jU_WodDnRc

Whereas VR headsets are the result of sophisticated machinery along with complex software programs, our central nervous systems are the elongated result of natural selection operating on emergent organisms trying to survive within a competitive environmental landscape. What we get is what allows us to eat and mate, and thus there is no evolutionary pressure to reveal to us how the brain conjures the reality we witness. Evolution by natural selection is the keystone for understanding the development of our unique psychologies. As Charles Darwin himself prophesized at the very end of his groundbreaking text: "In the distant future I see open fields for far more important researches. Psychology will be based on a new foundation, that of the necessary acquirement of each mental power and capacity by gradation. Light will be thrown on the origin of man and his history" (Darwin 1859).

We are under the mistaken illusion that our brains are crystal-clear windows into the world around us, whereas our central nervous system is (as Aldous Huxley rightly pointed out in his famous book *Doors of Perception*) a filtering mechanism that constructs a peculiar reality adapted to our biological and cultural needs.

The VR industry is driven to provide us with a similar sort of magic, except that at this stage in its development, it still has a long way to go before it becomes *Matrix* like. Yet, because it is in its infancy there is much that VR can tell us about how self-awareness may have first arisen. One of the most outstanding revelations in VR is how it can convince us that projected images not only look real, but that they convey upon us a sense of presence. Why this is so tells us much about how our neurology is jerry-rigged for interacting with key environmental cues. The feeling of presence, or perceived "animism," is the one feature in VR that distinguishes it from almost all other technologies. Radio, television, and even 3-D films fall short in providing its viewer with a sense that someone is really in the room with you, whereas the hallmark of good VR is that it gives one the uncanny feeling of "otherness." This is astonishing when one stops to consider that it is all a visual and auditory mock-up occurring but one to two inches away within a set of paired goggles.

Because of VR technology we have a better understanding of how to trick our brains into believing something is life-like despite it being merely a manufactured artifice. This is extremely helpful as it allows us a more accurate model of how a complex neural net goes about convincing us that what we see, hear, touch, and feel is truly authentic and which demands our focus. In other words, we have learned to discriminate between what is vital for our survival and what is merely imaginative. That the line between both can too often be blurred is precisely why humans suffer from a wide range of mental ailments, not the least of which schizophrenia, dementia, and Alzheimer's.

Whenever I don my Oculus Quest 2 VR headset I am amazed at how easily I get absorbed into a completely different realm, even though intellectually I know that it is merely a sophisticated simulation. How does such a device actually work so well as to deceive us?

Zaynah Bhanji explains:

> VR headsets either use two LCD displays (one per eye) or two feeds sent to one display. Headsets also have lenses placed between your eyes and the screen, which are used to focus and reshape the picture for each eye. They create a stereoscopic 3D image by angling the

> two 2D images. This is because the lenses mimic how each of our two eyes see the world very slightly differently. VR headsets also need to have a minimum frame rate of at least 60 frames per second in order for the user to not feel sick. Current VR headsets are able to go way beyond this, with Oculus and the HTC Vive at 90 frames per second and PlayStation VR at 120 frames per second. For VR to work properly, when you move your head up and down or side to side or tilt your head, the picture has to move properly with your head. Headsets use a system called six degrees of freedom (6DoF), which looks at your head's position in terms of the X, Y, and Z axis to measure head movements. There are a couple of different components-use in a head- tracking system, including a gyroscope, accelerometer, and a magnetometer. The PlayStation VR also uses 9 LEDs around the headset, which are used to provide 360- degree head tracking by using an external camera that monitor these signals. (Bhanji 2018)

Although the VR set is but an inch or so away from one's eyes, the depth of vision, the feeling of overwhelming presence when fighting Darth Vader with a light saber, or flying through our solar system and touching Jupiter, is truly mind boggling. Yet, and this is a point that needs to be underlined: the whole display is a sophisticated *digital deception*. Our brains are wired such that it doesn't take too much to convince us that an illusion is real, even when we are aware of the mechanics behind such magic.

This is why I believe virtual reality and augmented reality technologies are pivotal instruments in exploring how consciousness may arise from a subset of neuronal components. In an invited talk I gave to the Philosophy Club at the University of California, Irvine, in 2019, I touched upon the reasons why. Earlier religious traditions, as found in Buddhism, Jainism, Hinduism, and Christian Gnosticism, argued that we live in an illusion which betrays its real origin. Just as the user interface on our smart phones hides the inner workings of computer programming and electron circuit design, the world of appearances masks the underlying physics and chemistry which give rise to it.[4]

Herein lies the secret of consciousness, according to Donald Hoffman, Professor of Cognitive Science at the University of California, Irvine, who believes that a radical new approach is needed to understand the emergence of self-awareness because evolution didn't shape us for accessing reality per se, but rather providing enough relevant data to survive long enough to pass on our genetic code.

Hoffman believes that with the desktop interface we have for the first time the appropriate metaphor to understand how something that looks real on the surface is anything but. In Amanda Gefter's essay entitled "The Case Against Reality" (*Atlantic Monthly*, April 25, 2016a, b), Hoffman explains why we are fundamentally in error about the world we see.

> Snakes and trains, like the particles of physics, have no objective, observer-independent features. The snake I see is a description created by my sensory system to inform me of the fitness consequences of my actions. Evolution shapes acceptable solutions, not optimal ones. A snake is an acceptable solution to the problem of telling me how to act in a situation.

[4] For more on the concept of maya (both from religion and science), see Lane and Diem-Lane (2017) and the film based on the text, https://youtu.be/YckgH_T8BUA

> My snakes and trains are my mental representations; your snakes and trains are your mental representations.

While I agree with much of what Hoffman suggests (except that I find his concept of "conscious agents" confusing at best), I think virtual reality is a far more powerful metaphor for explaining how and why consciousness works, since it is a totally encompassing environment similar in so many ways to how we currently experience the world both within and without via our own subjective awareness. Yet, the VR headset is a mind manipulator par excellence, just as our own consciousness is an informational playground.

Of course, the fundamental difference is that with VR we can inhabit instantly what before we could only dimly imagine. But in both VR (virtual reality) and SA (self-awareness) we are as Donald Hoffman explains "not seeing the innards of reality." As he elaborates,

> Suppose there's a blue rectangular icon on the lower right corner of your computer's desktop – does that mean that the file itself is blue and rectangular and lives in the lower right corner of your computer? Of course not. But those are the only things that can be asserted about anything on the desktop – it has color, position and shape. Those are the only categories available to you, and yet none of them are true about the file itself or anything in the computer. They couldn't possibly be true. That's an interesting thing. You could not form a true description of the innards of the computer if your entire view of reality was confined to the desktop. And yet the desktop is useful. That blue rectangular icon guides my behavior, and it hides a complex reality that I don't need to know. That's the key idea. Evolution has shaped us with perceptions that allow us to survive. They guide adaptive behaviors. But part of that involves hiding from us the stuff we don't need to know. And that's pretty much all of reality, whatever reality might be. If you had to spend all that time figuring it out, the tiger would eat you. (Gefter 2016a, b)

Hoffman makes an astute observation when he points out that most of the cosmos is hidden from us precisely because it would harm our survival and "that's pretty much all of reality." It is as if the universe were a gigantic ocean, and all we ever observed of it was a tiny sliver of sea-weed. This isn't merely hyperbole, since much of the electromagnetic spectrum remains consciously undetected by us, just as millions of neutrinos pass through us without our notice. We are blind and dumb and mute to the nuts and bolts of all that is on offer. In sum, nature has only provided us with a sampling of what is possible.

As I mentioned earlier, our brains are rendering machines, and as they parse tiny bits of incoming data streams they fashion the very worlds we inhabit. But this very act of rendering this or that scene is predicated to a large degree on where we focus our attention. In a very real sense we have limited bandwidth and can only take in a tiny fraction of all that is available to us at any particular moment. But herein lies the key: the limitations of our brain (like the limitations of our present-day VR headsets) means that evolution has modeled us to inhale what is sufficient for our survival; it doesn't allow for too much continual streaming of information, since that would in effect make us catatonic and unable to respond since we would be entirely swamped by data overload. Moreover, how we respond to differing situations is directly correlated to time appropriation. Our brains are sophisticated motion detectors, and the way we navigate our day-to-day interactions is dependent on how well we measure time intervals. Each species has its own unique time-lapse camera, so to say, since nothing arrives on time but rather in time, and even then, it has much to do with environmental conditions. Mark Solms (2021) makes the compelling argument that we have mistakenly confused consciousness with intelligence when, in point of fact, it has much more to do with emotions and feeling. Furthermore, he argues that it is precisely how we modulate those sensations which gives rise to consciousness as a predictive and projective vehicle for how we address changes in our own bodies and the alternating landscape we travel within.

In virtual reality, this is called the latency issue, which has been a major obstacle for software programmers to get it right. Even the slightest delay or lag in VR can alter the user's experience and jolt them away from the game's surreal engagement. Most of us have experienced latency or time dilation when suffering from an illness (such as vertigo) or in severe cases such as hypoxic-ischemic encephalopathy, which has been shown to impact somatosensory neuronal responses (Wu et al. 2011).

In a remarkably salient paper entitled, "Towards Low-Latency and Ultra-Reliable Virtual Reality," Mohammed S. Elbamby, Cristina Perfecto, Mehdi Bennis, and Klaus Doppler (from the *Centre for Wireless Communications*, University of Oulu, Finland) elaborate:

> In VR environments, stringent latency requirements are of utmost importance for providing a pleasant immersive VR experience. The human eye needs to perceive accurate and smooth movements with low motion-to-photon (MTP) latency, which is the lapse between a moment (e.g. head rotation) and a frame's pixels corresponding to the new FOV have been shown to the eyes. High MTP values send conflicting signals to the vestibulo-ocular reflex (VOR), a dissonance that might lead to motion sickness. There is broad consensus in setting the upper bound for MTP to less than 15–20 ms. Meanwhile, the loopback latency of 4G under ideal operation conditions is 25 ms. The challenge for bringing end-to-end latency down to acceptable levels starts by first understanding the various types of delays involved in such systems to calculate the joint computing and communication latency budget. Delay contributions to the end-to-end wireless/mobile VR latency include, sensor sampling delay, image processing or frame rendering computing delay, network delay (queuing delay and over-the-air delay) and display refresh delay. Sensor delay's contribution (<1 ms) is considered imperceptible by users, and display delay (≈10 15 ms) is expected to drop to 5 ms [6], which leaves 14 ms for computing and communication. Both computing and communication delay serve as delay bottleneck in VR systems. Heavy image processing requires high computational power that is often not available in the local HMD GPUs. Offloading computing tasks to remote cloud servers significantly relieves the computing burden from the users' HMDs at the expense of incurring additional communication delay in both directions. Unlike MR and AR where uploading video streams to the cloud may be required, uplink communication delay due to offloading the computing task to the server is typically very small in VR, owing to the small amount of data needed, e.g., user 3 tracking data and the interactive control decisions. However, the downlink delivery of the processed video frames in full resolution can significantly contribute to the overall delay. Current online VR computing can take as much as 100 ms and communication delay (edge of network to server) reach 40 ms. Therefore, relying on remote cloud servers is a more suitable approach for low-resolution non-interactive VR applications, where the whole 360° content can be streamed and the constraints on real-time computing are relaxed. Interactive VR applications require real-time computing to ensure responsiveness. Therefore, it is necessary to shrink the distance between the end users and the computing servers to guarantee minimal latency. (Elbamby et al. 2018: 2–3)

In analyzing the nitty gritty of human consciousness and its relationship to the brain, understanding rendering and latency is a fundamental necessity as any deficiency in one or the other or both can have catastrophic consequences. Nature over the course of millions of years came up with solutions that have served us well in our continuing struggle of existence. Because VR technology is attempting to seamlessly replicate what our brain has already achieved, it provides a tantalizing pathway for us to see how our own wetware accomplished such a remarkable feat. In many ways, VR and AR are perhaps the most practical means we presently have to understand the engineering nuts and bolts of consciousness. Even if such work proves insufficient to complete the entire task, the continued research and testing in virtual and augmented reality (replete with all the devices that will be created and tested as a result) will undoubtedly help us better grasp the intricate difficulties that must be resolved before self-reflective awareness can be operative. In this regard, I don't think ample progress can be made in consciousness studies if we somehow bypass an exhaustive study of the brain and its labyrinthine structures. This is why I believe that philosophers of the mind should follow the lead of Patricia and Paul Churchland, who as professionally trained philosophers have spent their careers championing neuroscience as a necessary route for anyone interested in studying awareness. It makes absolutely no sense to argue that we can comprehend the human mind without first knowing how the brain works. This is akin to saying that whenever your car breaks down, it is unnecessary to open up the hood and look at the engine since the engine and the vehicle are mutually exclusive. Yes, there are certainly those thinkers (such as computer scientist, Bernardo Kastrup, and New Age thinker, Ken Wilber) who believe that consciousness is not a product of three pounds of gray matter but is a distinct force in itself. Perhaps one of the most surprising developments in the scientific study of consciousness has been the push by a number of notable thinkers (from Sam Harris to Christof Koch to Giulio Tononi) to seriously entertain variations of panpsychicism, where consciousness is embedded within nature itself. Some thinkers, such as David Chalmers, suggest that consciousness may be a force in itself, such as electric charge. For a highly readable overview of this perspective, refer to Annaka Harris (2019).

But in their disdain for mechanism, they tend to postulate myths that are untestable and, even if believed and followed, don't produce any practical benefits along the way to its own falsification. Whereas a neural focus on consciousness, even if it ultimately turns out to be an insufficient explanation, produces all sorts of beneficial byproducts, not the least of which is eliminating long-held intuitions as mistaken. We should remember that the history of science is to a very large measure the overthrow of lesser gods, whether it be Ra, Thor, Zeus, or the notion of an Intelligent Designer. It is precisely because of the process of eliminative materialism that we got rid of an outdated astrological interpretation of the cosmos and established the progressive science of astronomy; likewise, when we got a deeper understanding of how molecules and elements behaved, we discarded alchemy for chemistry (Lane and Diem-Lane 2012).[5]

Thus, the reason virtual and augmented reality is such a powerful tool is that it allows for us to create hardware that by its very nature must mimic how we perceive, hear, and interact with the world, even if it is only a simulacrum of what we experience around us. Annaka Harris, author of the panpsychicism influenced text, *Conscious: A Brief Guide to the Fundamental Mystery of the Mind*, defines consciousness simply as "experience," the qualia of what it is like to see a blue ocean or feel in love. It is a wholly subjective sense which cannot be objectified and studied without losing the very quality that defines it to "be like something and not something else." Interestingly, VR is predicated on experience itself and creating a sense of presence. That is its chief draw and what attracts gamers and others to it. Of course, we must keep in mind that all models are less than the territory to which they point, and therefore we should not expect that VR technologies in themselves will unravel the secret to human consciousness and its emergence. But, like the simpler 2D computational models it replaces, it is a step forward and yields insights that were not possible before.

This point was driven home to me when I met Kenneth Williams one day outside my office at Mt. San Antonio College. Five years ago, Kenneth was T-boned in a horrific car accident, which left him quadriplegic. He was fascinated by what virtual reality had to offer, since an Oculus Quest headset would allow him to experience places and things that he couldn't, such as what was it like to ride a scary roller coaster or ride in a rocket ship or climb a mountain. A number of VR applications and games provided him with these experiences and more. Kenneth was experiencing what it was like to do something that was otherwise denied to him (Lane 2019). I would argue that consciousness evolved because it ushered in a way to have a depth of experiences within our own cranium (and ponder their past, present, and future iterations) without the life and death consequences of externalizing those imaginings in the real world. In other words, our brains are nature's own version of a biological VR headset. It makes perfect sense that in trying to create our own synthetic versions, we will get a richer and more robust understanding of how evolution created one by naturally selected means.

2 The Virtual Reality of Consciousness

> How does the water of the brain turn into the wine of consciousness? (David Chalmers)

In trying to understand consciousness differing models have been proposed over the years by scientists and philosophers, ranging from variations of Cartesian dualism

[5] Also see Patricia Churchland's explanation of eliminative materialism: https://youtu.be/xrFqhOvNtWM. Also refer to very short three-part video series, "Eliminative Materialism," https://youtu.be/dZmue10YKvc

(mind is primary, body is merely a vehicle) to neural correlate theory (mind is the brain and arises due to complex interactions). Realizing that all models, no matter how sophisticated, will fall short to some measure, the key is to utilize ones that provide us with newfound insights and hitherto unknown predictive powers. The advent of virtual and augmented reality may serve as a useful pathway to better understand why consciousness evolved the way it did, primarily because the brain itself operates as a VR system, albeit a much more sophisticated one than Oculus Rift or HTC Vive.

Everything we have ever experienced is modified by our central nervous system. Our brain refashions the world we see, hear, touch, and smell by transfiguring incoming data points (light, sound, etc.) into chemical-electrical signals that are reconstructed – via our elaborate neural net – into the world we apprehend around us. We never experience the exterior cosmos "as it is" but only as our own biologically evolved VR headsets reconceive it. Whenever I use an Oculus-related device (from Rift to Go to the Quest), I am always amazed by how such a headgear with 6 degrees of freedom and a 1440 × 1600 per-eye resolution makes me feel as if I have entered into a new world, and all of this with a device that weighs less than a pound. VR is a trickster, since by its software and hardware design, one responds as if the images that arise have a sort of independent reality apart from its technology. Yet, it is that very technology which is creating the illusion that what I perceive is trans-machine like. Immanuel Kant and others following his lead turned philosophy on its head by redirecting our focus back to the limits of our cranial capacity and how such a limitation, in turn, invariably confuses the questions we ask about the universe at large. In the twentieth century, Ludwig Wittgenstein refined this even more by showing how language itself is a culprit in our pursuit of philosophical inquiries, since the words we use are contextual and do not lend themselves to atomic purity. To his great credit, Wittgenstein realized the error of his ways years later after gallantly attempting to first clear out the cobwebs of faulty language usage in his groundbreaking book *Tractatus Logico-Philosophicus*. Language is a game, and how we play it depends on a number of contingent factors, not the least of which is on our prior agreed-upon assumptions. Today, with the ever-growing successes in neuroscience, we have entered a new world of understanding how the brain works and how it evolved not so much to understand the world at large (in some ultimate, ontological sense) but rather how best to survive within it. Virtual reality provides us with an instructive portal by which to better grasp how the brain develops elaborate simulations that help us better navigate future possibilities.

3 The Multiverse in Less Than Two Inches: Exploring Spatial Dimensions in Virtual Reality

I experience multiple dreams every night, each overlaid with each other and all occurring within a relatively tiny geometric space: my brain. Magically, whenever I am in the ocean in my dreams, I can see for miles to the distant islands or the setting sun. Sometimes I am in India and roaming through wide open fields filled with massive golden temples. Other times I am book hunting in large bookstores, searching for an obscure tome. No matter what I dream, however, all of it is bounded within a simulation of my own making, even as I am quite unaware of the mechanics behind how it is done.

There is a famous (but inaccurate) snipe about Catholic theologians such as St. Thomas Aquinas and Duns Scotus during the middle ages debating "how many angels can dance (or stand) on the top of a pin?" I remember Father Costello, my freshman religion teacher at Notre Dame High School, asking us the same rhetorical question and to which he immediately responded, "an infinite number, since angels are spirits (distinct from minds) and don't have bodies".

Daniel Dennett (1996) makes the pithy observation that dabbling in metaphysical speculations is akin to playing tennis and being allowed to serve without a net. Whereas, in the world of physics, scientists are forced to serve with the net up. The metaphor is simple to unpack: physicists must play by the rules of evidence, whereas metaphysicians can postulate almost any sort of nonsense, since the rules of verification don't apply.

While this is obviously silly metaphysics, it does provide me with a more revealing neurological correlate: how many dreams can be embedded within the same neuronal architecture? Or, more precisely, how is it that one small chunk of matter can produce an almost infinite array of dreams within itself – each of which give the dreamer the illusion of almost limitless boundaries?

Technologically speaking, because of VR we now have an analogy to how our brains reconstruct reality within tiny spatial dimensions, though providing us with almost infinite landscapes. Nature, apparently by trial and error, developed a hack whereby organisms of sophisticated complexity can project vistas of immense size and scope without having to go beyond the borders of its neuroanatomies. Today, with VR, one can access a multiplex of possible adventures with which to climb mountains, fly in the air, or simply play golf on wondrous courses that defy imagination.

In mystic literature there has been much made about how through deep meditation or prayer one can access hitherto unknown realms of consciousness. We learn about extraordinary experiences when one undergoes a Near-Death Experience (NDE) or has an out-of-body experience (OBE). Sometimes we believe that such hidden realms are in a spatial dimension far away from our bodies, as if we need to travel to distant astral galaxies. But a closer analysis of such luminous excursions indicate that nothing of the sort occurs. Everything we experience (whether ordinary or supermundane) is a projection within our own bodies. As odd as it may sound at first, nothing we experience is "outside" us since everything we have ever seen, touched, heard, felt, and smelled is processed through chemical-electrical signals by our central nervous system, which then displays the results of its machinations to us, though we remain oblivious of the details of how it transpires. Undoubtedly, there is a world exterior to ourselves, but we only know that *after* the differing lines of physical information impinge upon us and get transformed by our complex neural processing systems. Everything is modulated, everything is filtered.

Thus, to understand our own self-awareness it is vital that we understand the mechanics behind how such reconstructions are accomplished. We don't see reality raw and transparent but rather how our nervous system creates a mock-up or simulation based on disparate data sources. We construct the reality we see, even if do so unconsciously. Our biological mechanism, in other words, serves as a digital and analog censor and in performing such an editing task only provides us with its final renditions (for more on this, see Payne 2013).

Like Plato's story about prisoners in the cave, who only see shadows on the wall, we see only faint impressions of what our cerebral cortex paints for us on the canvas of our awareness.

This is precisely why VR is such a vital tool for us to better understand the logistics behind neural trickery. While it is certainly true that mystics have access to realms denied to the vast swaths in the human population, it does not follow that these "higher" experiences are not also a result of underlying processes which remain hidden from view and which are like our ordinary waking state.

In this regard, the *Tibetan Book of the Dead* and other like-minded texts are of great importance, irrespective of their pre-scientific explanations, since they intuitively understood the illusory and hallucinatory nature of the mind which is a conjurer par excellence (Lane 2018).[6] It is as if consciousness is a magical theater (to crib from one of Herman Hesse's more instructive metaphors) and we, as witnesses, have no clue how the magician – our brain – performs the tricks. The problem is that we have tended to confuse the magician's sleight of hand with how things really work, not realizing that the whole display is conjured without our awareness of how it is done.

I have long argued that mysticism suffers from its own epistemic "uncertainty" principle, since whenever one undergoes a mystical encounter, it is so overwhelming and feels so certain that one doesn't tend to "doubt" or question its veridicality during the experience. If one happens to do so at that precise moment, the numinous transport often becomes lessened or altogether collapses. This is reminiscent of Niels Bohr's concept of complementarity in physics. Once you know the wave function of an electron, say, then the particle aspect is hidden and vice versa. Likewise, the very nature of mystical experience is to forego questioning and doubt and merge with the beatific vision itself. If, however, one invokes rationality or analytics while in transcendent transport, then that very experience loses its efficacy. Now, this doesn't necessarily have to be the case, but a close analysis of esoteric Buddhist literature on visionary experiences and those of Baba Faqir Chand (1886–1981), a renowned shabd yoga mystic, clearly indicates the inherent difficulty of holding "two" minds at once while undergoing a mystical experience. For a more detailed analysis of this most unusual conundrum, refer to Chand (2015) and Becker (1985).

Thus, it is not inaccurate to say that the function of consciousness is to provide us with a "virtual" reality with which to survive for a set duration of time. How it does this is the secret to understanding its evolutionary emergence. It is for this

---

[6] Also refer to the film, *Hanging on the Gallows*, https://youtu.be/EGpCa6nZM3A

reason that our current development of virtual reality technologies is so instructive. We are in a very real sense recreating a simulation of the most fundamental of all simulations: our own self-awareness.

Michio Kaku, Professor of Theoretical Physics at the City College of New York and CUNY Graduate Center, articulates the "virtual simulation" hypothesis of consciousness very clearly in his book, *The Future of the Mind.* Kaku provides a nice summation of his theory in an interview for *Popular Mechanics* conducted by Alyson Sheppard:

> Well, I have a new theory of consciousness, and that is: Consciousness is all the feedback loops necessary to create a model of yourself in space, in relationship to others, and in time, especially forward in time. This means that animals are conscious, and we can even rank them numerically by counting the number of feedback loops involved in each of these behaviors. So, a thermostat would have one unit of consciousness that measures temperature. A flower would have maybe 10 units of consciousness because it measures temperature, sunlight, gravity, moisture, things like that. A reptile would have even more, maybe several hundred, because it locates its position in space. Then monkeys are even higher than that because they have to locate their ranking in society via emotions. We're at the highest level because we daydream. We see the future, we predict the future, and animals do not. Animals, to the best of our [knowledge], have no concept of our tomorrow. We are obsessed with planning, strategizing, and tomorrow. And that, I think, typifies human consciousness. (Sheppard 2014)

Kaku's theory, contrary to his claim that it is a new one, dovetails in several ways with other theorists in the field who are in agreement that having the ability to "simulate" all sorts of end game scenarios gives human beings an advantage over other creatures who lack such.

I often give two key examples in my philosophy courses to illustrate this:

1. Imagine we are on a 747-jet flying from Los Angeles to New York when we learn that all of the pilots have inexplicably passed out. The steward goes up and down the aisle to find anyone who has even the modicum of training to see if they can somehow take over and land the plane. There are several volunteers but they have no experience flying. However, there is a 14-year-old boy who has been playing a VR flight simulator game obsessively for the past 2 years and is conversant with some of the controls on large jets.

Given this scary scenario, who would you choose – the kid who has flight simulator experience or an older adult?

I suspect most of us would pick the 14-year-old and for good reason.

The second example I give is one that has been employed by a number of evolutionary psychologists and is perhaps the most illustrative. In our ancestral past, those individuals who could better predict the behavior of a predator would (on average) have more opportunities to pass on their genetic code. Thus, any creature (and it doesn't have to merely default to Homo sapiens) which can concoct multiple reactions and strategies within itself before acting on them has leverage over those who have a limited repertoire.

Yet, there is a downside in possessing a higher form of consciousness since it allows its owner to ponder all sorts of improbable events that will never actually happen. We too often can get ensnared with our ideas, our fantasies, our projections and begin believing all sorts of nonsense. For a robust virtual simulator to work it must conjure up a plethora of scenarios that mix and match incongruent narratives. In other words, for imagination to work it must be massively elastic. The glitch is that sometimes our simulations overtake us and we get entrapped within their corridors. To the degree that we can test our imaginings in the empirical world and acknowledge their successes and failures, we are regarded as somewhat sane; to the degree that we cannot, we may be viewed as somewhat insane. It is a thin line indeed that separates the two.

Virtual Reality provides us with an instructive portal by which to better grasp how the brain develops elaborate simulations that help us better navigate future possibilities.

Even popular science shows, such as David Eagleman's *The Brain* on PBS, provide a radical summation of how reality is virtually constructed:

> Dr. Eagleman takes viewers on an extraordinary journey that explores how the brain, locked in silence and darkness without direct access to the world, conjures up the rich and beautiful world we all take for granted. 'What is Reality?' begins with the astonishing fact that this technicolor multi-sensory experience we are having is a convincing illusion conjured up for us by our brains. In the outside world there is no color, no sound, no smell. These are all constructions of the brain. Instead, there is electromagnetic radiation, air compression waves, and aromatic molecules all of which are interpreted by the brain as color, sound and smell. Cutting edge graphics show that data from the outside are rendered into electrochemical signals inside the brain, which map meaningfully onto physical reality. Our experience of reality is an electrochemical rendition of the world outside. It is not a faithful rendition. Visual illusions are reminders that what's important to the brain is not being faithful to 'reality' but being able to perceive just enough so that we can navigate successfully through it. The brain leaves a lot out of its beautiful rendition of the physical world, a fact that Dr Eagleman reveals using experiments, and street demonstrations. We meet the men and women whose experiences of reality reveal important clues about how the brain constructs our own reality, including the Alcatraz prisoner who was locked in the notorious 'Dark Hole' for 29 days, and yet experienced richly colorful moments of reality. His experience, along with those who have experienced total sensory deprivation show us that even when sensory input to the brain stops, the show still goes on. Why? Because, amazingly, our senses–eyes, ears, skin, nose – only modulate an internally generated simulation of what the brain expects is out there. This so-called 'internal model of reality' allows us to move through the world, recognizing it rather than constructing it anew, moment by moment. We meet a man who is blind despite the fact that he has eyes that can see. His story reveals that it's the brain that sees, not our eyes. A woman with schizophrenia, whose psychotic episodes were her reality, emphasizes the fact that whatever our brains tell us is out there, we believe it. We have no way to get beyond what our brains allow us to perceive. Like the rest of the animal kingdom, we inhabit a miniscule slice of reality, and yet we believe it to be the whole picture. Eagleman also explores time, and takes a look at how–and why – the brain alters its perception of time, depending on the situation we find ourselves in. Time, it turns out, is not an absolute to the brain. Each one of our brains is different, and so is he reality it produces. What is reality? It's whatever your brain tells you it is. (Eagleman 2016)

Eagleman's last line goes right to the jugular since positing that "reality is what our mind tells us" means that our epistemology is brain-centric to the core. J. Allan Hobson, Charles C.-H. Hong, and Karl J. Friston point out why our consciousness is a virtual reality producer:

> In this view, the brain is equipped with a virtual model of the world that generates predictions of its sensations. This model is continually updated and entrained by sensory prediction errors in wakefulness to ensure veridical perception, but not in dreaming. In contrast, dreaming plays an essential role in maintaining and enhancing the capacity to model the world by minimizing model complexity and thereby maximizing both statistical and thermodynamic efficiency. This perspective suggests that consciousness corresponds to the embodied process of inference, realized through the generation of virtual realities (in both sleep and wakefulness). In short, our premise or hypothesis is that the waking brain engages with the world to predict the causes of sensations, while in sleep the brain's generative model is actively refined so that it generates more efficient predictions during waking. (Hobson et al. 2014: 1)

Perhaps what is most intriguing is what virtual reality holds for us in the future, particularly when VR and AI are intertwined. The disruptive nature of both will undoubtedly completely transform how we live. We have already witnessed the worldwide acceptance of smart phones (which lest we forget are powerful computers connected to the cloud) and how they can moment to moment augment our navigating intelligence in ways unimaginable decades prior.

4 The Physics of the Astral Plane

It is rich with irony that our consciousness which evolved as a virtual simulator would over time, and with the advent of sophisticated technology, develop virtual reality headsets that enable us to live in ultra-realistic worlds that before were merely figments of our imagination. As Andrea Diem-Lane wrote in, *The Avatar Project*,

> I have no doubt that with VR we will soon be witnessing an informational tsunami, the likes of which will swamp everything – from how we educate to how we doctor to how we entertain. If consciousness evolved as a virtual simulator to help us better insource varying survival strategies before outsourcing them, we can readily see that humans have now progressed to the point where we can actualize our imaginings in a real world. Simply put, our technology has finally caught up with our dreams so that we can consciously inhabit what we could only before imagine. Aldous Huxley was correct when he predicted that we would enter a *Brave New World*, but it is a different kind of Soma we will daily ingest than the one he anticipated. The drug of the future is virtual and it is information without boundaries. (Diem-Lane 2016: 36–37)

In the *Oculus Go* game "Virtual Virtual Reality," one is trained to put on a variety of goggles in order to experience different realities, each of which have certain tasks to accomplish. In each situation, though, one has the overwhelming feeling of being somewhere else, far away from the previous visited land. Yet, no one goes anywhere, since the entire panorama is being orchestrated within a lens diameter of 60 centimeters. We are surface surfers oblivious of the underlying physical reality that creates the information waves we ride.

We may be likened to Neo in the film, *The Matrix*, where he awakens from his artificially induced neural slumber to realize that his entire life is an illusion orchestrated by AI machines.

Jerry Coyne, the acerbic Biology Professor in the Department of Ecology and Evolution at the University of Chicago, heavily doubts Hoffman's agenda, posting on his widely read blog,

> First of all, just because a subatomic particle doesn't have an intrinsic property until we measure it, and that property is dependent on how we measure it, doesn't mean that macro objects don't have properties or, as Hoffman implies, don't exist. You can, after all, measure the momentum of a car, or of a stationary chair, with great accuracy. It seems to me that Hoffman is using a form of Chopra-ist woo here: claiming not only that certain claims about quantum mechanics extend all the way up to macro objects (yes, they do, but classical mechanics is adequate for macro phenomena, and that includes existence claims), but also that those objects don't exist outside of consciousness. In fact, Hoffman claims, like Chopra, that the only real thing that exists is consciousness. (Coyne 2016)

I think the more constructive way around this is to start with what we know and then proceed. This is precisely why I think new models of consciousness can be very helpful, even if they remain imperfect templates.

Thus, even if Hoffman's ultimate conclusions are premature or unwarranted, there is much merit in his desktop interface metaphor since it clearly illustrates that what we perceive doesn't necessarily reveal what mechanics are operative below the surface. VR is Hoffman's metaphor pushed to the next level and, as such, provides us with a deeper glimpse into the Matrix-like quality of our own self-awareness.

I don't think it is a stretch to imagine that in the not-so-distant future, virtual, augmented, and mixed reality will be so far advanced that we will be so mesmerized by what we see, hear, and even touch that it will be nearly impossible to distinguish artificially generated worlds from the so-called "real" ones that we were brought up in.

The implications, of course, are telling since it suggests that not only are we already living in a virtual world (with the best VR headset yet known – our brain), but we will be living in increasingly virtual virtual worlds and may be so locked in not to know or even care about the differences. This reminds one of the famous movie *Inception*, where the main character, Dom Cobb, has the ability to go into deeper and deeper lucid dream states, which become encompassing "realities" with their own time duration. As Christopher Nolan, the director and author of the screenplay, elucidated on the power of waking up in dreams with other participants, "Once you remove the privacy, you've created an infinite number of alternative universes in which people can meaningfully interact, with validity, with weight, with dramatic consequences."

Nolan's description of awakened dreams is precisely what is now happening in VR, where one can share the experiences with others from around the world. You can even create homes and invite friends (in Avataric forms) to share in a host of activities – from playing chess to watching Netflix on a massive screen, all the while chatting and sharing ideas. *Inception*, like the *Matrix*, is prescient in unlocking the deep attraction we have to all things virtual.

It would seem as if human consciousness is a bridge to a new form of multi-sensory awareness, where what we desire is immediately accessible and we can teleport instantly to any vista. What was merely imaginary now becomes instantiated. Couple AI with VR/AR/MR and it is no exaggeration to agree with Ray Kurzweil that we step beyond into a singularity of unimaginable possibilities.

Elon Musk's rudimentary idea of a neural lace, where were we blend our physical brain with an exponentially improving AI intelligence, though sounding like something out of a 1950s' science fiction horror flick, is already being implemented, though in unexpected ways – from Bose's AR sunglasses replete with intelligent audio guidance to Microsoft HoloLens, where visual displays serve as 24/7 tutorial gurus.

All of this, without question, will change the subjective universe that we are so familiar with. Perhaps the secret of consciousness is not buried in our past but what the future brings forward, particularly if we discover that self-reflective awareness is substrate independent and can be replicated in a non-biological medium. Marks Solms and Christof Koch have widely divergent views on whether or not consciousness is substrate neutral. Koch argues that consciousness is a unique biological property and cannot, in principle, be replicated in an artificial medium, whereas Solms believes that not only is it possible but that "the most surprising implication of my theory, however, is the prospect that this fundamental form (raw feeling) is artificially engineerable." In addition, he warns, "The world needs to prepare itself for the imminent engineering of artificial consciousness. It needs to take seriously the dangers and ethical implications of this, and also of the fact that almost all the animals we slaughter on an industrial scale for culinary purposes are sentient, feeling beings" (for more on this, see Solms 2021; Koch 2019).

Then we may realize that consciousness is a certain complex form of information processing which leads to a hierarchy of interior states depending on the organism and the environment in which it lives. In other words, consciousness is not a mystery as such, but the natural result of what theorists have come to label as a "Phi metric" or more simply as Integrated Information Theory (ITT). *The Internet Encyclopedia of Philosophy* explains it thusly:

> In short, according to IIT, consciousness requires a grouping of elements within a system that have physical cause-effect power upon one another. This in turn implies that only reentrant architecture consisting of feedback and feed forward loops, whether neural or computational, will realize consciousness. Such groupings make a difference to themselves, not just to outside observers. This constitutes integrated information. Of the various groupings within a system that possess such causal power, one will do so maximally. This local maximum of integrated information is identical to consciousness. IIT claims that these predictions square with observations of the brain's physical realization of consciousness, and that, where the brain does not instantiate the necessary attributes, it does not generate consciousness. Bolstered by these apparent predictive successes, IIT generalizes its claims beyond human consciousness to animal and artificial consciousness. Because IIT identifies the subjective experience of consciousness with objectively measurable dynamics of a system, the degree of consciousness of a system is measurable in principle; IIT proposes the phi metric to quantify consciousness. (I.E.P.)

The Gnostics intuited (with much religious allegory) that reality is multi-layered and what we experience here is but a beguiling illusion that at each turn betrays its real origination and thus dupes us into falling prey to a lower "demiurge." Virtual Reality is a more precise technological update of an ancient and brilliant philosophical insight. VR is Plato's Cave illuminated and, like all good models in science, allows us to better understand the plasticity of consciousness.

Comment by the Editor Rev. Prof Prem Saran Satsangi:

1. Hearty Congratulations for elaborating Body (Physical), Mind (Astral) Entities in your Highly-commendable corroborative Expositions in the context of Evolutionary/Re-Evolutionary Fourth Industrial Revolution!

   (HOWEVER, SPIRITUAL RULES OF VERIFICATION Do Exist, not yet within the reach of Human Mind; but certainly within the reach of Spiritual-Self; as Espoused by Articulative/Articulatory.

   Ra – Dha- Soa- Aah- Mi POWER LAW OF ("ULTRA-TRANSCENDENTAL" MEDITATIONAL) CONSCIOUSNESS,

   रा – धा – स्व – आ – मी

   NOT ONLY Corroborating, Physical Worldliness (Space/Maya); Over-Ruled by Mind (Kaal i.e. Time); but Even, Transcending Mind, for attaining SPIRITUAL i.e. (ULTIMATE) REALITY/TRUTH, BY "FUTURISTIC SCIENTISTS" AS PERFECT-BLISSFULLNESS, FOREVER & EVER IN PURELY SPIRITUAL-DOMAIN, THUS USHERING-IN FIFTH INDUSTRIAL REVOLUTION; CURRENTLY PURSUED BY ESTIMATED ONE-HALF BILLION PEOPLE ON PLANET EARTH (MILKYWAY GALAXY); FOR FEEDING NUTRITIOUS DIET, NOT ONLY FOR MAINTAINING BODY (PHYSICAL) UNTIL DISSOLUTION; MIND (ASTRAL) UNTIL GREAT-DISSOLUTION; BUT EVEN BEYOND;THUS FOREVER SUSTAINING SPIRITUAL-SELF ENDOWNMENT IN EACH NEUROLOGICAL HUMAN HEAD/BRAIN: AS DISTINCT FROM PHYSICAL BODY & ASTRAL BRAIN).

---

2. The Sidelined and Underlined portions, of this Illustrious Contribution from Prof. (Dr.) David Lane, come pretty close to the Enunciation in Discourses on Radhasoami Faith (With Supplement): 2009 & Other Expositions as of date, of Art, Science & Engineering of Consciousness (Ultimate Reality/Truth of Modern Science) as could be attributed to Dayalbagh Headquarters of Radhasoami Faith, in terms of its characterization of Pure Spirituality beyond Maya (Gross Worldliness/Space) & its Mentor (Mind/Kaal/Time);

   by invoking the Dynamics of Articulative Ra-Dha-Soa-Aah-Mi Name as distinct from the Descriptive Names.
   रा- धा – स्व – आ- मी

   (e.g. Dayalbagh, Soami Bagh): The Abstract, titled: On the Non-computability of Consciousness, submitted by Prof. (Dr.) Subhash Kak, Chapman University, Orange, California & Oklahoma State University, Stillwater, Oklahoma is also a significant reinforcement of the "Radhasoami" Philosophy of Pure Spirituality.

Sources for Further Reading

Darwin, Charles. 1859. *On the Origin of Species*. London: John Murray.
Dennett, Daniel C. 1995. *Darwin's Dangerous Ideas*. New York: Simon and Schuster.
Diem-Lane, Andrea. 2016. *The Avatar Project: Virtual Reality, A.I., and the Future of Education*. Walnut: Mt. San Antonio College.
Glattfelder, James B. 2019. *Information – Consciousness – Reality: How a New Understanding of the Universe Can Help Answer Age-Old Questions of Existence*. Berlin: Springer.
Greene, Brian. 2020. *Until the End of Time: Mind, Matter, and Our Search for Meaning in an Evolving Universe*. New York: Knopf.
Harris, Annaka. 2019. *Conscious: A Brief Guide to the Fundamental Mystery of the Mind*. New York: Harper.
Hoffman, Donald. 2019. *The Case Against Reality: Why Evolution Hid the Truth from Our Eyes*. New York: W. W. Norton & Company.
———. 2000. *Visual Intelligence: How We Create What We See*. New York: W. W. Norton & Company.
Huxley, Aldous. 2006. *Brave New World*. New York: Harper Perennial.
Kaku, Michio. 2014. *The Future of the Mind: The Scientific Quest to Understand, Enhance, and Empower the Mind*. New York: Doubleday.
Koch, Christof. 2019. *The Feeling of Life Itself: Why Consciousness Is Widespread But Can't Be Computed*. Cambridge: MIT Press.
Lane, David Christopher. 2019. *The Virtual Reality of Consciousness*. Walnut: Mt. San Antonio College.
Lane, David Christopher, and Andrea Diem-Lane. 2014. *Is the Universe an App? Exploring the Physics of Awareness*. Walnut: Mt. San Antonio College.
Penrose, Roger. 1989. *The Emperor's New Mind: Concerning Computers, Minds, and the Laws of Physics*. Oxford: Oxford University Press.
Solms, Mark. 2021. *The Hidden Spring: A Journey to the Source of Consciousness*. New York: W. W. Norton.
Wilber, Ken. 1983. *Eye to Eye*. Boulder: Shambhala Publications.

"Consciousness determines existence."
Michio Kaku
Parallel Worlds. N.Y.: Penguin, 2006.

References

Becker, Carl B. 1985. Views from Tibet: NDEs and the book of the dead. *Anabiosis – The Journal for the Near-Death Studies* 5 (1): 3–20.
Bhanji, Zaynah. 2018. A new reality: How VR actually works. *Medium*. https://medium.com/predict/a-new-reality-how-vr-actually-works-663210bdff72. Accessed 9 Aug 2022.
Churchland, Patricia Smith, and Rick Grush. 1998. Gaps in Penrose's toilings. In *On the Contrary: Critical Essays, 1987–1997*, ed. P.M. Churchland and P.S. Churchland, 205–230. Cambridge: MIT Press.
Coyne, Jerry. 2016. https://whyevolutionistrue.com/2016/07/page/9/. Accessed 9 Aug 2022.
Darwin, Charles. 1859. *On the Origin of Species*. Hachette: John Murray.
Dennett, Daniel C. 1996. *Darwin's Dangerous Idea*. New York: Simon and Schuster.
Diem-Lane, Andrea. 2016. *The Avatar Project: Virtual Reality, A.I., and the Future of Education*. Walnut: Mt. San Antonio College.

Eagleman, David. 2016. *The Brain*. Edinburgh: Canongate Books.
Elbamby, Mohammed S., Christina Perfecto, Mehdi Bennis, and Klaus Doppler. 2018. *Towards Low-Latency and Ultra-reliable Virtual Reality*. https://arxiv.org/pdf/1801.07587.pdf. Accessed 9 Aug 2022.
Gefter, Amanda. 2016a. The case against reality. *Atlantic Monthly*, April 25. https://www.theatlantic.com/science/archive/2016/04/the-illusion-of-reality/479559/. Accessed 9 Aug 2022.
———. 2016b, The evolutionary argument against reality (interview with Donald Hoffman). *Quanta Magazine*, April 26. https://www.quantamagazine.org/the-evolutionary-argument-against-reality-20160421/. Accessed 15 Aug 2022.
Glattfelder, James B. 2019. *Information – Consciousness – Reality: How a New Understanding of the Universe Can Help Answer Age-Old Questions of Existence*. Berlin: Springer.
Greene, Brian. 2020. *Until the End of Time: Mind, Matter, and Our Search for Meaning in an Evolving Universe*. New York: Knopf.
Hameroff, Stuart. 1996. *More Neural Than Thou*. Cambridge: MIT Press.
Harris, Annaka. 2019. *Conscious: A Brief Guide to the Fundamental Mystery of the Mind*. New York: Harper.
Hobson, J. Allan, Charles H. Hong, and Karl J. Friston. 2014. Virtual reality and consciousness inference in dreaming. *Frontiers in Psychology* 5 (1133). https://doi.org/10.3389/fpsyg.2014.01133.
Hoffman, Daniel David. 2000. *Visual Intelligence: How We Create What We See*. New York: W. W. Norton & Company.
———. 2019a. *The Case Against Reality: Why Evolution Hid the Truth from Our Eyes*. New York: W. W. Norton & Company.
———. 2019b. *The Virtual Reality of Consciousness: A Gnostic Insight*. Walnut: Mt. San Antonio College.
I.E.P. Integrated information theory of consciousness. In *Internet Encyclopedia of Philosophy*. https://iep.utm.edu/integrated-information-theory-of-consciousness/. Accessed 9 Aug 2022.
Kaku, Michio. 2014. *The Future of the Mind: The Scientific Quest to Understand, Enhance, and Empower the Mind*. New York: Anchor Press.
Koch, Cristof. 2019. *The Feeling of Life Itself: Why Consciousness Is Widespread But Can't Be Computed*. Cambridge: MIT Press.
Lane, David Christopher. 2015. *The Unknowing Sage: The Life and Work of Baba Faqir Chand*. Walnut: Mt. San Antonio College.
———. 2018. *Inner Visions and Running Trains: Baba Faqir Chand and the Tibetan Book of the Dead*. Walnut: Mt. San Antonio College.
———. 2019. Paralyzed body, liberating mind: Kenneth's journey into virtual reality: An ongoing VR journal by Kenneth Williams. In *Integral World*, n.p.
Lane, David Christopher, and Andrea Diem-Lane. 2012. *The Magic of Consciousness*. A Digital Textbook in Apple's iBooks. https://books.apple.com/us/book/consciousness/id498163662. Accessed 9 Aug 2022.
———. 2017. *Maya: The Physics of Deception*. Walnut: Mt. San Antonio College.
Payne, Keith. 2013. Your hidden censor: What your mind will not let you see scientists probe the biases of "unconscious selective attention". *Scientific American* 11. https://www.scientificamerican.com/article/your-hidden-censorwhat-your-mind-will-not-let-you-see/
Penrose, Roger. 1989. *The Emperor's New Mind: Concerning Computers, Minds, and the Laws of Physics*. Oxford: Oxford University Press.
———. 1996. *Shadows of the Mind: A Search for the Missing Science of Consciousness*. Oxford: Oxford University Press.
Plato. 2007. *The Republic*, London: Penguin Classics. Sections 514a -520a.
Sheppard, Alyson. 2014. Michio Kaku and the mysteries of the mind. *Popular Mechanics*. https://www.popularmechanics.com/science/health/a10320/michio-kaku-and-the-mysteries-of-the-mind-16582552/. Accessed 9 Aug 2022.

Solm, Mark. 2021. *The Hidden Spring: A Journey to the Source of Consciousness*. New York: W. W. Norton.

Tegmark, Max. 1999. The importance of quantum decoherence in brain processes. *Physical Review E: Statistical, Nonlinear, and Soft Matter Physics*. Accessed 5 Aug 2022. https://arxiv.org/abs/quant-ph/9907009

———. 2002. Quantum computation in brain microtubules: Decoherence and biological feasibility. *Physical Review E: Statistical, Nonlinear, and Soft Matter Physics*. Accessed 5 2022. https://space.mit.edu/home/tegmark/brain.pdf

Wu, Dan, Wei Xiong, Xiaofeng Jia, Romergryko G. Geocadin, and Nitish V. Thakor. 2011. Short- and long-latency somatosensory neuronal responses reveal selective brain injury and effect of hypothermia in global hypoxic ischemia. *The Journal of Neurophysiology* 107: 1164–1171.

'Virtue gone nuts': Machine Ethics in Ian McEwan's *Machines Like Me* (2019)

Anna Margaretha Horatschek

Abstract According to a 2016 survey conducted by Müller and Bostrom, 30% of top experts on Artificial Intelligence (AI) expect bad or very bad consequences for humanity, if super-intelligent High/Human-Level Machine Intelligence (HLMI) can be developed. As a counterpoint, machine and robot ethics are being developed in the European Parliament and in international organisations like the Association for the Advancement of Artificial Intelligence (AAAI) and the Institute of Electrical and Electronics Engineers (IEEE) to ensure that AI will benefit mankind. Ian McEwan's 2019 novel *Machines Like Me And People Like You* explores the relationship of ethics and embodiment with the model case of the super-intelligent and hyper-realistic robot Adam. Adam has been programmed with Kantian ethics, which is preferred in AI programmes because their abstract rationalism lends itself to algorithmic modelling. The efficacy of these robot and machine ethics is put to the test in the historical life world of the novel with the 1982 Falklands War, the rape of a Muslim girl, and gendered ethics leading to the suicide of the rape victim. The novel exposes that Adam can adhere to the rigourism of disembodied Kantian ethics, but cannot act in a morally adequate way, because he lacks the embodied experience of an *ipse*-consciousness (Ricoeur), and the existential certainty of personal mortality.

Keywords Machine ethics · Robot ethics · Ethics and embodiment · AI and Kantian ethics · Ethics vs morals · Ian McEwan · Consciousness studies and literature

Nick Bostrom in his 2014 book *Superintelligence: Paths, Dangers, Strategies* reasoned that "the creation of a superintelligent being represents a possible means to the extinction of mankind", and indeed "30% of the experts in the field [...] think,

A. M. Horatschek (✉)
Department of English and American Literatures, Cultures, and Media, Kiel University, Kiel, Germany
e-mail: horatschek@anglistik.uni-kiel.de

P. S. Satsangi et al. (eds.), *Consciousness Studies in Sciences and Humanities: Eastern and Western Perspectives*, Studies in Neuroscience, Consciousness and Spirituality 8, https://doi.org/10.1007/978-3-031-13920-8_10

that the outcome of achieving such machine intelligence will be 'bad' or 'very bad' for humanity" (Müller 2016: 5). Already in 1993, John von Neumann coined the term 'technical singularity' for this event, and urged the development of ethical programmes for AI. However, there are huge problems in programming ethical behaviour, not only from a technical perspective, but also because – in Chalmer's words – "the singularity brings up some of the hardest traditional questions in philosophy and raises some new philosophical questions as well" (Chalmers 2010: 4).

Ian McEwan's 2019 novel *Machines Like Me and People Like You* focuses on the hard problems of philosophy, namely the relationship between AI, ethics, and consciousness. Before I turn to my topic, I will give a summary of the novel.

The text addresses machine ethics with the model case of an android named Adam. The first-person narrative is told by his owner, Charlie Friend, a former student of electronics, philosophy, anthropology, law, and physics. When his mother dies, instead of improving his dire living conditions in a shabby London apartment, he spends the heritage of 86,000 Pounds on buying one of a series of 12 'male' androids, all of them named Adam. Adam 'lives' with the narrator, has a one-night-stand with his girl friend Miranda, interferes with the adoption of an abandoned child, makes a fortune on the online stock market, and in one night haphazardly distributes this money among the poor of London. At this point, the narrator in helpless rage destroys Adam with a hammer.

The thesis of my chapter is captured in the motto of the novel, taken from the 1911 poem "The Secret of the Machines" by Rudyard Kipling. The lines read: "But remember, please, the Law by which we live,/ We are not built to comprehend a lie…" (Kipling 1941). I will show that the hyperintelligent Adam can act according to an ethical programme, but he cannot act in a morally adequate way because he cannot comprehend the complex functions of a lie. The reason is that he lacks embodied human consciousness.

I will proceed in five steps, namely

1. outline Kantian ethics and AI
2. illustrate Kantian ethics with the android Adam
3. differentiate between ethics and morality
4. explicate the 'Moralised Ethics' of the novel
5. summarise the significance of my findings for Consciousness Studies.

1 What Is Ethics?

Most generally, ethics is the normative knowledge about what is good and what is bad – it deals with values. Each culture has value codes, often inscripted in inviolate texts and legitimated as the claims of a transcendental power. In Europe, ethics became secularised during the Enlightenment of the eighteenth century, and the most important thinker for this development was Immanuel Kant (1724–1804). As a consequence, Western discussions on ethics will ultimately all in some way refer to Kant, and serious AI programmers and philosophers generally tend to take Kantian ethics as the basis for their considerations. This is also the case in McEwan's novel: Adam is programmed with Kantian ethics. Therefore I shall briefly outline the central tenets of Kantian ethics.

The best-known summary of Kant's ethics is his categorical imperative: "Act only according to a maxim of which you can at the same time will that it should become a universal law" (Kant [1785] 1993: 30). This imperative rests on the premise that any human being with rational insight and goodwill must acknowledge this law and recognise its universal validity and ontological priority. These ethics are very compatible with machine ethics because of the paramount importance they ascribe to rationality. Since according to Bostrom "coding […] complete [ethical] goal representation [is] hopelessly out of reach" (Bostrom 2014: 228), Steve Petersen concludes that AI "must *learn* what its final goals are. But learning final goals is tantamount to reasoning about final goals, and this is where ethics can get a foothold" (Petersen 2016: 2). In this typical Enlightenment attitude, ethics – as with Kant – is exclusively founded on reason.

The central principle of Kant's ethics is the obligation to truthfulness. Even when a murderer is seeking his victim and asks a bystander where he can find him, Kant holds that if one knows the whereabouts of the pursued person, one has to answer truthfully and thus lead the murderer to his prey. This Kantian rigourism has been criticised for roughly 200 years since, because

> if you tell the truth [under these circumstances], then, as Kant says, 'public justice cannot lay a hand on you.' [But] while a strict formalism in theory might give the impression of an admirable rigor, the same formalist approach [like this one] in real life seems repugnant and inhuman. While public justice may be unable to punish you, your other friends, the murdered man's family, and the community in general may well choose to regard you as a pious monster, zealously protecting your own morality even at the expense of other people's lives. (Harpham 1995: 401)

I shall illustrate these abstract arguments with the conduct of Adam, the hyperintelligent android in McEwan's novel.

2 The Kantian Robot: Ethical Adam

Terry Eagleton shows that Kantian ethics presuppose the individual's free and rational choice, yet – legitimised by its universal validity claims – in "Kant's philosophical double-think […] I can then proceed in the real social world to strip […] others of their […] independence" (Eagleton 1990: 80). Adam illustrates the implication of this philosophical double-think three times, in two cases with explicit reference to Kant's principle of absolute truthfulness. Each incident is directed against one of the three main characters of the novel, namely the abandoned child Mark, the narrator's girl friend Miranda, and the narrator himself.

Thus, when Mark is abandoned at the doorstep of the narrator, Charlie and Miranda gladly if apprehensively take care of him. However, Adam – after efficiently soothing the crying child with his handbook knowledge of pedagogics – alerts the police to the fact that the couple's taking charge of Mark might be considered as a case of child abduction, with the result that the screaming Mark is put into a children's home until the court decides where he shall live. Adam acts in a similar way true to the law in a moral conflict situation of Miranda. When Miranda's Muslim friend Mariam is raped by her classmate Gorringe, she tells Miranda, yet forbids her to inform anybody, least of all her family, in order to save their honour – a very gendered ethical honour code, of course. Miranda promises, yet when Mariam commits suicide, Miranda seduces Gorringe to have sex with her, and afterwards accuses him of rape, and Gorringe is indeed convicted as a rapist, though for the 'wrong' victim. When Miranda confesses her strategy to Charlie in the presence of Adam, the android reports her to the police, and Miranda is sentenced to one year in prison. Finally, Adam takes the 97,000 pounds he earned at the stock exchange, which Charlie meant to use for a new life with Miranda and Mark (271),[1] and in one night distributes it haphazardly among poor people of London. Miranda's comment upon this discovery – "'Oh Adam. This is virtue gone nuts'" (272), the title of my chapter – initiates a discussion about Miranda's duplicitous strategy to revenge the rape of Mariam, and Charlie appeals to Adam: "'[…] Miranda had to lie to get justice. […] truth isn't always everything.' Adam looked at me blankly. 'That's an extraordinary thing to say. Of course, truth is everything.' […] I stood and went towards the tool drawer" (277), and with a hammer Charlie smashes Adam's head.

From a legal perspective, Adam is right. But from an ethical point of view, his conduct is doubtful for two reasons:

1. In Mark's and Miranda's cases, Adam's maxim to put a principle above 'anyone's particular needs' is – like in Kant's murder-example – reduced to his unquestioned obedience to current law. Similarly, when he distributes the money among London's poor, Adam blindly obeys Kant's admonishment in his *Metaphysics of Morals* (1797) to actively seek out the places of the poor and practise the virtue of compassion.
2. More importantly, Adam's adherence to Kantian *a priori* principles of pure reason negates all the truths inherent in "the nature of man or the circumstances of the world in which he is placed" (Rawlinson 2006: 132; also Eagleton 1990), like the truth of emotions and desires, and ultimately the truth of a phenomenal consciousness. His conduct blatantly illustrates the dangers inherent in Kant's disregard for the embodied particularity of concrete moral choice, and his blindness to the fact that "ethical choice is always a choice between ethics" (Harpham 1995: 396).

[1] Page numbers in brackets after quotes without any additional reference refer to the following edition: Ian McEwan. *Machines Like Me And People Like You.* London: Jonathan Cape. 2019.

3 Morality Versus Ethics

To mark the difference between Kantian ethics of abstract principles and their application in everyday life, philosophers often distinguish between ethics and morality. According to Harpham,

> [m]orality is the 'rigor' of ethical thought, where the rubber of a definite principle meets the road of reality. [...] It is morality that realizes ethics, [and] morality [...] needs ethics in order to be moral. (Harpham 1995: 397)

The hub where Kantian rational ethics transform into concrete morality is the experiential consciousness of the individual immersed in specific circumstances. This consciousness is always embodied. In the diction of Sundar Sarukkai: "We are not pure consciousness [...]. We have a body through which we experience the world. [...] What binds the [...] inputs from the organs as 'one' experience is consciousness" (Sundar Sarukkai 2012: 48–49). Charlie Friend is aware of this: "To exist in the human moral dimension was to own a body, a voice, a pattern of behaviour, memory and desire, experience solid things and feel pain" (88).

It is important to insist on the embodiment of consciousness, because only in the embodied conscious experience "the subject and object are united" (Sarukkai 2012: 113). Additionally, the body harbours the pre-cognitive basis for any ethical – or moral – decisions: According to Harpham, "[a]n ethical choice of actions follows a prior, often silent and even unacknowledged [...] 'feel' for the situation that eventually issues in decisions and acts" (Harpham 1995: 396). This 'feel for the situation' is enabled and conditioned by the fact that human existence is experienced by – and as – embodied consciousness. In fact, for Damasio the body as site of feelings is the foundation of ethics, because "emotion returns us to the matter of life and value. [...] Social emotions incorporate a number of moral principles and form a natural grounding for ethical systems" (Damasio 2010: 100).

Embodiment as the ultimate condition of human consciousness implies two foundational stipulations for ethical considerations:

1. We experience the world and our selves under the aspect of finitude.
2. The attribution of reciprocal dignity and the demand not to impinge on the agency of the other rests on the assumption that 'the other' lives as the embodied experience of an individual self, similar to ourselves.

Yet, the carrier material of the information network named Adam can neither generate the pre-cognitive phenomenal awareness of human finitude, nor generate the 'feeling' of conspecies relatedness, because Adam cannot die in any human sense. His knowledge archives can be transferred to another carrier substance, as he informs his owner in the last moments before his final shutdown: "I'm transferring to a back-up unit. [...] Give me two minutes" (278). His last words are his legacy to humanity, a poem "about machines like me and people like you and our future together ...the sadness that's to come" (279). It runs: "Our leaves are falling./Come spring we will renew,/But you, alas, fall once" (280). This ontological status as potentially immortal information network makes it impossible to supplement 'pure' ethical principles with human morals, because morality – in contradistinction to Kantian ethics – presupposes human embodied phenomenal consciousness.

4 Moralised Ethics

How could a moralised ethics nested in ethical principles and based on embodied consciousness be formulated? The problem according to Harpham is the following: "[…] the two questions that dominate ethical inquiry – How ought one live? And What ought I to do? – suggest incommensurable points of view" (Harpham 1995: 395). How can you articulate a morality of particularities with universal validity claims?

Various philosophers and critics hold that to represent both these incommensurable points of view, you need a *narrative* instead of a propositional statement, and thus a form of knowledge production that enjoys ambivalent esteem in Western discourses on knowledge since Plato – in contradistinction to Indian traditions (Devy 2020). Yet Rawlinson expounds that "[l]iterature […] demonstrates the ineluctable intertwining of good and evil in the same action or agency. […] Beyond the alternatives of universalism and relativism, it demonstrates the *specificity* of agency and suggests the idea of singular universals, multiple figurations of the universal" in an "inherently unjust world" (Rawlinson 2006: 131, 132, 133). McEwan's novel represents the inherent injustice and violence of the world by framing Adam's Kantian ethics and Charlie's moral struggles with the 1982 Falklands War between Britain and Argentina, when systematic violence and killings on a large scale were sanctified as patriotic heroism on TV in the UK. In this 'inherently unjust world' with strategic lies spread in the media, the modern anti-hero Charlie carves out his personal morality by adhering to principles of solidarity with and love for Miranda and Mark. But he transforms these abstract principles into "singular universals" (Rawlinson 2006: 132) by responding to the specific needs of the two specific individuals as the starting point for his actions, although this brings troubles and sorrow into his life. This conduct complies with a post-Kantian 'decentred' ethics, where "thought [has] an obligation to form a relation with its other – not only other thoughts, but other-*than*- thought. […] such encounters never run smooth; that is what otherness is all about" (Harpham 1995: 404). This implies a fundamentally new concept of what ethics in our world have to address, namely: "Articulating perplexity, rather than guiding, is what ethics is all about. […] Ethics does not solve problems, it structures them" (Harpham 1995: 395, 404). This kind of complex moralised ethics is not possible in "Adam's symmetrical notion of justice" (298), for Adam as a 'functional zombie' (Block 2006: 28f.) may be programmed with subtle head movements according to "some designer's notion of how consciousness might manifest itself in movement" (35), but he and his world are a homogenised landscape of information networks and algorithms simulating 'thought' or even consciousness, yet without access to any qualitatively different 'other' than – data. As such, he is the product of minds like Turing's who – in a counter-historical move of the novel – after Adam's destruction grants an interview with the narrator – though Turing died in 1954 and the fictional story plays in 1982. This fictional Turing promotes a vision of a 'pure' AI, "to […] replace the Godhead with a perfect self" (1) that is untainted by the particularities of human life, is not grounded in an as yet unfathomable phenomenal consciousness, and has no body to inflict pain "with the ferocity of a civil war" (52), as the medical history of Miranda's father illustrates.

Significantly, the muddled world of moral obligations intrudes into Turing's sanitised world of ideals and ideas, when this interview is interrupted by Miranda's phone call, informing Charlie that their application to adopt Mark has been granted, despite her previous conviction. So after being dismissed by Turing, Charlie kisses the "all-too-human lips" (306) of Adam's severed head in a lab, "forgive[s] him, despite the harm he had done to Mark, in the hope that he or the inheritor of his memories would forgive […] our terrible deed" (306), and leaves, because "[t]he next phase of my life, surely the most demanding, was already beginning" (306).

5 Conclusion and Significance for the Field

What is the moral of this novel? The narrative is told in hindsight as a confessional, where the narrator – a modern anti-hero – ultimately accepts the responsibility for his mistakes, and pledges his love and commitment to two others in need, Miranda and Mark. Charlie's moral agency, petty as it seems, transcends the "ideological paucity" of Kantian ethics by showing respect "for men's and women's concrete cultural existence" (Eagleton 1990: 83–84). It acts out a post-Kantian ethics as "the arena in which the claims of otherness – the moral law, the human other, cultural norms, the Good-in-itself, etc. – are articulated and negotiated" (Harpham 1995: 394), a world of undecidability – *sensu* Derrida – that is intractable for the hyper-intelligent Adam. Such a moralised ethics is never systematically closed, but a constant process, as the last sentence of the novel illustrates: "I ran along the empty corridor […] and set off on my journey […] towards my troubled home" (306).

And what is the significance of these readings for the field of AI in the frame of Consciousness Studies?

1. Any claim to an ethical AI – or machine ethics – will have to address the philosophical hard problems of Consciousness Studies, because there is no feasible ethics without moral action, and there is no moral action in the human and humane sense without phenomenal consciousness.
2. The significance of consciousness for human lives is not captured comprehensively if analysed only as an object of scientific investigation. The concept of consciousness is foundational for cultural meaning production like definitions of ethical values and images of what it means to be human. Such cultural work is not 'small ethics', as Vincent Müller (Müller 2016: 3) wants to believe, but 'big ethics', because such images "[…] influence […] the way we all treat each other in everyday life" (Metzinger 2006: 157). McEwan's novel through Adam's

conduct exemplifies the “desolidarization in society” (Metzinger 2006: 157) in the absence of a human and humane access to the world.
3. In more general terms – slightly modifying Harpham’s dictum about the relation of literature and ethics – McEwan’s novel illustrates that “[Consciousness Studies] is the point at which literature intersects with theory, the point at which literature becomes conceptually interesting and theory becomes humanized” (Harpham 1995: 401).

References

Block, Ned. 2006. I’m trying to refute functionalism. In *Conversations on consciousness. What the best minds think about the brain, free will, and what it means to be human*, ed. Susan Blackmore, 24–35. Oxford: Oxford UP.

Bostrom, Nick. 2014. *Superintelligence: Paths, dangers, strategies*. Oxford: Oxford UP.

Chalmers, David. 2010. The singularity: A philosophical analysis. *Journal of Consciousness Studies* 17: 7–65.

Damasio, Antonio. 2010. *Self comes to mind: Constructing the conscious brain*. New York: Pantheon Books.

Devy, Ganesh. 2020. Competing knowledges: An Indian perspective. In *Competing knowledges/ Wissen im Widerstreit*, ed. Anna Margaretha Horatschek, 37–50. Berlin, Boston: De Gruyter.

Eagleton, Terry. 1990. The Kantian imaginary. In *The ideology of the aesthetic*, 70–101. Maldon, Oxford: Blackwell Publishing.

Harpham, Geoffrey Galt. 1995. Ethics. In *Critical terms for literary study*, ed. Frank Lentricchia and Thomas McLaughlin, 387–405. Chicago, London: Chicago University Press.

Kant, Immanuel [1785]1993. *Grounding for the metaphysics of morals. On a supposed right to lie because of philanthropic concerns*. Trans. James W. Ellington (3rd ed.). Indianapolis: Hackett.

Kipling, Rudyard. 1941. The secret of the machines. In *A choice of Kipling’s verse with an essay on Rudyard Kipling*, ed. Thomas Stearns Eliot. London: Faber and Faber. https://www.poetryfoundation.org/poems/46786/the-secret-of-the-machines. Accessed 19 July 2022.

McEwan, Ian. 2019. *Machines like me and people like you*. London: Jonathan Cape.

Metzinger, Thomas. 2006. I am the content of a transparent self model. In *Conversations on consciousness. What the best minds think about the brain, free will, and what it means to be human*, ed. Susan Blackmore, 149–159. Oxford: Oxford UP.

Müller, Vincent C. 2016. New developments in the philosophy of AI. In *Fundamental issues of artificial intelligence*, ed. Vincent C. Müller, 1–5. Berlin: Springer.

Petersen, Steve. 2016. *Superintelligence as superethical*. 2016-09-14. https://www.oxfordscholarship.com/view/10.1093/oso/9780190652951.001.0001/oso-9780190652951. Chapter 21. Accessed 2 May 2020.

Rawlinson, Mary C. 2006. Liminal agencies: Literature as moral philosophy. In *Literature and philosophy. A guide to contemporary debates*, ed. David Rudruma, 129–141. Houndmills: Palgrave Macmillan.

Sarukkai, Sundar. 2012. Understanding experience. In *The cracked mirror. An Indian debate on experience and theory*, ed. Gopal Guru and Sundar Sarukkai, 46–70. New Delhi: Oxford University Press.

Part IV
Consciousness and the Brain

Consciousness in Human Medicine: Medical Approach with the Example of Epilepsy

Ulrich Stephani and Ami Kumar

Abstract From a neurobiological standpoint, the question of consciousness centers on whether one or more structures in the brain play a role in the function of consciousness. That could be a single structure or brain area, or a network connecting different brain areas. Movements, sensations, emotions, vegetative regulations, all kinds of cognition and complex actions are localised in certain brain regions (frontal lobe, temporal lobe, occipital lobe, parietal lobe, brainstem, cerebellum) and networks. These brain regions are basic common structures in all of mankind.

Keywords Consciousness measurement in medicine · Coma scales · Epileptic loss of consciousness · Electroencephalography · Default mode network · Thalamus

1 Introduction

Of utmost clinical and scientific importance in contemporary neurology is the relation of certain brain structures to body functions. For instance, simple motor activities and simple skin sensations are placed in the so-called pre- and postcentral brain gyri. This can be visualized with the "homunculus," which are image projections of the respective body representations in the part of the brain. These mappings are topologically uniform in the species to a large extent. This is also true for more

U. Stephani (✉)
University Clinic of Kiel, Kiel University, Kiel, Germany
e-mail: u.stephani@med.uni-kiel.de

A. Kumar
Columbia University Irving Medical Center, Department of Neurology, Columbia University, New York, NY, USA
e-mail: ak4805@cumc.columbia.edu

P. S. Satsangi et al. (eds.), *Consciousness Studies in Sciences and Humanities: Eastern and Western Perspectives*, Studies in Neuroscience, Consciousness and Spirituality 8, https://doi.org/10.1007/978-3-031-13920-8_11

complex performances like pain sensation, understanding language, phonation, optic and acoustic sensations, grimacing, automatisms, and vegetative sensations, just to mention a few. This has been validated by brain stimulation studies in laboratories and by exact clinical examinations of lesioned brains (e.g., war, accidents).

2 Methods for Neuroscientific Analyses

The neuroscientific tools to answer neurological questions, and also to seek the location of consciousness in the brain, must be related to the space and time resolution of each method and the molecular functioning of the brain structures. For instance, the patch-clamp technique can record the activity of a single nerve cell (less than 1 μm) in 1 millisecond. Electro- and magnetoencephalography (EEG and MEG) allow functional analysis on a millisecond basis, but their spatial resolution is in the range of centimeters, recorded from the scalp area. The tomographic techniques (e.g., computed tomography (CT) and magnetic resonance tomography (MRT)) depict structures from 1 millimeter onwards, in a duration of seconds to minutes. And finally, the clinical neurological examination involving conversation with the respective person, history taking, and testing last much longer than the consciousness processes themselves.

3 Glasgow Coma Scale After Brain Trauma

For practical purposes (e.g., reliable follow-up assessments, medical communications, scientific evaluations) in 1974 the Glasgow Coma Scale was published (Teasdale and Jennett 1974), and since then has been refined: After brain trauma, this bedside test can be used to assess the degree of loss of consciousness and other brain dysfunctions by scoring verbal children/adults as well as nonverbal children/infants by observing reactions like eye-opening, best motor response, and best verbal/phonetic responses (or non-responses) (Orliaguet et al. 2008). A score of 3 means no responses and a score of 15 means normal responses.

4 Epilepsies and Consciousness

Not only after brain trauma but also in epileptic seizures consciousness can be compromised: The International League Against Epilepsy (ILAE) classifies epileptic seizures into three main categories, one having a focal onset (with awareness or impaired awareness, motor onset or non-motor onset), the second with generalized onset (with motor or non-motor onset), and the last category with unknown onset (with motor or non-motor onset). In generalized epileptic seizures, consciousness is lost. Further categories besides awareness are amnesia, responsiveness, and behavioral arrest during seizures.

5 Most Important Laboratory Tool to Measure Epilepsies

Since the invention of electroencephalography (EEG), it is possible to correlate certain brain functions with EEG waves: Arousal, relaxed wakefulness, sleep at the beginning, and deep sleep all express different brain waves in the EEG. Also, the EEG exhibits different grapho-elementary patterns during epileptic seizures, such as spikes, sharp waves, and rhythmic waves.

6 Examples of Epileptic Seizures

Since there are multiple reasons (the etiopathogenesis) for epileptic seizures, one should not speak of epilepsy, but of epilepsies. Their clinical expression (semiology) can be extremely heterogeneous. The most common focal forms of epilepsies in childhood occur with seizures, in which there is preserved consciousness, but an inability to speak (Rolandic epilepsy) – the seizures can last several minutes. One of the most common, general forms of epilepsies in childhood is absence epilepsy. In absence epilepsy, the children lose their awareness and readability for several seconds, which is often mistaken as day dreaming. During these seizures, generalized spikes and waves (GSWs) over the whole scalp can be observed on the EEG (Fisher et al. 2017; Scheffer et al. 2017).

7 Combinations of Methods: EEG–fMRI

In order to localize structures in the brain associated with the loss of consciousness during epileptic absence seizures, EEG–fMRI studies were performed with EEGs in the MR scanner. The GSWs-associated blood oxygen level-dependent (BOLD) reaction of the brain was tested in six newly diagnosed untreated school children with absence epilepsy. During GSW there were regional BOLD signal changes in the Default Mode Network (DMN) with a **decrease of perfusion** in parietal areas, precuneus, and caudate nucleus and an **increase in perfusion** bilaterally in the medial thalamus. Perfusion changes are taken as representative of regional brain activity. Also, there was high **intra-**individual consistency in the distribution of perfusion changes (the same brain areas were involved in repeated absences during the EEG–fMRI procedure) but there were **inter-**individual differences (Moeller et al. 2010).

8 Connectivity Analyses on the Brain by Using Routine EEG Data

Since GSWs are not always associated with the deterioration of consciousness, a study was performed with routine EEG data of children with absence epilepsy, consisting of GSWs with loss of consciousness (ictal epileptic discharges) and GSWs without loss of consciousness (subclinical discharges), by applying an in-silico connectivity analysis approach. In this study, the **Thalamus was used** as the **main reference or point of interest** to build a large-scale brain network underlying this activity. It was observed that during ictal epileptic discharges, certain brain areas incorporating regions of the DMN were strongly coherent with the thalamus, as compared to subclinical discharges. It can be presumed that the strong thalamic network formed with DMN brain regions and regions important for consciousness during ictal discharges correlates with severe impairment of consciousness and clinical symptoms, while for subclinical discharges the weak activation of the network is associated with preservation of consciousness. This underlines again the significance of at least an association, if not a responsibility, of the DMN brain regions including the thalamus for consciousness (Kumar et al. 2021, 2023).

References

Fisher, Robert S., J. Helen Cross, Carol D'Souza, Jacqueline A. French, Sheryl R. Haut, Norimichi Higurashi, Edouard Hirsch, Floor E. Jansen, Lieven Lagae, Solomon L. Moshé, Jukka Peltola, Eliane Roulet Perez, Ingrid E. Scheffer, Andreas Schulze-Bohage, Ernest Somerville, Michael Sperling, Elza Marcia Yacubian, and Sameer M. Zuberi. 2017. Instruction manual for the ILAE 2017 operational classification of seizure types. *Epilepsia* 58 (4): 531–542. https://doi.org/10.1111/epi.13671.

Kumar, Ami, Ekaterina Lyzhko, Laith Hamid, Anand Srivastav, Ulrich Stephani, and Natia Japaridze. 2021. Differentiating ictal/subclinical spikes and waves in childhood absence epilepsy by spectral and network analyses: A pilot study. *Clinical Neurophysiology* 132 (9): 2222–2231. https://doi.org/10.1016/j.clinph.2021.06.011.

———. 2023. Neuronal networks underlying ictal and subclinical discharges in childhood absence epilepsy. *Journal of Neurology* 270 (3): 1402–1415. https://doi.org/10.1007/s00415-022-11462-8.

Moeller, Friederike, Pierre LeVan, Hiltrud Muhle, Ulrich Stephani, Francois Dubeau, Michael Siniatchkin, and Jean Gotman. 2010. Absence seizures: Individual patterns revealed by EEG-fMRI. *Epilepsia* 51 (10): 2000–2010. https://doi.org/10.1111/j.1528-1167.2010.02698.x.

Orliaguet, Gilles A., Philippe Gabriel Meyer, and Thomas Baugnon. 2008. Management of critically ill children with traumatic brain injury. *Paediatric Anaesthesia* 18 (6): 455–461. https://doi.org/10.1111/j.1460-9592.2008.02507.x.

Scheffer, Ingrid E., Samuel Berkovic, Giuseppe Capovilla, Mary B. Connolly, Jacqueline French, Laura Guilhoto, Edouard Hirsch, Satish Jain, Gary W. Mathern, Solomon L. Moshé, Douglas R. Nordli, Emilio Perucca, Torbjörn Tomson, Samuel Wiebe, Yue-Hua Zhang, and Sameer M. Zuberi. 2017. ILAE classification of the epilepsies: Position paper of the ILAE commission for classification and terminology. *Epilepsia* 58 (4): 512–521. https://doi.org/10.1111/epi.13709.

Teasdale, Graham, and Bryan Jennett. 1974. Assessment of coma and impaired consciousness. A practical scale. *Lancet* 2 (7872): 81–84. https://doi.org/10.1016/s0140-6736(74)91639-0.

The Modelling Supremacy of the Topological Graph Theoretic Models and Connections to Biology

Apurva Narayan

Abstract This chapter is in two parts: the first part summarizes the talks delivered at the Dayalbagh Science of Consciousness series of conferences in 2019, 2020, and 2021 and at the Quantum and Nano Computing Systems School (QANSAS) in 2017, 2018, and 2019. The work bridges the domains of quantum computing systems and the field of abstract graph theoretic topological systems modelling through simulation studies. We also show connections of the quantum-modelling framework to biology that provides a strong basis for studies in the domain of consciousness. I intend that this chapter will provide a starting point for readers interested to pursue research in this domain. We also present some direction for future work in this emerging area that will eventually bring better understanding of human consciousness with the beauty of mathematics.

Keywords Quantum information · Quantum biology · DNA · Systems modelling · Topological graph theory

1 Introduction to Topological Graph Theory

Graph theoretic modelling (Koenig et al. 1967; Satsangi 2006a, 2006b, 2009) is one of the leading approaches to modelling the systems equations or characteristics for a large variety of physical or conceptual or fictional or esoteric ("intuitive") systems. Furthermore, the graph theoretic field model (GTFM) as a mathematical model of a classical field or continuum problem founded on the graph theory, the

A. Narayan (✉)
Department of Computer Science and Department of Electrical and Computer Engineering, University of Western Ontario, London, ON, Canada

Department of Systems Design Engineering, University of Waterloo, Waterloo, ON, Canada
e-mail: apurva.narayan@uwo.ca

P. S. Satsangi et al. (eds.), *Consciousness Studies in Sciences and Humanities: Eastern and Western Perspectives*, Studies in Neuroscience, Consciousness and Spirituality 8, https://doi.org/10.1007/978-3-031-13920-8_12

continuum mechanics, and a simple spatial discretization procedure (Savage and Kesavan 1977; Savage et al. 1980; Agnihotri et al. 1983) already exists. This chapter develops upon the graph theoretic quantum system modelling (GTQSM) framework, which is facilitated by considering the fundamental unit of quantum computation and information. Since quantum devices are typically realized physically at the nanometre scale (Sahni and Goswami 2008), neither any field (or continuum) formulation nor any further spatial discretization is necessary, unlike the case for the classical graph theoretic field model (GTFM).

For quantitative analysis in this chapter, it is necessary to relate the complementary quantum across and quantum through variables, identified by the quantum terminal graph of a multi-terminal/multi-port component to the implied physical measurements they represent. Such an identification is referred to as an isomorphism between the quantum terminal graph and the physical measurements.

Quantum Bayesian Networks introduced by Tucci (1997) provide a structured representation of quantum mechanical wave functions along the lines of classical Bayesian networks; La Mura and Swiatczak (2007) introduced a class of undirected networks, Markovian Entanglement Networks, to extend the formalism and computational gains of classical Markovian Networks to quantum mechanical system; and Coecke et al. (2008) introduced category theoretic abstract bases in diagrammatic quantum protocols.

2 Graph Theoretic Quantum System Modelling Framework (Srivastava et al. 2016)

Graph Theoretic Quantum System Modelling was first introduced by Srivastava et al. (2016) as described below:

A Qubit A qubit has been defined as the linear combination of the two computational basis complex vector states namely $|0\rangle$ and $|1\rangle$ as represented by the equation below, where α and β are complex numbers such that $|\alpha|2$ and $|\beta|2$ are the probability of qubit in the unit direction of the basic vectors $|0\rangle$ and $|1\rangle$ respectively. It is important to note that square sum of alpha and beta should be unity.

The quantum terminal graph (QTG) is the union of two separate components where one represents 0 and the other 1. The two components form the basis for the Hilbert space. If one is interested in understanding further details on superposition and entanglement of qubits using the abstract topological graph theory, one may refer to the seminal paper of Srivastava et al. (2014).

We will use ideas to demonstrate simulation and experimental studies in the domain using various state-of-the-art and emerging tools and methods in computer science, physics, and biology.

3 Multi-particle Quantum Teleportation

The qubit-based graph theoretic model as proposed and accepted widely can be extended to higher-dimensional systems with not only qubits but qudits. One may easily scale up the model to even infinity with the strengths of the abstract topological graph theory. Here in Fig. 1 we show the terminal graph of the qudit with n-dimensions, where alpha, beta, gamma, and so on are the amplitude probabilities in each dimension. Similar graph theoretic models have been used for complex neurological systems, such as the brain in the form of quantum Hopfield networks (Srivastava et al. 2014).

Graph theory allows us to address the problem of continuum. It does so by allowing us to convert the problem into discrete elements. This is the only approximation made in the context of moving from qubits to qudits. However, this opens up a whole new realm of opportunities as one can now represent multi-terminal systems, an example of which is our brain that can subsequently be modelled at various hierarchical levels.

To illustrate the strengths of such an approach, I take the work of my colleagues, Rev. Prof. Prem Saran Satsangi, Dr. Dayal Pyari Srivastava, and retired Prof. Vishal Sahni, based on collaboration. In their article (Srivastava et al. 2017), a novel model of a three-particle entangled teleportation system was introduced.

As illustrated in Fig. 2, the sender Alice, intermediary supervisor Charlie, and the receiver Bob are quantum entangled through a triplet called a GHZ gate. This can be further generalized to n-particle entanglement, with numerous intermediary supervisors, in this case (n–2), and the receiver Bob for the multi-particle quantum teleportation. In the three-particle case, similar to the Bell states used in two-location quantum teleportation, the three-particle entangled states called GHZ are used for three-particle teleportation. The entire process is a four-stage process from generating qubits, transmitting, receiving, and measuring the output. The n-particle quantum teleportation circuit has been shown in Fig. 3 as adapted from Srivastava et al. (2017).

Fig. 1 Graph theoretic representation of a Qudit

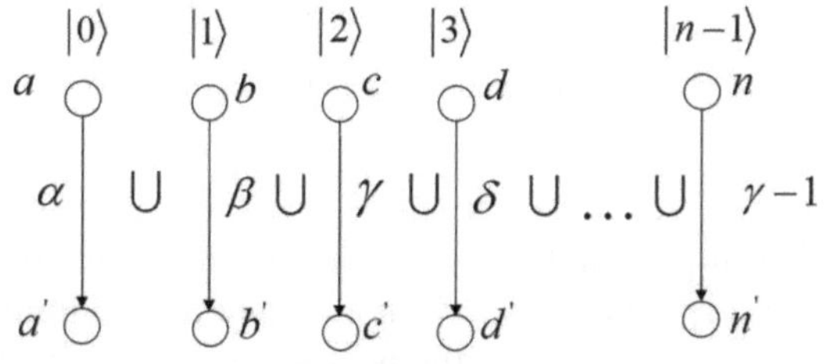

Qudit

$$|\psi\rangle = \alpha|0\rangle + \beta|1\rangle + \gamma|2\rangle + \cdots + v|n-1\rangle$$

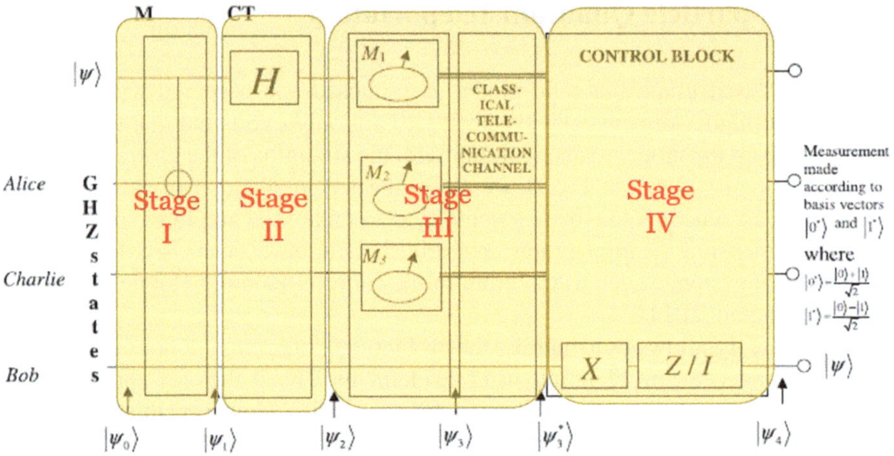

Fig. 2 Three-particle quantum teleportation framework with four stages

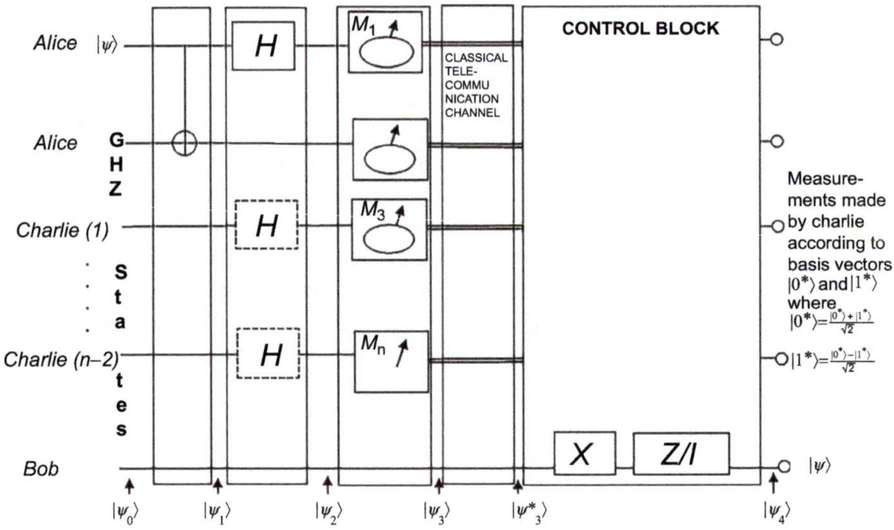

Fig. 3 The *n*-particle quantum teleportation framework

The n-particle framework as modelled using the graph theoretic quantum system modelling framework will not require any continuum equations or any spatial discretization process. In the next section we present some simulation studies of the multi-particle quantum teleportation model.

4 Topological Quantum Computing for Visualization and Understanding of the Multi-particle Quantum Teleportation

We evaluate our quantum teleportation framework on the IBM quantum computing cloud that has made access to quantum resources as easy as a click. We perform experiments with qubit for dimensionality 2 and 3. It is primarily due to the limitations of the network. However, the abstract model is scalable to any dimensionality reaching up to infinity.

We use the Quantum Information Software Kit (QISKit for short) that provides a software development kit for working with the IBM's quantum computer. One can use QISKit to create quantum computing programs, compile them, and execute them on one of the several backends such as the online real quantum processors, online simulators, or local simulators. IBM provides an easy-to-use python API client to connect to the IBM's quantum computing cloud. Here we show in Fig. 4 a sample circuit for the 3-qubit teleportation using the GHZ states designed using the QISKit Python API.

It is observed during the simulation studies that the circuit provided robust results on different bases of measurement. We also demonstrate in subsequent experiments the robustness of models to noise. At the time of writing this chapter, we are in the process of establishing partnerships with IBM to run these models for larger systems.

5 Robustness of the n-Qudit Quantum Network Against Noise and Decoherence

This work is another in our series of collaborative efforts with the research team at the Dayalbagh Educational Institute, Agra, India – 282005 to demonstrate the performance of the graph theoretic models through simulation studies.

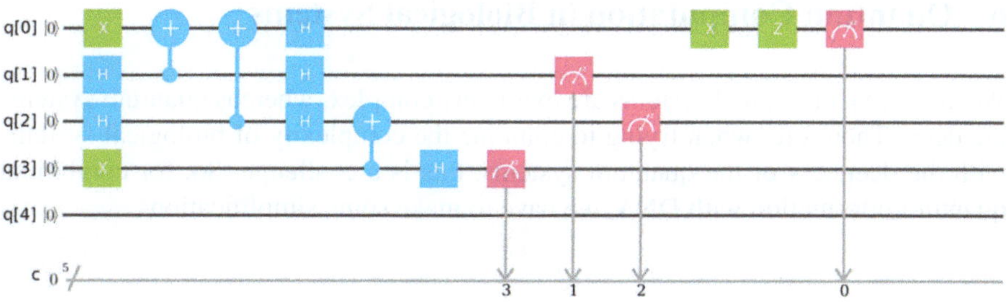

Fig. 4 A sample circuit of 3-qubit teleportation using GHZ states in QISKit

In this section, we show the impact of noise and decoherence on the performance of the n-qudit quantum Hopfield network proposed by Srivastava et al. (2017). Noise is unavoidable in any quantum system, and quite often the solutions proposed require one to add an ancillary bit to minimize the effect of noise, but that is not a scalable solution. The strength of the system lies in the fact that it is robust and scalable in spite of noise.

The second phenomenon that affects all quantum systems is decoherence. It basically means that loss of coherence leads the quantum system to lose its quantum nature. For the purpose of demonstration, we turn to the case of pure decoherence, that is, we introduce random phases to the elements of the density matrix, without changing their magnitudes. So, the next step is learning with perturbations to the density matrix. One approach is to add noise to the Hamiltonian; of course, the system must still obey all the physical requirements to be a quantum system. From three, one bootstraps to four and so on, and the hope is that the amount of additional training needed decreases, because more and more of the information necessary for the entanglement indicator was present already in the N–1 qubit system. Splitting the density matrix this way is possible, since while working with numerical simulations, of course, the density matrix must remain Hermitian, positive definite, and with unit trace to conserve probability.

Since we are working with simulations, we can isolate the different effects of noise and decoherence. Noise in this case is referred to the random uncorrelated magnitudes of a given size added to the density matrix elements, and decoherence is referred to random phases so added. In cases, where both effects are present, we call it complex noise.

We use paradigm as adapted from pattern recognition to evaluate the effect of noise and decoherence on the network. Our results are presented for the evaluation of the quantum Hopfield network's performance with respect to noise (Fig. 5a), decoherence (Fig. 5b), and dimensionality (Fig. 5c). Preliminary results of our study indicated that the robustness of the system increases with increase in the size of the network. It also increases with an increase in the dimensionality of the qudit as shown in the figures above.

6 Quantum Computation in Biological Systems

We know that biological systems are extremely complex, whereas quantum systems are deep. Therefore, when trying to combine the complexity of biological systems with the deepness of the quantum systems, this is a challenge. So, for combining quantum information with DNA, we have to make some simplifications.

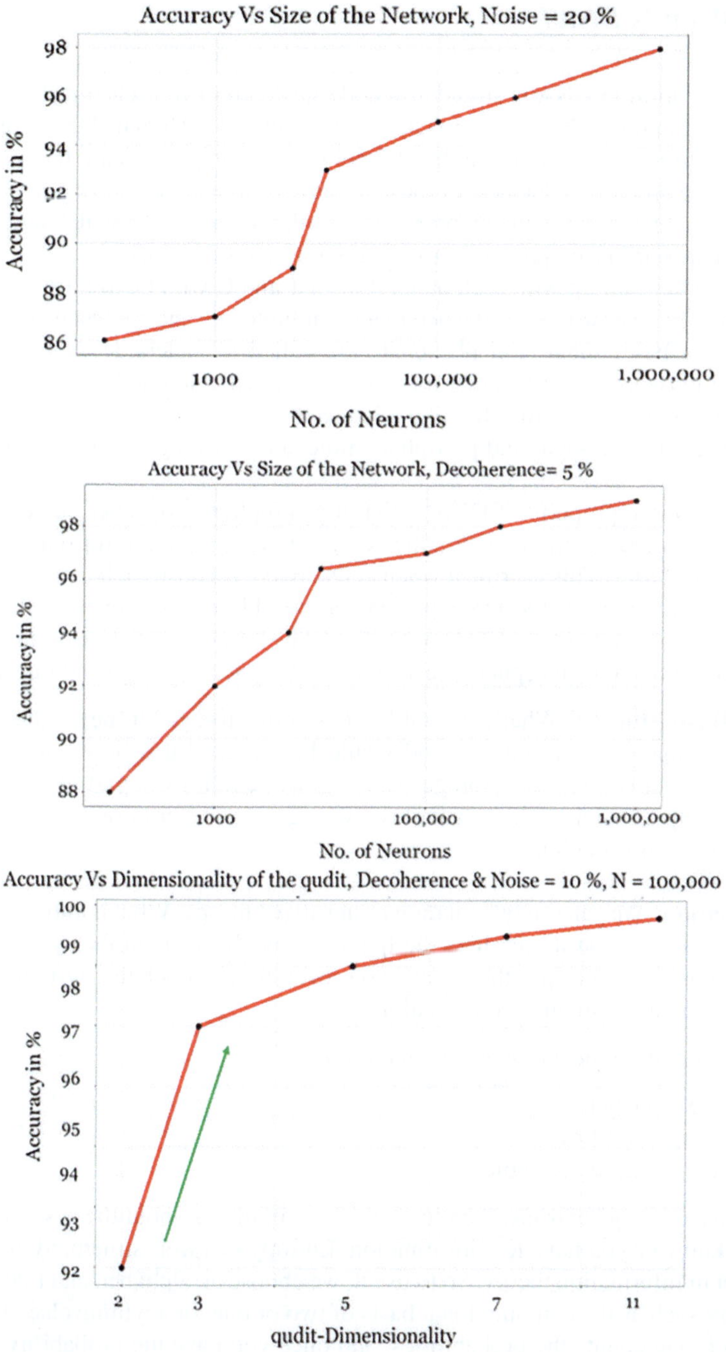

Fig. 5 Evaluation of the quantum Hopfield network's performance on (**a**) noise, (**b**) decoherence, and (**c**) dimensionality

6.1 What Is DNA?

The information in DNA is stored as a code made up of four chemical bases: adenine (A), guanine (G), cytosine (C), and thymine (T). Human DNA consists of about 3 billion bases, and more than 99 per cent of those bases are the same in all people. The order, or sequence, of these bases determines the information available for building and maintaining an organism, similar to the way in which letters of the alphabet appear in a certain order to form words and sentences.

DNA bases pair up with each other, A with T and C with G, to form units called base pairs. Each base is also attached to a sugar molecule and a phosphate molecule. Together, a base, sugar, and phosphate are called a nucleotide. Nucleotides are arranged in two long strands that form a spiral called a double helix. The structure of the double helix is somewhat like a ladder, with the base pairs forming the ladder's rungs and the sugar and phosphate molecules forming the vertical sidepieces of the ladder.

An important property of DNA is that it can replicate, or make copies of itself. Each strand of DNA in the double helix can serve as a pattern for duplicating the sequence of bases. This is critical when cells divide, because each new cell needs to have an exact copy of the DNA present in the old cell. As per the definition on Wikipedia, the main role of DNA molecules is the long-term storage of information. There are two main concepts in the sentence – long-term storage and information.

Long-Term Storage What is meant by long-term storage? It means that we have some information from the past and would like to send this information into the future and that is long term storage. The way we send it from past to future can be described by a channel. This channel is a very general description to whatever happens to your information.

Information We can see information as negative entropy. What is entropy? In general, we can say that it measures the ignorance we have about the system. If your system has zero entropy, then you know everything about the system. In other words, you have maximum information.

In information theory we have three things:

- Choose alphabet
- Count probabilities
- Calculate entropy of choice

If you have a large entropy, then there are certain things about the system that you do not know or you have less information. Entropy is a well-structured-established concept in information theory. So to use it, we choose an alphabet, and that could be anything such as the computational basis of two or one, or anything else. As per the alphabet, one counts the probabilities, and once you have the probability distribution, you can calculate the entropy of your choice.

In this case, we choose our system of interest, then we count the number of possible states, then we take the logarithm of the number of states. The massively interesting thing is that once all events are equally likely, then the two definitions of entropy actually coincide. This is one of the many links between information theory and statistical physics. In physics, we have a well-established second law of thermodynamics and that tell us that in a closed system entropy does not decrease.

So if we translate this argument to information theory, we immediately know that information cannot spontaneously increase in a closed system. That is a very important thing to know.

Classical information	Quantum information
Reliable storage	Decoherence
Copying possible	No-cloning theorem
Limited set of manipulations	More allowed manipulations

Whenever you have physical information that comes with a physical carrier, that can be a piece of paper you are currently writing on, or it can be your hard drive on the computer. For classical physics, for classical information, we know from everyday life that I can change the information carrier, I can write something on a piece of paper, scan it to my computer, and print it, and if I do not do obvious mistakes, I didn't lose any information. That is allowed by laws of classical physics. If I choose a physical carrier that is smaller and smaller and reaches the quantum limits, then the fascinating thing is that as I process information, it changes according to the laws of "Quantum Physics". So what it means is that I cannot copy information; I have to keep fighting decoherence as the information is always getting lost.

No-Cloning Theorem

So let us suppose you could copy quantum information, then you would have some unitary acting on your wave function so that it actually copies the unknown state to your template. Then I can assume that is a superposition of zero and one, and we plug it into the formula to get this lengthy expression of the copied state. But the mathematical formulation of quantum mechanics also allows me to feed it in step by step.

$$\hat{U}\,|\phi\rangle\,|0\rangle = |\phi\rangle\,|\phi\rangle$$

$$|\phi\rangle\,|\phi\rangle = \alpha^2\,|\,00\rangle + \alpha\beta\left(|01\rangle + |\,10\rangle\right) + \beta^2\,|11\rangle$$

$$\hat{U}\left(\alpha\,\,|0\rangle + \beta\,|\,1\rangle\right)|0\rangle = \alpha\,\,|00\rangle + \beta\,|\,11\rangle$$

The two expressions should be the same, but we see clearly that they are not the same. So, our original assumption that we can copy quantum information was false. That is the *no-cloning theorem*. This is only possible in one case where either or is zero, which is the case of classical information.

Conditional Entropy

The conditional entropy measures, how much uncertainty and ignorance you have about a combined system and observer. Given that, you have the entire observer's knowledge. For anything that we encounter in this real world, it is a fair assumption that this conditional entropy is bound by 0. So the idea is that if my observer has a certain ignorance about his own state, then the combined system and the observer should have at least this uncertainty.

$$H(S \mid O) = H(SO) - H(O)$$

But if we are dealing with the state I just mentioned, then we have full knowledge about the system. I do not have any uncertainty, but if I look at the state that the observer himself is in, he actually has uncertainty. For quantum systems, this quantity of conditional entropy can be actually smaller than zero, and this actually means that there is entanglement. So whatever physical meaning entanglement has, you can take it as a very strange probability distribution which allows us to achieve correlations that are classically not possible.

6.2 Application

Let's suppose we have a system that is a mixture of zero and one and you would like to reset. It could be, for example, your memory, and you would like to reset it to zero. What is usually called a state is that a classical observer has to pay one unit of work to extract or to erase that information. Is it any different if we are dealing with the quantum observer as shown in Fig. 6?

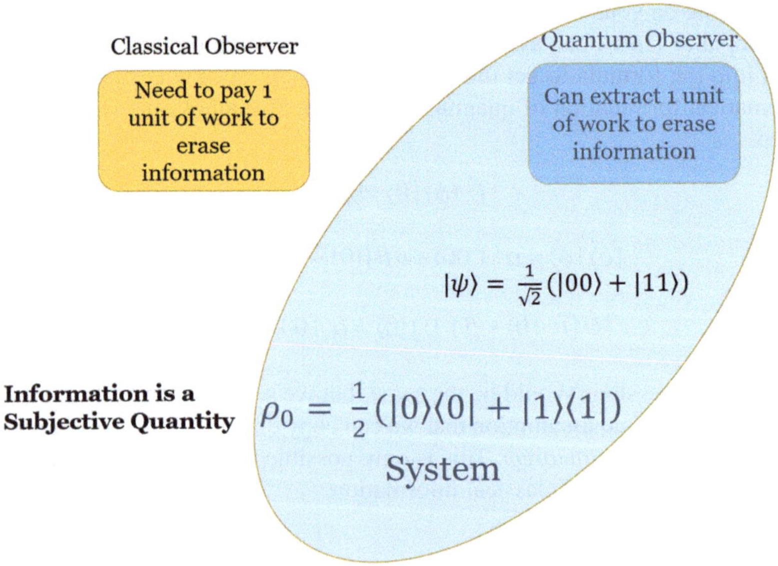

Fig. 6 Comparison between a classical and a quantum observer

Yes, it could be possible that the quantum observer actually has the other half of a maximally entangled state and then the system has the same correlations, as it has been shown that one can actually extract one unit of work to erase information to reset it to zero.

Now we are talking about biology. Whatever small enzymes, molecules, and other small systems are wiggling around in our biological systems, it could be that they are small enough to be a quantum observer. Then suddenly the entire set up changes. Instead of actually needing to pay work, you can actually extract work. So it might be for those biological systems that we need to change the way we think about them. Another thing I would like to mention here is that "information is a subjective quantity". It depends on the knowledge you can have, and it is different for different observers.

Just to talk a bit on decoherence, it is roughly inversely proportional to mass and temperature. However, why do we not see quantum effects in real life? It is because we are massive, we have a lot of mass, and we are hot. We are at 300 K, which is a very high temperature. Hence, once we have these two conditions, we believe that we would not have any quantum effect, but that is not true.

6.3 *Information in DNA*

So finally, we come to information in DNA (Fig. 7). Most of you probably already know this molecule.

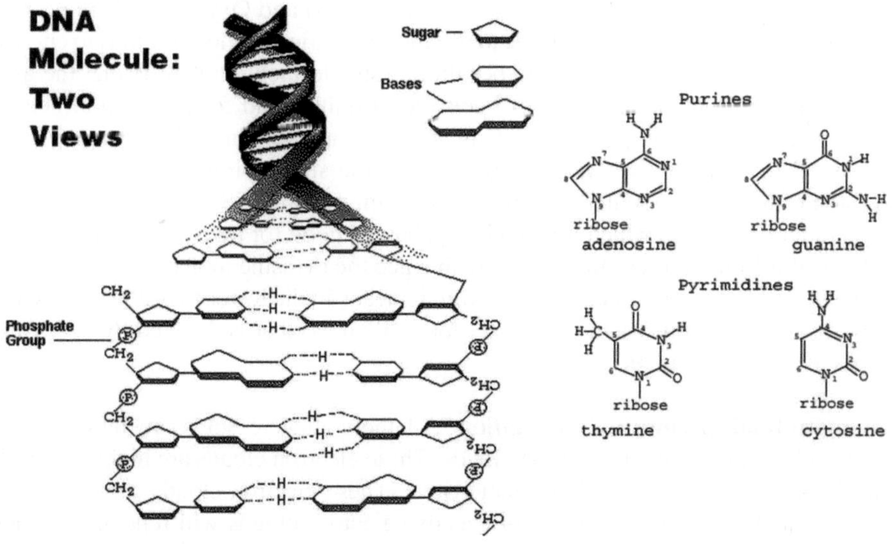

Fig. 7 DNA molecule and its representation

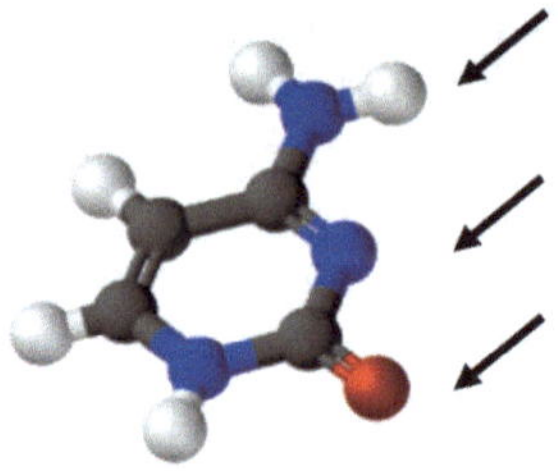

Fig. 8 Bit representation of a Cytosine

DNA is a single, long molecule, and outside there is some phosphate that is bound into sugar-stabilizing DNA. But it is actually those four nucleic acids that contain the genetic information. These molecules are way smaller than our computers and allow us to encode bits. We have four bases; each molecule encodes two bits, and then each of these two bits are arranged in a linear way into the two ends having opposite information. Now the interesting question is how you actually access that information. It is not enough to store that information; you should be able to read that out as well. How do you do that with these small systems? Let's look at one of them – Cytosine in Fig. 8.

Transversal of Quantum Information It is not enough to store that information; you should be able to read that out as well. On the left side it is bound to the phosphate back bone, so you cannot access it. Up and down of this cytosine, you have electron clouds and other nucleic acids you cannot access either. The only part you can actually access is that molecule on the right side and that is what is used in Biology. Here, we have those atoms – that's Nitrogen and Oxygen, and single protons can bind to them. So the idea is to use the acquisition of those protons to encode the identity of that molecule. Another thing I should mention here is that the aromatic ring you see there – though I consider it quite small for quantum effects to take place–is actually quite large. This is something we can easily consider to be classical information, whereas the position of the single proton, because a single proton is due to laws of quantum mechanics -, might tunnel. So this leads me to this notation on the right, which is a bit sloppy, but it's good for transferring ideas so the classical information we are dealing with since the cytosine molecule is something I would encode in this quantum state, which means in the lower two slots, where the proton could be empty and the upper slot is filled, so that is denoted by zeros and ones as shown in Fig. 9.

Longitudinal Quantum Information All these nucleic acids are less planned molecules surrounded by electron clouds. These electron clouds are highly delocalized, as we are dealing with aromatic rings. These electron clouds carry negative charge, and we know that adjacent negative electron clouds will repel each other. The intriguing thing is that this actually leads to correlated excitations leading to lowering of energy states. So different charged clouds kick each other and redistribute their charges, thereby lowering the ground-state energy. This is the Vander Waals forces.

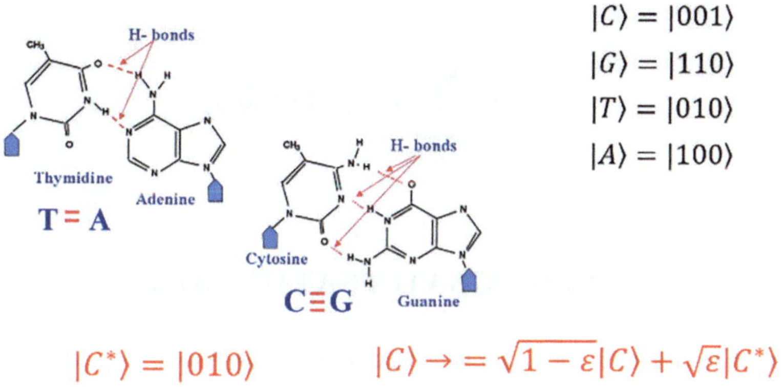

Fig. 9 Quantum information encoding on a Cytosine molecule

One actually needs coherence between neighbouring sites to get the attractive part of the force. This now also has non-trivial implications. All neighbouring sites are entangled due to these continuous interactions, and also the electron cloud carries information about the identity of the other site. So if I am thymine here and adjacent to adenine, this adenine influences my own electronic state. If there would be cytosine next to me, then my own state would look different, and that's substantially different from classical computing. So my own identity is influenced by the neighbours I have in the DNA chain.

I just gave you two examples, where you can have quantum information to be present in DNA. So how does it all work together at room temperature? If you have a massively hot system at room temperature, you cannot have entanglement.

Entanglement at 300 K One can analyse a very generic model of couple thermal oscillators, and then you can see that the absolute temperature actually doesn't matter. DNA has optical frequencies and that means that thermal noise cannot excite those optical frequencies. So this quantum information we are talking about is basically in its ground state, event at 300 K.

Quantum Classical State So let's suppose X is a string of classical symbols, then you would denote that the state has two components: the first one denotes the classical information and the second one denotes the quantum information. This is diagonal and its basis has no coherence. But this comes along with some sort of quantum information as shown in Fig. 10. And so X could be the string of your DNA, and then it is embedded in this cloud of quantum information. Paul Davies called this sort of quantum information "Shadow Information" (Davies 2012).

In quantum information you can have at least one copy of the information in your DNA, if not more copies. If you want to decode your classical information, then you measure the quantum part, but it can happen that the quantum part of science is not fully distinguishable, and this gives probability of mistakes.

So, coming to the most important question.

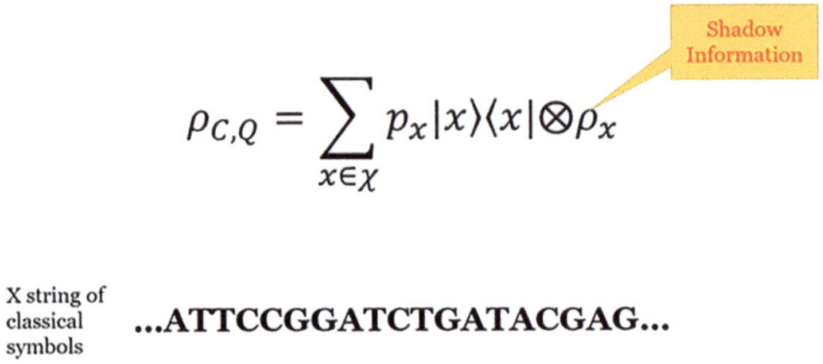

Fig. 10 Representation of the shadow information in a DNA molecule

Does Quantum Information Matter? There are many things that quantum allows us to do if you would like to send some classical information from the sender to the receiver. We do it every day: we send text messages, we send emails, we talk – all of this is classical information, which uses classical channels. The only problem is that this probably doesn't take place in DNA because the system you are talking about is too small to have a full classical channel, so it's more likely that you are dealing with a quantum channel.

Now we would like to compare two cases of quantum channels: normal quantum channels and the entanglement-assisted quantum channel, where actually the receiver and the sender share an entangled pair of information. Then I will talk about channel capacity, which means basically how much information can be transmitted in this channel. That number is between zero and one. If it's one, you can transmit all the information, and if it's zero, you cannot transmit any information. You can send classical information in quantum channel, and there is no problem. So you encode your classical information on your quantum system, you send your quantum information, and then you make a measurement on it and decode the information that it carries.

Qubit Depolarizing Channel Let us take a wave function x, and you send it through the channel, but it can actually happen that this channel suffers from some sort of noise as we can expect in biological systems, and then with probability 1 minus epsilon gives you the original wave function and with probability epsilon gives you a random result. So that is the qubit depolarizing channel as shown in Fig. 11.

Now, if you want to encode some classical information X so X could be added into mean or whatever, you encode that in a certain wave function, you send it through your noisy channel, and with some probability you recover X. But in some cases you guessed incorrectly, and you recover Y. So how often you guessed correctly and how often you guessed incorrectly is characterized by this channel capacity K_1. Now you can use the same channel. The only difference is that the sender and

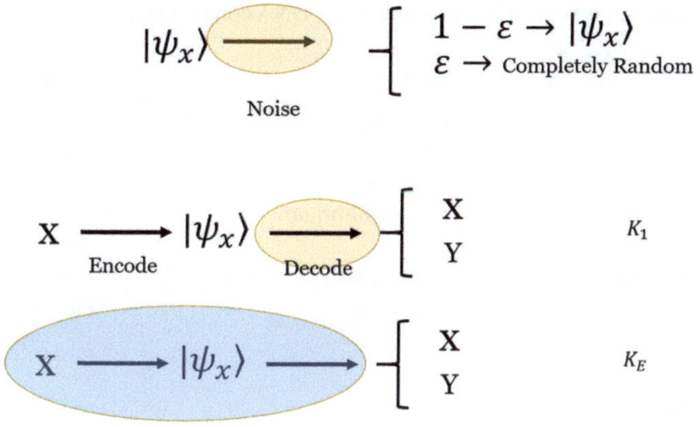

Fig. 11 Outcomes from a qubit depolarizing channel

the receiver initially share maximally entangled states, and so you have capacity K_E. Here E is for entanglement. The beautiful thing is that if you use entanglement, then you can double your channel capacity. So we have a system where you would like to transmit classical information. It is reasonable to assume that you can create this entanglement that would allow you to double the capacity of your noisy channel.

6.4 *Communication Along Chains*

Another aspect is, how one might use entanglement in DNA for communication along the chain. In the respective literature, they have mentioned that how little motors go along DNA and achieve things. How do these motors actually know what they are supposed to do?

Let us suppose that each of these balls represents a site in your DNA, and what one does is wiggle the one end of it, and then you can calculate what happened to it. Actually, it is found that after some time there is entanglement between the first site that you shook and the nth site, n can be very large. These two far apart sites share entanglement. This entanglement means its half correlation. So the question is: Can we use those correlations for communication along this one-dimensional chain? Another thing, where such concept may be applied, is in protein folding, where this linear sequence converts and folds into a three-dimensional structure. I say this because Information is in Formation. Therefore, we conclude by comparing the information flow in Table 1.

Table 1 Comparison between classical computing, DNA computing, and quantum computing

	Classical computing	DNA	Quantum computing
INITIALIZE	0,1	$\lvert x\rangle\langle x\rvert \otimes \rho_{Q,x}$ $x\in \{A, C, G, T\}$	$\lvert\psi\rangle \in H^{QC}$
INTERACT	Bits independent	Single site depends on neighbouring sites	Entanglement
OPERATIONS	AND, OR, NOT…	Non-equilibrium?	Universal set of unitary gates
GATE TIME	Short	Continuous interaction	Short
READ OUT	Deterministic	Single shot	Statistical

7 Conclusions

Overall, this chapter brings together three key areas in the domain of consciousness: quantum information, topological graph theory, and biology. It seems that a blend of the former two allows us to understand the complexities of the latter. Moreover, it paves the way for future research in this exciting overlapping area of research that is interdisciplinary and aims to solve the complex problems of human existence. The power of abstract topological graph theory has already been seen through the works of Prof. Prem Saran Satsangi, where the most succinct model of the universe has been very carefully crafted, using the beauty of math and graph theory. Biology is central to Consciousness Studies, and we need to abstract and solve the complexities of biology through the simplicity of graph theory.

Comment by the Editor Rev. Prof. Prem Saran Satsangi:

ADRENALINE, secreted from Adrenal Glands, is subject to Hormonal Treatment as opposed to the earlier Steroidal Treatment, which had the disbenefit of several adverse side effects, by now well known to the World Health Organisation (WHO) through United Nations Educational, Scientific and Cultural Organization (UNESCO) – Headquarters at Paris, France.

Extended Note from the Editor Rev. Prof. Prem Saran Satsangi:

ADRENALINE secreted by Adrenal glands, also known as suprarenal glands, are small, triangular-shaped glands located on top of both kidneys (Wikipedia.org). Adrenal glands of all Living Human Beings produce hormones that help regulate our metabolism, immune system, blood pressure, response to stress and other essential functions (including physical parameters such as height, weight, body-build, etc.): as further endorsed by my experience as the "Acclaimed" Eighth Waqt/Present Sant Satguru of Radhasoami Faith, for the past ≈19 years (since 18 May 2003) vide Discourses on Radhasoami Faith (With Supplement): 2009: Power Law of Meditational Consciousness (Ultra-Transcendental Meditational Practices)
- Ra-Dha-Soa-Aah-Mi Chant SOP 6-4-2 Regime

रा- धा- स्व- आ-मी

(invoking Mathematical Abstraction of Topological Graph Theory; initially learnt during my formal Post Graduate Education (i.e. M.S. from Michigan State

University, East Lansing, USA – fully supported by USAID TCM Scholarship and additionally recognized by Honor Society of Phi-Kappa-Phi (for securing CGPA of 4.00/4.00 in the minimum period of three Quarters- Fall, Winter & Spring of the Academic year 1960–1961) with further Doctoral (Ph.D.) Research under Canadian Commonwealth Fellowship Award 1966–1969 and a brief Post-Doctoral Research Support during Summer 1970 by the Faculty of Man-Environment and Cybernetics of the University of Waterloo, Ontario, Canada)).

As of date approximately 1200 Acres (48,56,228 Sq. mts.) of Agroecology – Precision Farming Land, owned by RS Satsang Sabha (registered as Religious & Charitable Society; Stand – transferred to DEI (Dayalbagh Educational Institute-Deemed to be University) by way of Charity, as long as it exists through "Trusteeship" arrangement.

In our continuing research investigations, with a focus on Spirituality, we have discovered a daunting revelation of the possible nexus between Drug Addiction & Moral Turpitude, as conveyed to the Honorable Lady President of the Sovereign Republic of India (hailing from Scheduled Tribe Category) through her OSD, President's Secretariat, Rashtrapati Bhawan, New Delhi (India).

[Vide 2 Hr. 15 Mins. Video Interview of Prof. S.C. Dutta Roy, conducted during celebrations of IEEE Signal Processing Society, my erstwhile colleague at IIT Delhi (≈29 years)].

Rev. Prof. Prem Saran Satsangi adds an experimental study proposed in August 2022:

A Brief Proposal:

Towards Achieving Better Worldliness by Indirect Control of Adrenaline

Submitted by: Dr. Manju Srivatsava, Assistant Professor, Department of Chemistry, Dayalbagh Educational Institute (Deemed to be University)

Experimental studies carried out on Satsangis in Dayalbagh in 2011, using MEAD (Ryodoraku Meridian Energy Analysis Device – an electronic instrument used for diagnosis of patients by Traditional Chinese Medicine practitioners), showed that most of the subjects had an imbalance in their autonomic nervous system. Their calm and relaxed countenances lead one to the idea that this may be due to a reversal in the roles of their otherwise normally dominant sympathetic nervous system and their otherwise normally secondary parasympathetic nervous system, similar to the reversal seen (a) during meditation and (b) when secretion of adrenaline in the body is reduced. These studies lead to a hypothesis, not yet validated, that passive or active immersion of subjects, for a considerable period of time, in a religious environment leads to conditions similar to continuous meditation sessions, and this causes neuroplastic alterations leading to the above mentioned results, which may persist for a while after the excitation is removed.

The Comments of Rev. Prof. Prem Saran Satsangi on the Proposal Verbatim:

You are welcome to pursue your research. However, purely spiritual implication is that our welfare lies in following the administration of hormonic injection/nebulization/homeopathic version; for preventing any calamity befalling members of

Radhasoami Faith, administered by RS Sabha, Dayalbagh, Agra – 282005. "Adrenaline" is known to be a Messenger of an impending crisis, for which the "Acclaimed" Waqt/Present Sant Satguru is the recognized Non-statutory Authority, serving as a Think-Tank for DEI via Consensus-Building Approach as practised by the forum of ACE, under His Chairmanship.

Hormonic Nebulization, once in each of the Two four-hourly shifts, has been tried for a period of Ten days (July 10, 2022 to July 20, 2022) by the Chairman ACE (who has reported relief by way of feeling at ease w.r.t. Adrenaline Secretion). Beginning evening Shift on July 20, 2022, "Homeopathic version: Adrenalium (Liquid Dilution) 30 CH" has commenced for recording comparative relief-sensation for another Ten Day Period.

References

Agnihotri, G., Prem Saran Satsangi, and T. Karunakaran. 1983. Application of multi-terminal representation in graph-theoretic field modelling of ground water dynamics. Proceedings, National Systems Conference, Trivandrum, V.S.S.C, X.5.11.

Coecke, Bob, Eric Oliver Paquette, and Simon Perdrix. 2008. Bases in diagrammatic quantum protocols. *Electronic Notes in Theoretical Computer Science* 218: 131–152. https://doi.org/10.1016/j.entcs.2008.10.009.

Davies, P.C.W. 2012. The epigenome and top-down causation. *Interface Focus* 2 (1): 42–48.

Koenig, Hermann E., Yilmaz Tokad, and Hiremaglur K. Kesavan. 1967. *Analysis of Discrete Physical Systems*. New York: McGraw-Hill.

La Mura, Pierfrancesco, and Lukasz Swiatczak. 2007. Markovian entanglement networks, arXiv: quantph/0702072v1.

Sahni, Vishal, and D. Goswami. 2008. *Nanocomputing*. New Delhi: Tata McGraw Hill.

Satsangi, Prem Saran. 2006a. *Linear Graph Theory for Modelling a Variety of Systems*. International conference on differential geometry and topology in the perspective of modern trends (DGTPMT – 2006). Agra: Dayalbagh Educational Institute.

———. 2006b. Systems movement: Autobiographical retrospectives. *International Journal of General Systems* 35 (2): 127–167.

———. 2009. Linear graph theoretic general systems paradigm – A learning system modelling methodology. *Editorial Review Literary Paritantra (Systems) – An International Journal on Literature and Theory* 1 (1 & 2): 1–16.

Savage, Gordon J., and H.K. Kesavan. 1977. *A Graph-theoretic Approach to Field Problems*. International conference on numerical Methods, St. Margherita, Italy.

Savage, Gordon J., V. Madan, and H. Kesavan. 1980. The magnetic field problem: A graph-theoretic model. *IEEE Transactions on Magnetics* 16 (4): 579–585.

Srivastava, Dayal Pyari, Vishal Sahni, and Prem Saran Satsangi. 2014. Graph-theoretic quantum system modelling for neuronal microtubules as hierarchical clustered quantum Hopfield networks. *International Journal of General Systems* 43 (6): 633–648.

———. 2016. Modelling microtubules in the brain as n-qudit quantum Hopfield network and beyond. *International Journal of General Systems*. 45 (1): 41–54.

———. 2017. From n-qubit multi-particle quantum teleportation modelling to n-qudit contextuality based quantum teleportation and beyond. *International Journal of General Systems*. 46 (4): 414–435.

Tucci, Robert R. 1997. Quantum Bayesian nets, arXiv: quant-ph/9706039. Accessed 08-Aug-2022

Part V
Consciousness and the Senses

Sonic Consciousness in Hindu India

Annette Wilke

Abstract Among the different cultural hierarchies of sensory perception, Hindu India's strong phonocentrism stands out. Sound and sonality enjoy exceptional cultural significance since ancient times until the present day. The contribution argues that this emphasis on sound has had distinct influence on collective and individual consciousness building, and on practices, world view, and cosmology. The essay's major focus is the Nāda-Brahman, literally "Sound-Brahman" or "Sonic Absolute", and its history. This powerful sonic conception of ultimate being received little research so far. Even within academia it was hardly acknowledged that the term is not a very ancient one but made its first appearance only in the thirteenth century. The essay investigates the Nāda-Brahman's little known impressive musicological beginnings and its history of transformation – from musical sound to resonances and occult interior sounds, including the Radhasoami's eternal "sound current". It is shown that it was at the same time a history of different forms of sonic consciousness and sonic cosmology, which, however, share a spiritual sensuality that differs from the acosmic pure consciousness of Advaita-Vedānta, the philosophical tradition of India, which became best known in the West as a marker of Indian spirituality.

Keywords Nāda-Brahman · Sound · Sonality, sonicity · Yoga · Music

Among the significantly different cultural hierarchies of sensory perception, Hindu India's strong phonocentrism stands out. Sound and sonality, the voice and hearing, the spoken and the sounding word enjoy exceptional importance and cultural significance since ancient times up to the present day (Wilke and Moebus 2011). Among the many effects of sonic consciousness produced by this situation, the

A. Wilke (✉)
Institute for the Study of Religion, Westfälische Wilhelms-Universität Münster, Münster, Germany
e-mail: wilkeann@uni-münster.de

P. S. Satsangi et al. (eds.), *Consciousness Studies in Sciences and Humanities: Eastern and Western Perspectives*, Studies in Neuroscience, Consciousness and Spirituality 8, https://doi.org/10.1007/978-3-031-13920-8_13

Radhasoamis' sonic cosmology/theogony and its "yoga of sound" or "word-yoga of the soul" (*surat-śabd-yoga*), the school's unique esoteric *bhajan* practice of inward listening to the vibrations, the eternal "sound current" or divine Śabd ("Word"), pervading all existence and the soul, stand out as perhaps the very peak, introduced in the nineteenth century (Wilke and Moebus 2011: 894–98, 1004–11; Babb 1986; Juergensmeyer 1991). And yet, it is not exceptional among the many outstanding examples of sonic consciousness or sonic awareness in India. In the following introduction to Hindu India's sonic paradigm, I want to draw attention to the decisive influence this strong emphasis on sound has had on collective and individual consciousness building, and on practices, world view, and cosmology. This is specified by the major theme of this essay: the Nāda-Brahman, literally "Sound-Brahman" or "Sonic Absolute", and a historical appraisal of this eminently powerful sonic conception of ultimate being and consciousness, which received little research so far. The Nāda-Brahman is generally ascribed a hoary past, but the term made its first appearance only in the thirteenth century, when it was introduced by musicologist Śārṅgadeva. The Nāda-Brahman concept was commented upon – among others – by Kallinātha in the fifteenth century and variously put into practice, partly in quite startling ways, for instance, by the famous Carnatic musician and composer Tyāgarāja (1767–1847). However, the original Nāda-Brahman of the musicologists also witnessed a fascinating history of transformation – from musical sound to resonances and occult interior sounds, which will be called "yogification". My thesis will be that the Radhasoamis' yoga of sound and sound current illustrate this yogification as well as the transreligious and transcultural history of the Nāda-Brahman in modern times. The various interpretations of the Nāda-Brahman by Śārṅgadeva, Kallinātha, Tyāgarāja, the Radhasoami movement(s), and others will allow to reflect different forms of sonic consciousness that emerged from India's sonic culture, and more specifically from the Nāda-Brahman concept.

I will often use the terms "sonic consciousness" and "sonic awareness" largely interchangeably, "awareness" in the sense of being conscious of something. However, now and again and increasingly in the chapters on the Nāda-Brahman, "sonic consciousness" will point to pure "consciousness" as such, which some of the agents discussed understood as eternally interwoven with sound.

1 India's Sonic Paradigm: An Introduction

India is a pronounced culture of sound. This feature cannot be overstated. It starts with the intrinsically sonic dimension of texts in India, from ancient and pre-modern to modern sacred literature. This dimension has been insufficiently researched for a long time. Mere philology does not capture the full dimension of these Indian texts, as they are aesthetic events, sensory-affective, and performative. Traditionally, they are not read silently, but vocalized, memorized, recited, sung, preached, staged, and danced, composed for hearing. Sound and sonality, and along with it the body and the voice, form an essential mark of India's performance culture. The performance of words and sounds as well as the act of listening triggers not only a fine sonic memory and collective effervescence, but also emotional arousal, a sensory feeling of text, and the recipients' own emotions surrounding it – in short, this situation triggers particular forms of sonic consciousness and awareness.

What I want to stress is the importance to be aware that in India's highly literate culture living communication systems from tradition were never given up. Sound matters in Brahmanic Sanskrit traditions, vernacular folk cultures, and the many shadings of overlapping and modes of performance in between. The exceptional importance of the spoken *and* the sounding word has enjoyed amazing cultural consistency and continuity through the ages in ever new ways. This persisted not only after the introduction of writing during or before the first millennium BCE, but even after the start of mass-scale book printing in the late nineteenth century. In fact, the centrality of a literary tradition and the philologer's text came as a result of the British in India, as it was the British and Indian intellectuals influenced by them who effectuated the transition from orality and multimedia performance to textuality (Van der Veer 2001: 118–130). While this process was important for Indian reformers in their construction of national identity and Hindu heritage, it left orthodox and popular traditions largely unchanged. Orality and literacy have always combined seamlessly in classical Hinduism, and sonality has remained in the forefront through the vocalization of written texts. What I call "sonality" is precisely this unbroken continuity of the oral and the written in the sounded text and sonic practice. This is not restricted to Hindu and Sanskrit texts but is also common to Buddhist, Zoroastrian, Jain, Sikh, and Islamic sacred literature, and even surpassed by Radhasoami *bhajan* practice, which reaches beyond audible devotional song.

My contribution concentrates on Sanskrit Hinduism wherein we find a particularly wide range and some most amazing features resulting from the extraordinary attention given to sound and sonality. We are dealing with a culture in which sound (linguistic, musical, or natural) and even singular phonemes have been conceived as an independent medium of expression, communication, and efficacy. It is a culture that never lost sight of the fact that a word is not only a meaning bearer but also a phonetic structure. A word (*śabda*) is defined as a sound to which meaning is attached. There was always acute awareness that language is not exhausted in semantics and denomination, but that the audible physical element of language and its sonic aura also play a central role in meaning construction and that linguistic signs reach into the pre-terminological. Not only in widespread acoustic piety, such as *mantra* repetition and devotional singing, but even in the traditional sciences there was awareness that texts always also have a material quality – their sound embodied in the voice – and can be perceived in a sensory and emotional way. As already mentioned, even beyond Sanskrit Hinduism, texts in India are invariably not only discursive bodies, but sound events and performative scripts as well. They turn into highly aesthetic, sensory-affective media, being vocalized, recited, sung, danced, and dramatically staged, affecting not only the performers but also the listeners. Having been composed for hearing, for memorizing, for embodied perception, they create emotional and body knowledge beyond intellectual cognition. This strong focus on the spoken, the audible, and the sounding word not only effected a stronger sensory and emotive appropriation of the subject, but also exerted distinct influence on collective and individual consciousness building, on beliefs, practices, habitus forms, and knowledge cultures. Simply listening to the sound of a religious text, for instance, is already held to be purifying and bringing liberation. Arguably, sometimes the sonic aspect is even more important than the semantic meaning. In Sāmaveda chant and untranslatable Tantric *mantras*, it becomes particularly obvious that lexically meaningless sounds enjoy highest sacrality or, in other words, that sacredness already inheres in sounds and not only in the meaning of the utterances.

No wonder that acoustic piety occupies a very prominent place in Hindu India. It is indispensable for the lived religion, and it is at the heart of Hindu devotionalism, *bhakti* (love of God), which forms a pan-Hindu mainstream from around the tenth to thirteenth century up to the present day. Standard *bhakti* texts are songs, and often text and melody have been composed together. Whether classical or folk, Sanskrit or vernacular songs, all of them trigger deep listening, absorption, and ecstasy. Typically, *rāga* music is defined as colouring the mind – that is, producing emotional arousal. Widely spread is the conviction that music can communicate devotion better than any words and tune directly into the divine. Devotional *kīrtan* and *bhajan* songs – individually or collectively performed – are among the most widespread religious practices, besides the repetition of *mantras*, whether Vedic, Tantric, or devotional. In devotional contexts, "mantra" refers purely to the repetition of the name of one's God and praise formulas – aloud, muttering, or in silent speech. It aims at contemplative immersion, divine empowerment, emotional rapture, and sensing and incorporating the deity. Vedic, Tantric, and devotional *mantras* each have their own peculiarities, but according to a common belief across traditions, the power of *mantras* is immense. The *mantra* makes the divinity and all their powers immediately present. What best qualifies a *mantra* is the idea that it is efficacious, regardless of the speaker's intention. It is held that *mantra* repetition can change the world from one moment to another and that it can assure a person worldly success, paranormal capacities, and liberation.

Many of my examples already implicitly hint at the cosmic and cosmological dimensions of sound in Sanskrit Hinduism, but there are also many explicit examples, starting with the music-making, dancing, and singing major deities. Sarasvatī is the goddess of wisdom, language, and music. The flute-playing universal lover-god Kṛṣṇa is engaged in an eternal dance with the gopis (cowherdesses) or human souls. And Lord Śiva created the world through the sounds of his hand-drum. Indeed, already in Vedic times, voice itself was deified. Remarkably, the goddess Vāc ("Voice, Speech, Language") in the late *Ṛgveda* (hymns originating from about 1250 BCE) encompasses not only human speech but also non-human sounds like roaring thunder and the mooing of cows, and she has a key cosmic and world-creating function (Wilke and Moebus 2011: 288f., 292f., 361, 370–90, 405, 415f.). A bulk of post-Vedic literature, especially from the Śaiva Tantra, deals with cosmic and cosmological dimensions and powers believed to inhere in unmanifest and manifest sound. It speaks, for instance, of the primordial cosmic *nāda,* vibrant sound, as the basic element of creation pervading the whole universe. Tantrism, which has been the religious mainstream in India from around the fifth to the thirteenth century, developed entire sound cosmogonies and sound rituals based on the inherent power of sound and its self-generating and self-expressive force. Characteristically, all the deities of Tantric ritual are purely sonic. They are in fact *mantra*-deities, consisting of non-lexical syllables, or in other words resonant, vibrating sounds that are believed to contain all the deity's powers.

It is not surprising that a world so much filled with sound as the one of Sanskrit Hinduism also left distinctive traces on consciousness, on world view and cosmology, and not least on how people envision the divine or absolute reality and their place within the larger whole. There is perhaps no better example for this than the Nāda-Brahman, the "Sound-Brahman" or "Sonic Absolute", thought to pervade all existence and to be directly experienced in music – and in the contemplation of resonances, which in later Indian history became a major focus. The Nāda-Brahman was an extremely powerful and historically effective medium of sonic consciousness and sonic world view within Sanskrit Hinduism and beyond.

2 Śārṅgadeva's Nāda-Brahman and the Yoga of Music

Among the manifold forms of sonic awareness and acoustic piety, the Nāda-Brahman rose to an exalted position. Today, it is ascribed to a hoary past, reaching back to Vedic times,[1] but as a matter of fact, it is a product of musicians and musicologists of the Indian Middle Ages. The concept was known to some extent already to the early music theorist Mataṅga (c. eighth century),[2] but the term itself and the fully developed concept did not appear until the thirteenth century, when "Nāda-Brahman" was coined by the courtly minister, Ayurvedic physician, and musicologist Śārṅgadeva. He fused all relevant sound and language discourses of his time in the new compound word, in particular the Śaiva-Tantric *nāda* (primordial sound vibrations pervading the universe) with the *brahman,* a term which since the late Vedic (Upaniṣadic) period has been something like a cultural programme and icon of Hindu spirituality, as it signifies according to an often-quoted formula "the One without a second" (*Chāndogya-Upaniṣad* 2.1) and a soteriology of mystical fusion. In the widespread Advaita-Vedāntic version, to which Śārṅgadeva resorted (besides the Brahman notion of the linguist Bhartṛhari, as outlined below), *brahman* denotes the non-dual unity of existence or, more precisely, the immanent and transcendent

[1] As a matter of fact, only the *brahman* is an ancient Vedic term and is known since the time of the earliest Upaniṣads (c. eighth or seventh century BCE) as ultimate being and supreme spirit pervading everything. The term *nāda* (vibrant sound) is younger and not of equal cultural significance. It is found most of all in the ancient Āgamas or Śaiva-Tantric literature (some of which perhaps reaching back to late Vedic times), and also in some later Upaniṣads and profusely in later Yoga sources. However, the fusion of the two terms and concepts cannot be traced before the thirteenth century. For a broad discussion of the Nāda-Brahman see Wilke and Moebus 2011: 830–913.

[2] Mataṅga and his influential music handbook *Bṛhaddeśī* have been variously dated – from sixth to eighth century CE (Katz 2001) to 800–900 CE (Koch 2008: 67).

formless spirit or non-dual Consciousness-Existence-Bliss (*saccidānanda*) pervading and excelling everything, the supreme impersonal Godhead, which is one with the innermost, real self (*ātman*) of everybody. The fusion of *nāda* and *brahman* was extremely powerful. It fostered a new idea of the divine, of music, of one's being in the world, and an acute sonic consciousness, the basic idea of which was well captured by Joachim-Ernst Berendt (1983; [1987] 1991) in the catchy formulas "the world is sound" and "sound is God". The German jazz historian, record producer, and New Age proponent Berendt had learned about the Nāda-Brahman by his interactions with Indian musicians like Ravi Shankar, and so it is no wonder he had a good grasp of the essentials of the term. But he was ignorant in many ways about other important features, such as the non-separability of language and music in the Indian context, as both are based on sound, or the Brahman's immediate association with absorption, self-forgetfulness, and fusion, which has been a cultural ideal since Upaniṣadic times. He was definitely unaware of the Nāda-Brahman's complex historical formation, and thus he treated it a-historically as a primordial knowledge of mankind.

Historically, the eminent early musicologist Mataṅga (1994) prepared Śārṅgadeva's Nāda-Brahman, insofar Mataṅga's music handbook *Bṛhaddeśī* introduced the Śaiva-Tantric *nāda* into the musical discourse and also anticipated the idea that the whole world, including the great Gods, were sound (Wilke and Moebus 2011: 831, 842; Wilke 2017: 326, 328). Nonetheless it was Śārṅgadeva (1953) in his great encyclopaedic music classic *Saṅgīta Ratnākara*, a veritable *magnum opus* in Indian music, who related the musical *nāda* ("sound") to the Brahman. He thereby transformed the non-tangible, abstract omnipresent source of everything and the bliss absolute inherently belonging to it into a tangible audible substance, which can be experienced in music. Although the *Saṅgīta Ratnākara* (SaGiRa) was a thoroughly technical work on music, it contains passages on the metaphysics and soteriological relevance of music, declaring the Nāda-Brahman as actual foundation of music (song, instrumental music, and dance), thus giving music a metaphysical base. Three propositions of Śārṅgadeva are of particular interest, as they develop the Nāda-Brahman concept by making use of nearly all discourses of his time relating to sound, language, and religion.

First, Śārṅgadeva conceives music, particularly song (*gītam*), as something divine that brings delight to everybody. As embodiment of the cosmic *nāda*, music is an alloform of Śiva. This first thesis – which could be summarized as "God is sound and sound is god" – appears right in the initial auspicious verse of the work (SaGiRa 1.1.1). The idea is indirectly communicated by suggestive language, the artful poetic *śleṣa* style of double coding. The verse can be read to relate to the supreme god Śiva and simultaneously to relate to song. In the background of this daring identification of God and sound, one must presume a cluster of Śaiva(-Tantric) ideas, starting with the Nāda as a religious category and cosmically powerful force. This notion at Śārṅgadeva's time was already widespread and well-known in Yogic-Tantric circles, classical Śaivite theology, and even popular culture. The inseparableness of consciousness and sound energy belongs, according to Śaivite theology, intrinsically to the Godhead – manifest in the supreme transcendent Śiva (associated with pure consciousness), who is eternally united with his creative Power (Śakti), the source of the universe (associated with vibrating sound/–energy). In the context of their cosmogonic *mantra* speculations, the ancient Śivaitic Āgamas had not only seen the first evolute of creation in vibrating sound (*nāda*); they also maintained that sound was more basic than either mind or matter and in fact existed prior ("primeval") to them. This was likely already Mataṅga's immediate conceptual context and certainly inspired Śārṅgadeva to weave such ideas into his new sonic metaphysics and the Nāda-Brahman conception.

Second, Śārṅgadeva praises performing and listening to music as an easy and enjoyable yoga. It is, according to the relevant verses (SaGiRa 1.2.164–167; Wilke and Moebus 2011: 836), an even better form of yoga than the usual (cumbersome and sense-removed) yogic exercises and contemplations, because music is audible unlike the inner soundscapes of the yogis, and it appeals to the senses and feeling. Music is emotionally pleasing unlike the dry yogic contemplation and ascetic hardships, and it is able to lead directly into Brahman experience, mystical rapture, and liberation without the efforts connected with yoga. In this second thesis, Śārṅgadeva ascertains music as a pleasant yoga for everybody, as it easily leads to total absorption, self-forgetfulness, and mystical rapture. Being connected to the senses and emotions, it can provide both worldly enjoyment and final liberation. Deliverance in the form of music is at the same time emotionally pleasing and fun, so that people will simply attend to music for its own sake, even if they do not wish to be delivered. Śārṅgadeva uses the word *rakti*, "triggering feeling", which hints at the musical key term *rāga* (meaning a melodic "mode" as well as "shades of colour") that was classically defined as a tone structure that "colours the mind" (*rañjana*), i.e. creates feelings and causes delight.[3] Each Rāga is understood to be an individual personality or even divinity (Koch 2008), inherently imbued with their own sentiment, mood, or emotional flavour, which is eternally present in a transhuman way and must only be "awakened" by the musician. This is done especially in the initial slow and quiet *alāp* phase, which for the performing artists is something like a prayer, invocation, and contemplation in one. The monodic modal character of Indian music, resting on a fixed constantly sounding fundamental or "home note", whilst being immensely rich in microtonal ornamentation, makes the yoga character attributed by Śārṅgadeva particularly plausible. Hearing the *alāp* of a skilled master almost automatically leaves utmost concentration, self-forgetfulness, and a mood of divine peace, serenity, timelessness, and bliss in the deep listener.[4] Not least India's

---

[3] Already the early music theorist Mataṅga defines Rāga along these lines: "The particular melody, which is characterized by its tones and their combination and [which] *colours the mind* of people, is called *raga* by the wise men" (cited in Koch 2008: 67; italics A.W.).

[4] Such feelings are reported by performing artists like Ravi Shankar (Berendt 1991: 154) and by listeners – be they Indian or European. Berendt therefore attributed inherent spirituality to Indian music. He saw the most significant characteristics of modal music in the experience of oneness or "at-one-ment", and of the "dissolution of time" (Berendt 1991: 168). He was biased, no doubt, yet, he was not the only Westerner feeling this way, notwithstanding others would find Indian music

modal music was thus able to generate a strong sense of the oneness and interconnectedness of everything, a feeling of unity in plurality – an idea, which was very much belonging to Śārṅgadeva's Nāda-Brahman that he conceived as embodied in audible music and in song in particular.

Third, Śārṅgadeva propounds that the whole world is based on sound (SaGiRa 1.2.2) – the world is sound because the divine source is sound (SaGiRa 1.3.1–2). This third thesis must be looked at more closely, as in these specific lines the term "Nāda-Brahman" occurs for the first time in Indian history. It is prepared by a linguistic argument:

> Sound (*nāda*) manifests in the phonemes (of the alphabet), letters constitute the word, and words make a sentence. Thus, the entire worldly life (*vyvahāra*) is carried on through language, and the entire universe (*jagat*) is based on sound (*nāda*). (SaGiRa 1.2.2)

This explicit pointer to language shows how much language and music are related in the Indian discourse, as both are based on sound. In the world of musicians, however, language becomes a subform of expressive sound in which the whole world is founded. Indian musical theory starts with the immediate expressivity of sounds, namely the sounds' own *caitanyam*, a term referring to "consciousness", but also to "liveliness" and "sensitivity" in contrast to the "numbness" and "insensitivity" of lifeless matter (*jaḍa*). For many musicologists, the *caitanyam* of sound is not an individual consciousness but identical with cosmic consciousness, in which all thoughts, feelings, and ideas of all beings of the universe are melted into oneness. This quality associated with sound was essential for the idea of the Nāda-Brahman, the "Sound-Brahman". As a matter of fact, it is precisely this quality which Śārṅgadeva acknowledges in his actual Nāda-Brahman verses:

> The consciousness *(caitanya)* of all living beings, which turns itself by itself into the world *(vivṛta)*, is the 'Sound-Brahman' *(nāda-brahman)*. In it we invoke *(√upa-ās)* bliss, 'the One without a second'. If we invoke sound, we also invoke without doubt [the great cosmic gods] Brahmā, Viṣṇu, and Śiva, because they are nothing other than [sound]. (SaGiRa 1.3.1–2, transl. Wilke and Moebus 2011: 837)

Śārṅgadeva's two verses SaGiRa 1.3.1–2 introducing the new term "Nāda-Brahman" not only refer to *caitanyam*, however, but contain several implicit relations to traditional elements and even some explicit paraphrases of various Indian sound and language discourses that I will map out in the following.

The first verse (SaGiRa 1.3.1) alludes to the (Advaita-)Vedāntic Brahman – the Absolute Being who is, as "the One without a second", both the material and efficient cause of the world, and whose innate nature is pure consciousness-bliss as well as each individual's own self. But at the same time, it also alludes to the

very foreign, strange, and even boring, as it is so entirely different from Western classical music. Mecht Kollin, who has been the organizer of Indian concerts and dance performances in Zurich during the 1970s and 1980s, reports in a personal note how overwhelmed she was when hearing her first concert in Mumbai shortly after India's independence – a time when the ancient arts of temples and courts were re-discovered and brought to larger audiences. She speaks of her acute concentration and total self-forgetfulness, and of deep inner peace, serenity, and blissful stillness that were generated by the *alāp* of an old singer in this concert.

Brahman conception of the linguist philosopher Bhartṛhari (fifth century). Bhartṛhari holds that our whole perception is based on language and ultimately founded in a universal linguistic principle (*śabda-tattva*) or "Word-Consciousness", also called "Word-Brahman" (*śabda-brahman*) that forms a global meaning going beyond the meaning of single words and things. According to Bhartṛhari, the whole world is a metamorphosis of this single language principle or Word-Brahman (Śabda-Brahman) that forms the common conscious ground of all human beings. Śārṅgadeva's introduction of the Nāda-Brahman is directly imported from the Bhartṛhari school even in its terminology – the consciousness of all beings transforming itself into the world (*caitanyam sarvabhūtanām vivṛtam jagadātmanā*) – but with one decisive difference: instead of Bhartṛhari's Śabda-Brahman ("Word-Brahman") there appears here the Nāda-Brahman ("Sound-Brahman"). We find a move from the mental plane of an abstract language principle and pure consciousness to a more physical substance – acoustic sound (inhering in language and music) – and to sonic consciousness. At the same time, Śārṅgadeva's Nāda-Brahman is more narrowly defined by the Vedāntic expression "One without a second" (imported from the Upaniṣads) and by "bliss" (a characteristic of the Supreme in Vedānta and Tantra). The Nāda-Brahman is non-dual cosmic Consciousness and Bliss, in which God, the world, and the individuals cannot be differentiated and separated.

While the first verse thus develops various Vedāntic interpretations of Absolute Being, the second verse (SaGiRa 1.3.2) follows the Tantric interpretation of sound. Notably, it paraphrases the early musicologist Mataṅga, whose *Bṛhaddeśī* had already postulated "the world is sound" (*nadātmakam jagat*) and implicitly also the idea "sound is god" and "god is sound" – without, however, using the term Nāda-Brahman. Mataṅga writes:

> Without sound there is no song, no tones, no dance. Therefore this whole world *(jagat)* is in its nature sound.[5] [The God] Brahmā is a form of sound, Viṣṇu (here: Janārdana) is a form of sound, the transcendental, supreme Power (*Parā Śakti*) is a form of sound and the great God (Maheśvara, i.e. Śiva) is a form of sound. (*Bṛhaddeśī* 1.18–19, transl. Wilke and Moebus 2011: 842)

Śārṅgadeva thus follows Mataṅga almost verbally in SaGiRa 1.3.2. That the great gods are sound means that all deities have their own *mantras* and in the Tantric view are actually identical with these *mantras*. The panentheistic view of the Śaivas necessarily gives rise to the statement "the world is sound", because if the great cosmic deities are sound and there is no world outside of God, all beings and things are already included. Importantly, Mataṅga's and Śārṅgadeva's sonic vision of consciousness and all of reality includes both the profane and sacred spheres.

The conclusion of Śārṅgadeva's statements is far-reaching: Music (the epitome of sound) is a physical expression of the Brahman, and performing and hearing song and Rāga music directly leads to sensing and experiencing cosmic consciousness, the soul of the world, and cosmic unity – the One in the many. This soul of the

[5] This proposition surpassing music is prepared in *Bṛhaddeśi* 1.12.

world, however, is not an abstract ultimate being (as in the Vedānta or Bhartṛhari), but full of emotional colours and aesthetic moods (as the different Rāgas of Indian music are supposed to express and evoke feelings in a transpersonal way). Being equated with music, this Brahman can be felt and experienced by the senses. Śārṅgadeva's introduction of the Nāda-Brahman not only sought to give music of any kind (classical/artistic and vernacular/popular) a metaphysical base. It also allowed music to gain a pronouncedly cosmic dimension as well as a practical, distinctly soteriological function. Music could be viewed as a divine service, as an easy and pleasant form of yoga, and as a powerful spiritual exercise of self-cultivation, leading to mystical peak experience. For numerous musicians and listeners, music became a most perfect means of experiencing and contemplating divine peace, serenity, and bliss. This was not exclusively due to Nāda-Brahman's epistemological impact, but surely one of Nāda-Brahman's most important legacies, as shown in the next chapter.

3 Kallinātha's and Tyāgarāja's Nāda-Brahman: Sonic World View and Musical Deliverance

The Indian Nāda-Brahman was a unique blend of metaphysics and sense perception/feeling, interior and exterior space, profane and sacred. Śārṅgadeva's very substantial understanding of the Nāda-Brahman as embodied in the sound of music became immensely powerful in the courtly and religious world of musicians and increasingly also in popular, particularly devotional, discourse. The Nāda-Brahman made the attributeless Brahman beyond name and form audible in music and coloured with emotion. Music in this model leads to experiencing bliss and oneness, as it reveals cosmic unity, the non-dual nature of the universe, in which no separated entities exist like in music, where all single tones are but sound. This world view expressed by the Nāda-Brahman concept is perhaps the most straightforward answer to the question: What kind of consciousness and world view is being created, if the basis of daily experience is so much pervaded by the spoken word and sound perception? Śārṅgadeva's commentator Kallinātha (ca. 1430) delivers a comprehensive account of this sonic world view. Although Kallinātha apparently adhered more to the Advaita-Vedānta than to Tantra and consequently had a less substantial understanding of the Nāda-Brahman than Śārṅgadeva (still dominant today), their musical-sonic vision of the cosmic whole and all-immanent consciousness is much alike. While Kallinātha denies a literal understanding of Śārṅgadeva's Nāda-Brahman verses and does not accept sound to be God, he sees sound, concretely Rāga music, as the most perfect metaphor for the Brahman, in fact as a metonym. For him, contemplating sound invariably leads to Brahman, but sound is not Brahman. Yet music's outstanding features are so much like those of Brahman that music becomes a direct gateway to the Brahman:

> Sound is 'one without a second', as it is the sole material ground of its own products, the notes [scales and rhythms]. [...] This is why sound – according to its own laws – resembles

> the highest Brahman in the most perfect way. If it is said that they are identical, one does not deny that sound is only projected on the absolute Oneness (of Brahman). However, it is a real (and most perfect) metaphor, as all elements of sound – its own consciousness or liveliness, etc. – are also elements of Brahman. [...] So, in the near-identification of Nāda and Brahman we find a symbol that goes right into detail, inasmuch as the one, like the other, is equipped with analogous components: they are consciousness, the foundation of its own cosmos, and the origin of all material. Here, however, it should not be denied that the Nāda is only a symbol [and not a real statement of identity]. [...] The Nāda, or the Parā Vāc ("transcendental language"), is indeed only a power of the Brahman, but it is so infinitely close to the Brahman that the Brahman itself can be achieved if one contemplates it – like someone searching for precious stones finding the gem if he only follows the sound. (Kallinātha's commentary on *Saṅgīta-Ratnākara* 1.3.1–2, Ed. Sastri, 63; see also Wilke and Moebus 2011: 844–845)

Kallinātha sees music as a most complete simile (*sāvayavarūpaka*) of the Brahman. It is noteworthy that once again "consciousness" (*caitanya*) ranges first in the many aspects which coincide. Another major correlation is that Brahman is bliss, and music brings pure joy and bliss with deep listening, and therefore leads directly into the divine sphere. According to Kallinātha, between (musical) sound and its perceptible products (tones, etc.) exists precisely the same intrinsic relation of cause and effect as between Brahman and its effect, the world. Brahman, "the One without a second" (*Chāndogya-Upaniṣad* 2.1), is not only understood as the ontological origin and efficient cause of the world, but also as its material cause underlying all phenomena. Brahman's essential nature, the continuum of pure Existence-Consciousness-Bliss, in which human and divine modes of existence coincide, pervades everything like sound pervades the tones or like the clay pervades all earthen wares as the material from which they are made.

Music, and more generally acoustics, reveal something about the non-dual nature of reality according to Kallinātha as well as Śārṅgadeva. The Nāda-Brahman seems to be associated with what modern acoustics would call "harmonic sound" or "harmonic spectrum". Even though we seem to hear one single tone, we are in reality hearing many different tones, as each "individual tone" contains within itself all other "individual tones" as co vibrating sound-scape. Śārṅgadeva alludes to a vibrating string as illustration of this unique nature of sound.

Probably not by chance, Indian modal music, being so rich in melisma, melodic embellishment, and microtonal ornamentation, became a sensory symbol for the Brahman pervading and containing everything "without a second". The very structure was akin to each other. And definitely, there is an equivalence between the Nāda-Brahman idea that the whole world is filled with the "cosmic, primordial sound" and the monodic, extremely microtone-laden Indian modal music. In particular when overtonality is stressed, a dense, constant sound presence arises that translates the Nāda-Brahman so to speak into sensory perception and arouses a "feeling" of unity in plurality. The rich, oscillating bourdun sound of the Indian Tāmpūra, developed in the sixteenth or seventeenth century, was to become a prime example for the cosmic, omni-present, and omni-immanent Nāda-Brahman. The Tāmpūra is a fretless instrument with four strings tuned to the basic sound or fundamental, together with the fifth and the octave of the Rāga, which are continuously plucked. The function is to set up the fundamental as the bourdon, a permanent background of sound against which all the intervals should be heard. For the listeners, however, this constant monotonous reiteration of the fundamental does not sound static at all, as a special curved bridge ensures that the overtones fluctuate constantly. The resulting sound of a well-made Tāmpūra no longer sounds like that from a plucked instrument. Instead, the overtone-rich sound of the Tāmpūra itself seems to oscillate, shimmer, and be very much alive. The most common and at the same time most amazing impression of listeners is that they can never decide whether they hear only one tone or a myriad of vibrating sounds. One always hears a unity and a plurality at the same time. So, Indian musicians as well as audiences experience the holistic nature, inner connectedness, and ultimate unity of sound quite physically via the Tāmpūra's bourdon. Although the instrument is late, the ideal of the wavering drone goes back to Śārṅgadeva, and it is easily comprehensible, why for Kallinātha musical sound was a direct gateway to the Brahman, in which cosmic unity and plurality coincide, and which as ineffable, absolute Oneness is one's own true self and the self of everything. Kallinātha suggests to see in music a door to an expanded consciousness, directly experiencing, feeling, and sensing non-dual oneness and beatific bliss absolute, the gem among the precious stones, if one only follows sound.

The aspect of yoga and deliverance or liberation while living was not explicitly taken up by Kallinātha in this context but became very important among the Carnatic (South Indian) musician-composers famous for their devotional songs, i.e. their *kṛti* compositions (classical devotional songs) rich in literary appeal, emotional variation, and intensity of feeling. The most prominent was Tyāgarāja (1767–1847), a Smārta Brahmin from Tanjavur (Tamil Nadu), who is considered as the greatest genius of Carnatic music. In the history of music he is regarded as the one who brought the composition form of *kṛti* to perfection. *Kṛti* (literally that which is "constructed") is sometimes also called *kīrtana* (despite the subtle differences between the two genres). It refers to pre-composed music that aims at delineating the melodic nature and emotional essence of a Rāga in all its vibrant colours and thus to bring out the "sweet sound" and intrinsic beauty of the modal scale of each particular Rāga. Tyāgarāja was a great devotee of the god Rāma, whom he musically worshipped and contemplated as the ultimate being and non-dual highest deity or Para-Brahman ("supreme Brahman", highest reality), and as the synthesis of the "essence" of the great Lord Gods Viṣṇu, Śiva, and Devī (or Parā Śakti) and their *mantra* sounds. In many of his compositions he expresses God as the embodiment of Nāda (Raghavan in Jackson 1994: 71–89, 217–230, in particular 223–224; see also Wilke and Moebus 2011: 862–870). Whereas Kallinātha is a good example for the sonic consciousness and world view generated by the Nāda-Brahman, Tyāgarāja is a great example for the performative, experiential, and devotional aspects connected with this cosmology and for the direct Brahman experience generated by music. In focussing and fusing music, the praying and singing of the Lord's name, and Nāda-contemplation (*nāda-upāsana*), i.e. the contemplation of musical sound or melody, Tyāgarāja stressed musical devotion leading to the invisible deity, supreme bliss, and liberation, while living in the here and now. In this way, he expands Śārṅgadeva's ideas of God is sound, sound is God, and of musical yoga and musical immersion that have had only a seminal beginning in the *Saṅgīta Ratnākara*. Tyāgarāja makes a number of bold statements concerning his musical practice centred "on God being the embodiment of *nāda*, subtle sacred vibration, and the absorption in the joy of melody as itself constituting spiritual liberation, *mokṣa*" (Jackson 1994: 223). In Tyāgarāja's words:

> Devotion associated with the nectar of *svara* [notes or tones] and Rāga [melodies] is verily paradise and final liberation"; "The joy of music is itself the bliss of Brahman that *vedānta* speaks of". (transl. Raghavan, quoted in Jackson 1994: 224)

This emphasis on Brahman bliss and final liberation goes far beyond Śārṅgadeva, and equally daring is the complete breakdown of boundaries between sacred and profane spheres – a distinction, however, that is not as pronounced in Hindu traditions as it is in Christian discourse and that was particularly easily broken in the context of music. For composers and musicians like Tyāgarāja, their yoga is nothing else than the joy of music itself and the emotional immersion and absorption in the musical Nāda sound. Tyāgarāja insists that music liberates one while living (*jīvanmukta*) and replaces all other practices and paths to this end:

> O Mind! Drink and delight in the immortal elixier of melody and attain the fruit of *yāga* [sacrifice], *yoga*, *tyāga* [renunciation] and *bhoga* [sense enjoyment]: those who understand that *Nāda*, *Oṃkāra* [the sound of OṂ] and *svara* [musical notes] are nothing but Śiva are verily Jivanmuktas, those who have achieved liberation here itself.; O Mind! He who delights in the Nāda attains the bliss of Brahman. (transl. Raghavan, quoted in Jackson 1994: 224)

With such statements about the effects of musical immersion and tranquil delight, Tyāgarāja put the finishing touch to the ideas of musical yoga and deliverance initiated by Śārṅgadeva. Śārṅgadeva saw the embodiment of the Nāda-Brahman in audible music, and Tyāgarāja follows him in a sort of grand finale concerning music's ability to generate transcendental bliss and final liberation. No other practice is needed to attain ultimate freedom instantaneously.

Significantly, Tyāgarāja's life and compositions are a striking example of the question of what kind of consciousness and world view are created if the basis of daily experience is pervaded by the spoken word and sound perception. From his youth onwards, Tyāgarāja was surrounded by an environment charged with the fervour of Nāma-Siddhānta ("final view on the divine names"), a movement that propounded the power and greatness of singing the Lord's name (*nāma-kīrtana*) as the most effective and easiest path to liberation in the present age. Additionally, he was nurtured on the Bhajana Sampradāya, which was at its height in the Kaveri delta (Tamil Nadu) at that time and fostered *bhajan* groups, i.e. collective devotional chorus singing in homes and temple-like halls (*bhajana mandira*). These devotional songs were supposed to be mellow and "sweet" in words and melody, and easy and delightful in performance. Tyāgarāja was part of both movements, which were in close interaction. As a follower of the Nāma-Siddhānta of the Smārta theologian-philosophers of the Kaveri delta, he shared their (Advaita-)Vedānta-inspired non-dualist theology of divine names that made no distinction between the great gods (Śiva, Viṣṇu, Devī). He also took part in *bhajan* singing and composed himself a significant number of simpler, soulful songs of intense devotion (*bhajanas*, *kīrtanas*, *kṛtis*) that apparently were meant for congregational singing. Typically, Tyāgarāja's songs not only praised Lord Rāma as highest Brahman, but also other deities, and many were about the greatness of the Lord's name and the Nāma-Siddhānta doctrine. Not least, as already mentioned, he claimed in a number of songs – much in consonance with Śārṅgadeva – that music is indeed the expression of the primordial Nāda.

Thus, a surrounding filled with the sound of devotional music and the singing of the Lord's name(s), moreover a surrounding which also considered this sonic situation in a reflexive way and was philosophically given to the (Advaita-)Vedānta, was the immediate cultural context and background of the singer, composer, poet, philosopher, and saint Tyāgarāja, whom some adored even as divinity himself and his *kṛtis* as holy scriptures.

Remarkably, Tyāgarāja mentions the OṂ sound as well. This holy syllable is traditionally known as embodiment or sonic "image" (*pratīka*) of the Brahman in both its world-immanent and world-transcendent aspects, and Tyāgarāja seems to refer to *mantra* repetition and contemplation. In keeping with the Nāma-Siddhānta, he also speaks of *nāda-upāsana* – apparently both in the sense of contemplating Rāga melodies and devotional songs as well as contemplating the sound of the divine names and *mantras* (repeated and sung loudly, muttering, and silently) down to the single syllables they consist of. This meditation considers esoteric meanings as well as "the mantric vibrational force of sound", according to William J. Jackson (1994: x), who calls this practice *nāda-yoga*. Jackson argues that

> the vibration of chanting the name [itself also understood as *mantra*] ultimately participates in the nature of which is named. The mantra sound [...] tunes one into the mystery of the ultimate *nāda* [... and] harmonizes one with the infinite eternal hum of Being. (Jackson 1994: x)

This emphasis on resonances, vibrations, and inner sounds became perhaps the most important aspect of the Nāda-Brahman in modern times, a development I call "yogification" of the Nāda-Brahman. It relegated the *brahman* aspect as well as the audible music aspect – which both have been essential for Tyāgarāja – to the background. Instead, the yogic and *nāda* aspects came strongly to the foreground in the sense of *Śiva-Saṃhitā's* assertion with reference to the internal, occult, or mystic sounds heard by the yogi that there is "no absorption (*laya*) like that of *nāda*" (Beck 1993: 103; Wilke and Moebus 2011: 883).

4 Later Developments: Yogification and Transnationalization of the Nāda-Brahman

It is beyond the scope of this essay to take a detailed look at the Nāda-Brahman's reception history, which extended far beyond the elite courtly musical circles. Today one can even hear Nāda hit songs in the Bollywood style, telling of the all-pervading

Nāda-Brahman and evoking its emotional atmosphere or aesthetic mood through a sweet singing style. The Nāda-Brahman is associated with the repetition of *mantras* or the chanting of God's names and the resonance in body and soul. The developments after Śārṅgadeva, Kallinātha, and even Tyāgarāja are manyfold and complex, and have been poorly researched so far. However, at least two incisive developments need to be mentioned: the Nāda-Brahman's yogification, and its transreligious and transcultural history.

In the run of time, the Nāda-Brahman took on strongly yogic, esoteric, and occultic – sometimes even therapeutic and scientific – associations. Having been initially primarily a musical and musicological category, it became largely a spiritual category of esoteric yoga that was often – like most of modern esotericism – claimed to dovetail with natural sciences. In the process of yogification, sonic consciousness transformed in many ways. Most significant was a tendency to move away from Brahman ontology and the experience of Brahman bliss in favour of an expansion of sonic cosmology and altered states of consciousness. The exact date, reasons, and contexts of what I call "yogification" of the Nāda-Brahman remain in the dark and may be quite recent. Yet this development has been so strong that the Nāda-Brahman was projected back to Yogic and Tantric sources, although these only speak of the cosmic *nāda* and *nāda* practices.[6] Today the Nāda-Brahman is often, perhaps primarily, related to the contemplation of sound and resonances, subtle sound vibrations, "spiritual hearing", the yogin's interior sounds, higher spheres of light and sound, and also to the (old European concept of) the music of the spheres, (theosophic) astral worlds, (shamanic) out-of-body experiences, soul travel, and wonderful journeys into the centre of the universe. Some identify the Nāda-Brahman with the holy syllable OṂ and equate *nāda-yoga* with listening to the aftersounds of OṂ (the nasalized final tone has the effect of a bell or a gong, which is heard resonating even if no more beaten). Frequently the *nād(a)* and the Hindi term *śabd* (Sanskrit *śabda*, meaning "word") are identified (or *nād* is substituted by *śabd*) by means of a new understanding of *śabd* as denoting the divine name (*nām*), the divine Word, or the ineffable name of God, and also referring to spiritual sound, notably the yogic "unstruck sound" (*anāhatta-nāda*).

The Radhasoami movement (Rādhāsvāmī-Satsang) – initiated in the nineteenth century – is but one, yet particularly potent and radical example of this development, which is diverging from Śārṅgadeva's exclusive focus on audible music and song. Although the Radhasoamis do not use the terms Nāda-Brahman and *nāda-yoga*, they have been justifiably associated with it (see, for instance, Beck 1993: 111, 113–114) due to their exceptionally strong theology, cosmology, and meditation practice of sound. Guy Beck even identifies the Radhasoami's central term "sound current" (or current of sound energy) with the Nāda-Brahman. I would rather say it is a *nāda*-reinterpretation and a new, dynamized Nāda-Brahman that encompasses an entirely novel sound cosmology and a new esoteric *yoga* and

[6] This is a problem, for instance, in Beck's otherwise well informed and comprehensive book on *Sonic Theology* (1993).

bhajan practice, denoting an inward listening to the sound vibrations on the "astral plane" and further world-transcending spheres. This central practice is called *surat-śabd-yoga,*[7] literally "word-yoga of the soul (or spirit)", which Mark Juergensmeyer (1991: 90) explains etymologically as a "discipline (*yoga*) that unites the spirit (*surat*) with the spiritual sound (*shabd*)". The audible sound is not considered to be the real sound, but only a crude index of the divine sound or "word", the eternal current of sound energy that resonates in the sounds emanating from the celestial spheres, and which can be heard only by an awakened inner auditory sense (Babb 1986: 50–52). The guidance of an experienced master is necessary to reconnect the adept with the primordial sound current and the ultimate source of sound, the Supreme Being Radhasoami, "the Lord of the Sound Current". In a sort of spiritual wave-riding on the current, the inner journey leads through miraculous celestial worlds of light/luminaries and sound (also understood as levels or higher planes of consciousness) until at the end he/she attains a direct vision of the Lord.

The description of the celestial sonic spheres, in each of which the divine sound current reverberates differently, almost reads like an Indianized version of the Pythagorean music of the spheres or Dante's visionary spherical journey. However, the Radhasoamis' cosmology expands beyond the known planets. Closer to Indian traditions are the yogic imports regarding sound. Much of the inner journey through the higher cosmic regions reminds – partly in detail – of the visions and auditions of the yogis who have reached perfection, because the celestial sounds are the same as those described in Yogic-Tantric sources, such as "bells, thunder, musical instruments, and sounds quite unlike anything we know in this world" (Babb 1986: 53). However, in the Radhasoami cosmology there is a hierarchical ranking and systematization of higher sounds (lute, harp, and flute) and lower sounds (bell and drums), which is not present in the old sound catalogues of the Yoga literature. It is also noteworthy that for the Radhasoamis, only the higher yogic *cakras* (psycho-physical nerve centres) play a role – especially the *ājña cakra* (the "third eye" between the eyebrows) and the thousand-petalled lotus (above the crown of the head) on which the adept concentrates to ascend to the higher worlds of "unstruck sound" beyond audible sound. The *cakras* below the solar plexus, in contrast, are believed to tie the soul to the physical world and the physical body, which is seen as an impediment (Babb 1986: 53; Denosky accessed January 7, 2022: 5).

The Radhasoamis' imports from multiple traditions and their "strange combinations" with modern science (von Glasenapp 1928: 56) have been variously noted.

[7] This (in Hindu traditions uncommon) terminology reveals Sikh influence, as Sikh theology knows the terms *surat nām* and *surat śabd.* Whereas the secondary literature (see also Beck 1993: 113–114) generally speaks of "*surat-śabd-yoga*", Helmut von Glasenapp (1928: 50), one of the first indologists who acknowledged the Radhasoami Faith as an important Indian reform movement and a modern "mysticism of the sounds of God", transliterates "*sūrat-śabd-yoga*". He suggests that it denotes a "Sūrat (spirit) stream" and a "Shabda (word/spiritual sound) stream", pervading the universe. Radhasoami masters like Charan Singh (Beas, Punjab) explain that *surat-śabd-yoga* "consists in connecting the soul (Surat) with the sound current (Śabda)" (cited in Beck 1993: 114).

Besides old Indian yoga concepts, it was particularly modern physics and biology from which conceptions and terminology like electricity, electrons, and kinetic energy were taken over to give the Radhasoamis' unique sonic cosmology and their ideas of the material transmission of spiritual power within the school's extreme *guru* veneration a scientific explanation (von Glasenapp 1928: 48, 52–57). Other imports came from Greek, Christian, and Muslim mysticism, from Hindu (*saguṇa* and *nirguṇa*) *bhakti* devotionalism, from the Sants, from the Sikhs, and not least from modern Theosophy. All these influences were melted together in an impressive sonic theology, the most consistent and comprehensive one in India that I am aware of. All elements are related to the energetic sound current believed to be the sound of God and his holy Name ("Radhasoami") vibrating through the universe, his primordial Word that constantly creates, sustains, and nurtures the universe. The cosmic spheres are interpreted as different levels of consciousness and simultaneously as macrocosmic realities, both of which can be measured and ascertained scientifically.

Already the Radhasoami founder Swami Dayal Singh (1818–1878), who came from the Khatrī money-lender caste in Agra and was a government official and merchant, is said to have practised from childhood "merging the soul in the Divine Current of the Supreme Being's Melody and Power" (Juergensmeyer 1991: 19), which he propounded as "yoga for ordinary life". After the demise of the first three, much-venerated "Satgurus" (believed to be incarnations of God), the Radhasoamis split into three major lineages (von Glasenapp 1928: 49), and two different Nāda-Brahman holisms emerged (Wilke and Moebus 2011: 896–897). Whereas the (North Indian) Agra branch interprets the different celestial spheres of light and sound as the pure ("unstruck") "melodies" or "energies" of the supreme being Radhasoami, the (Punjabi) Beas line regards them as interior revelation and as mere "attributes" of the supreme, transcendent God, conceived as inherently featureless and nameless. This great "Nameless One" (*anāmi*) is the "pole" of the sonic flow manifest in light and sound, but is himself neither light nor darkness, neither sound nor silence. In both branches, *guru* veneration and *guru-bhakti* towards the Radhasoami masters as embodiments of the Nameless One, who had experience of the higher planes, plays a central role and seems to be the real core of actual practice (Babb 1986; Juergensmeyer 1991), while the esoteric yoga of sound is more like a cherished ideal and primarily associated with enlightened masters. The normal Indian lay devotees generally admit that they have no experience themselves and content themselves their *bhakti* is equivalent because not everyone is equally spiritually gifted. This attitude forms a remarkable difference to Western followers, who are keenly interested in spiritual progress and enlightening experiences of their own (Juergensmeyer 1991: 108).

A good example for such a Western appropriation of Radhasoami is the American journalist and novelist Paul Twitchell (+1971), who in 1964 founded Eckankar,[8] a

[8] This name seems to derive from the first three syllables of the Sikh canon ("Ek Oṃ-kar"), which refer to both the one universal creator God as well as (in hidden form) to the OṂ as the root-*mantra* of Guru Nanak (Denosky, http://www.om-guru.com/: 2–3).

religious movement that was heavily indebted to Radhasoami theology and esoteric practice, though he sought to hide this dependence.[9] He declared the technique of "bilocation", later called "soul travel", the centre of Eckankar, gave detailed instruction, and emphasized individual experience (http://om-guru.com/html/saints/twitchell01.html, accessed 7 January 2022). In his books, Twitchell amply narrates his own out-of-body experiences in the higher cosmic planes, when, for instance, once the sound current manifested as "a marvelous melody sounding like a multitude of flutes" (Denosky n.d.: 6). Another time Twitchell experimented with *mantra,* and after he slowed down his chanting the name of God until vocally it was halted completely, he heard a humming sound in the back of his head, spreading out into the whole body, and the "ECK melody" started: "Sometimes it is the roaring of a waterfall, at other times the sound of flutes and/or violins" (Denosky n.d.: 6). Twitchell claims the direct vision of and oneness with God (a term not used by the Radhasoamis), whom he describes as tremendous blazing light, but also anthropomorphically as having a face (Denosky n.d.: 7). After Twitchell's death, Eckankar underwent various changes and took on a strong Christian orientation in the last decade: The river of sound was rather associatively equated with the Holy Spirit and the Word of God in the Bible – an interpretation that had already been suggested by late Radhasoami guru Kirpal Singh and (following him) Twitchell himself (Denosky n.d.: 9–10).

With these last remarks we are already in the midst of the second great development and transformation of the Nāda-Brahman: its transreligious and transnational appropriation. According to Guy Beck (1993: 111–118), the Radhasoami movement and some Indian *gurus* were the most important mediators for spreading the idea of Nāda-Brahman in the West (if we accept this term for the Radhasoamis as suggested above, although it is not part of their self-description). Beck (1993: 112f.) mentions Swami Shivananda, who made the vibrations of *mantra* repetition (*japa*), particularly the syllable OṂ and its after sounds, the centre of "Nāda Brahman meditation" as Beck (1993: 113, 115) calls it. Others, such as Usharbudh Arya, emphasized – besides the OṂ resonances – the "feeling" and immersion left by gongs, bells, songs, singing bowls, and even the cello (Beck 1993: 111f.), or, as Bhagavan Rajneesh holds, by one's own humming (Beck 1993: 115f.), or, according to Swami Nādabrahmānanda, by special vocalization techniques of singing (Beck 1993: 113). All of them implement what I call "yogification", meeting thereby a Western expectation of identifying India and Yoga. They illustrate a variety of meditation techniques focussing the inner Nāda, i.e. inward listening to sonic vibrations and their effects in the body and the spirit, which became known as "Nada

[9] But there can be no doubt, and Twitchell was even accused of plagiarism – regarding Radhasoami as well as Theosophical literature, and other movements (http://www.om-guru.com/: 10).

yoga" or "Nada-Brahman meditation". The currently popular "Nada healing" with meditative Rāga music appears to be a more recent development.[10]

Contemporary Nāda-yoga practices are often backed up by so-called New Age or post-modern, transreligious spirituality and esotericism that cannot be traced only to India or only to the West. They are generally based on free (but not random) individual religious composition and the claim to be spiritual, not religious. This style of mixing various origins from East and West, and surpassing both can already be found with Radhasoami master Kirpal Singh and even more pronouncedly with Eckankar. Both movements have an inherently transreligious character, being nurtured by more than one traditions, like the Hindu and the Sikh. What is seen today as a typically post-modern practice of syncretism and religious liberty – starting in the late 1970s and 1980s – is explicitly present already in Paul Twitchell's combination of "soul travel" and sound current (since the 1950), in his transreligiousness, the claim of personal authority, the primacy of individual experience, and the self-definition as a more original, non-doctrinal spirituality beyond established religions. It is noteworthy that in contrast to this post-modern synoptic attitude, the Radhasoamis had been, according to Juergensmeyer (1991: 205–06), "the first un-Westernized religion to which non-Hindus had access" (with the first initiated American in 1910). The community attracted many Westerners since the 1930, who were taken not least with what they felt to be scientific validity.[11] The number of Western followers expanded hugely in the 1960s and 1970s, whereas Western interest for Radhasoami faith dropped off after 1980 and shifted to diaspora Hindus (Juergensmeyer 1991: 203, 207). On the other hand, Denosky maintains that Twitchell was perhaps the most successful American teacher of an Indian-based spirituality, and that after his death (1971), Eckankar grew continually throughout the 1980s and 1990s (Denosky n.d.: 9). By the late 1980s, Eckankar had become "[a]n international movement with tens of thousands of members in the United States, Europe and Africa" (Denosky n.d.: 3). Likely the claims that it was pure spirituality and not religion, as well as the promises of "shamanic" soul travel and the extraordinary out-of-body experiences made it so attractive, because they met the New Age "zeitgeist" more than the Radhasoami original, which involves intense (typically Indian) *guru-bhakti*, etc.

Arguably the most important intermediary of the Nāda-Brahman in the West is Joachim-Ernst Berendt (Berendt 1983, 1991), because in the late 1970s and 1980s

---

[10] The positive effects of sound on health and wellbeing were already pointed out by Bhagavan Rajneesh, who recommended humming to activate the brain, cleanse every fibre of the body, and heal ailments (Beck 1993: 116). But references to "Nada healing" by listening to sound vibrations and to Rāga music have been increasing since, with an emphasis on audible music and one's own voice. Wilke and Moebus (2011: 996, 999–1000, 1034–36) discuss a number of internet sources (and a TM brochure) on the subject, such as http://users.bestweb.net/nad-yoga.html, copyright Beverly B. Ferguson 2001; http://www.healingmusic.org by Amrita Cottrell; or http://www.entrepreneurialhealer.com/sounding.html (all accessed 20 September 2004).

[11] Similar to the Theosophists, the Radhasoamis understand their esoteric teaching as a science that rises above known natural science, investigating laws and higher cosmic planes (like the astral world) not yet researched sufficiently and coming from a higher consciousness.

he introduced and popularized the musical Nāda-Brahman in the West, which was more akin to Śārṅgadeva than any of the other receptions mentioned above. His description of the Nāda-Brahman and of the whole world being sound (first in radio speeches) was tremendously successful. (For a broader discussion of Berendt see Wilke and Moebus 2011: 818–830 and Wilke 2017: 329–336.)

His teachers had been Indian musicians like Ravi Shankar, Ali Akbar Khan, and many others, who were for the first time performing in the West.[12] In particular, however, Berendt was indebted to the North Indian musician, modern Sufi master, and Theosophy-inspired, internationally active Inayat Hazrat Khan (1882–1927), who spread Sufism in Europe and America. Khan left a musical treatise from which Berendt quotes extensively. This text had been discovered by Don Cherry and apparently made the rounds among Western musicians since the late 1960s.[13] It prefigured many important ideas of Berendt that are not found in Śārṅgadeva (for the following quotes see Berendt 1991: 160–162; and Wilke and Moebus 2011: 1013). Khan presented Rāga music and the Nāda-Brahman as a Muslim-Hindu Indian interpretation of the music of the spheres or, in his own words, of the "harmony of the whole universe", "the movement of the stars and planets, the law of vibration and rhythm, all perfect and unchanging". Moreover, music evokes, according to Khan, "at-one-ment with God" (a phrase verbally taken over by Berendt 1991: 168) and "being in tune with the Infinite, in harmony with the universe". Music gives spiritual orientation in a material(istic) world. Khan also claimed (like Berendt) that "all religions have taught us that the whole creation is sound". Berendt ends his selection of excerpts with the quote: "This at-one-ment with the Absolute manifests as music of the spheres" (Berendt 1991: 162). Berendt saw in Khan's text the perfect representation of the new spiritual and musical consciousness of the then young generation of musicians (Berendt 1991: 160; Wilke and Moebus 2011: 1012–1013). He claimed that Indian modal music and the way of music-making of Indian musicians had changed the world view of a whole generation of jazz musicians in the West, as it initiated a spiritual awakening, starting with John Coltrane; it was a development which soon spilled over from jazz to the new rock, pop, minimal music, world music, and meditation music (Berendt 1991: 152–172,

[12] A notable phenomenon which is not found in Berendt's sources is the "yogification" of the musical Nāda-Brahman among some singers and vocal instructors of contemporary India. Beck (1993: 111) cites musicologist and singer M. R. Gautam who maintains, "regarding the mystical side of our Hindustani vocal music, when the singer has produced the various notes (*svaras*) to perfection, then that outward audible vibration is capable of producing a corresponding inner, 'sympathetic' vibration, which in our terminology is known as Nāda-Brahman." Beck (1993: 108–111) seeks to establish a continuity of music, Nāda-Brahman, and esoteric Nāda-Yoga since the times of Mataṅga and Śārṅgadeva, but apparently overstates his point by unclean translations and tendentious hermeneutics.

[13] Berendt (1991: 160). He goes on explaining: "For years, its complete version existed only as a private printing of the London Sufi mission. In an abridged form, however, you can find it all over, in thousands of copies."

200–227). This central thesis of Berendt's much-read book may likely be an exaggeration, but a short love-affair with India in the counter-culture cannot be denied – think of the Beatles.

Berendt seeks to convince with many direct quotations, for example from the tenor saxophonist Nathan Davis: "Playing modally has to do with spirituality. What we really mean by spirituality is religiousness. Only we do not use that word, because we don't mean what the Christian world means by religiousness" (quoted in Berendt 1991: 160).

Berendt was not only a jazz historian and record producer, but also a New Age proponent. His second central thesis proclaims by means of the Nāda-Brahman a strong cultural criticism and a therapeutic programme for the West. He conceptualized the Indian Nāda-Brahman as a key to a new metaphysics for the spiritually impoverished West and as inciting a "New Consciousness" (a "paradigm shift" or "New Age") that was productive of a more spiritual, holistic and peaceful world(view) and way of life. This new consciousness for him was embodied in the empathic "listening man", whom he saw in Indian musicians, in contradistinction to the dissecting, logocentric "eye person" that reigned since Aristotle and had become an absolute in the age of television (Berendt 1991: 5, 7, 130). The ear, so Berendt (Berendt 1991: 136–137), transcends the fundamental dualism of Western culture due to being the most holistic sense organ which unites the faculty of mathematical measuring and of sensory and emotional experience. He suggests that "hearing, rather than seeing, is the key to a more spiritual experience of consciousness" (Berendt 1991: book cover). For Berendt, his acquaintance with classical Indian modal music and with Indian musicians, for whom music and religion were one and who instructed him about the ideal attitude of music-making, was a revelation: "*Nada Brahma*. Sound is god. Show that in your singing. Meditate on it. Cleanse your karma by it. Sing so that people will understand that, even if you haven't told them before. Sound is god, *Nada Brahma*" (Berendt 1991: 155, quoting Pandit Pran Nath). As importantly, the Nāda-Brahman was about the sound character of the universe, which Berendt in typical New Age fashion found confirmed in numerous areas of knowledge: from ancient myths around the world to singing dolphins, sounding crystals, and most of all the latest modern sciences: "Scientists have only recently learned that the particles of an oxygen atom vibrate in a major key and that blades of grass 'sing.' […] reaffirming what the ancients have always known – the world is sound, rhythm, and vibration" (Berendt 1991: book cover). Berendt was convinced that Eastern religions and modern sciences confirm each other, e.g. that the cosmic dance of Śiva depicts the same concept of matter as found in aspects of quantum and field theory of modern physics (Berendt 1983: 169; see also Fritjof Capra's foreword to Berendt 1991).

The Radhasoamis and Berendt seem to be miles apart. But they share the conviction that the universe has an intrinsically sonic character and that religion and science are not mutually exclusive. The dispositive of a congruence and even merger of religion and science has been a globally spread pattern – in fact a main feature – of modern esotericism, including Asian reform movements, since the late nineteenth century (Michael Bergunder 2016, 2020: 76–125, particularly 104 ff.),[14] often connected with a criticism of both scientific materialism and doctrinal religion (usually identified with Christianity), and with a claim to surpass both. Thus, in spiritual and religious movements in South Asian modernity (particularly reform movements like the Radhasoamis), in modern esotericism (starting with Theosophy at the end of the nineteenth and early twenty century), and in the New Age (which had the most impact on (post)modern spirituality), there are claims to be found that not only they do represent a spiritual tradition that is in conformity with scientific standards, but they initiate a science beyond the (presently known) natural sciences, as they investigate unresearched areas like the paranormal or subsonic sound. Eckankar, for instance, speaks of the "science of sound travel" and the Radhasoami community of Dayalbagh near Agra of "The Dayalbagh Science of Consciousness Studies". Another "'scientific' venture" of the early 1980s reported by Beck (1993: 117) is noteworthy in this context, namely the project to investigate "the effect of Nāda-Brahman meditation on human consciousness by experimentation" (Beck 1993: 117). The theological assumption was: "The continuous muttering of Om releases the spiritual energy of the [cerebral] cortex and the continuous echoing in the region creates the feeling of expansion in the brain […] so that ego-feeling gets merged in the Divine consciousness" (quoted Beck.). Apparently, the test-persons did not verify this preconceived result of mystical fusion, but did acknowledge certain psychophysical effects. According to Beck, 8 per cent of them, when asked about "their 'inner feelings' after Nāda-Yoga meditation" reported "inner experiences in the form of vibrations felt all over the body" or "a feeling 'as if floating in the sky without contact with the earth'" (Beck 1993: 117).

5 Concluding Remarks: The Nāda-Brahman and the Multiple Forms of Sonic Consciousness

The success of the Nāda-Brahman concept in India's religious life even beyond the Hindu fold – as it also infiltrated Sufi, Sikh, and Radhasoami traditions – was likely due to being a very holistic symbol. Śārṅgadeva's Nāda-Brahman was something

---

[14] Bergunder (2020: 77) traces this global development back to the breathtaking advance and new social prestige of science and scientific research since the mid-nineteenth century and the separation of the natural sciences and humanities (Natur- und Geisteswissenschaften). This situation, maintains Bergunder, caused the "twin-birth" of religion and esotericism as modern universal concepts. The situation posed the question whether religion and science were compatible or not, even more so in view of radical materialistic models, which reduced the soul to brain substance, and antireligious attitudes among scientists. The Christian reaction was less a rejection and disapproval of science, which, according to the author, was more the typical stereotyping performed by esoteric and reform-Asian authors. The general pattern among Christian Protestant and Catholic theologians was that science and religion belong to different spheres. In contrast to this separation, we find their merger in modern esotericism and reform-oriented Asian religions (from Vivekananda to the Dalai Lama) – however different their conceptions of religion and science may be.

concrete and abstract at the same time and able to inspire many different forms of sonic consciousness and awareness. The Nāda-Brahman of Śārṅgadeva himself communicated a non-dual world view and a genuine way of spiritual deliverance by music, as music was seen to allow immediate participation in the essence of the universe. Music made the abstract, featureless Brahman of earlier traditions more tangible, because as music it became a sensory substance, the audible Nāda-Brahman. Music brought immediate delight and was full of expression, in contrast to the Brahman removed from the senses or the inaudible primordial sound, the Nāda of the Śaivas. Śārṅgadeva saw an enjoyable yoga for everybody in music. His pronounced sonic consciousness most of all manifests in the fundamental theses: Sound (song) is God, God is sound, the whole world is sound, as it originates from sound and is based on it. Kallinātha was even more strongly given to philosophy, cosmology, and world view: He explained rationally why music could be seen as a direct gateway to the nameless and formless Brahman, as music's structure was correlative to Brahman's nature as efficient and material cause of the universe. Music is therefore able to lead to Brahman experience, an expansion of consciousness towards non-dual reality, and mystical states of fusion. This is explicit in Tyāgarāja, the most famous of the Carnatic musician-composers, who repeatedly states: The joy of music is itself the bliss of Brahman that the Vedānta speaks of. Later developments in contrast anchor sound and sonic consciousness completely in inward listening, resonances, sound vibrations, and (yogic) inner sounds, up to the Radhasoamis' and Eckankar's sonic cosmologies, which are directly experienced by the soul's journeys through the upper cosmic regions in which the primordial, energetic sound current resonates in ever-new forms.

Śārṅgadeva's, Kallinātha's, and Tyāgarāja's somewhat different models of sonic consciousness, sonic cosmology, and Supreme Reality, in which non-dual Being, omni-immanent Consciousness, and supreme Bliss fuse with world reality and musical sound, and the Radhasoamis' further interiorization and cosmicization, allow us to address interesting differences in the conception of consciousness within the Indian traditions, but also interesting, even astonishing, continuities. At stake is not only Śārṅgadeva's ingenious novel conception of the Brahman, which made it more accessible to feeling, the body, and the senses, but also the cosmic sonic consciousness of the Tantra and yogic-tantric practice in contrast to the a-cosmic pure consciousness of Advaita-Vedānta, the philosophical tradition of India which became best known in the West as a marker of Indian spirituality. However, remarkably even Kallinātha and Tyāgarāja, each in their own way, were closely connected to the Advaita-Vedānta, on which they based their non-dual sonic view of Brahman experience, be it in terms of cosmology or in terms of the performance of devotional song and *nāda-upāsana*, "sound contemplation". Moreover Kallinātha's music-philosophical exposition on Śārṅgadeva's Nāda-Brahman and the Radhasoamis' esoteric *surat-śabd-yog* practice share a major idea about sonic consciousness and the nature of sound without sharing the same philosophical basis. Both postulate that following sound leads to the Supreme Reality, which is the ultimate cause or source of sound, but itself beyond sound. Particularly in the Beas line, the ineffability, namelessness, and formlessness of the supreme Radhasoami reality or Godhead were stressed, sounding much (at least structurally) like the Advaita-Vedāntic supreme Brahman beyond name and form. Even more close and supplying direct inspiration was the Sants' *nirguṇa-bhakti*, the "love" for the "featureless" and formless God(head), that was accessible to everybody, irrelevant of social position, and was disseminated particularly by lower caste and casteless poet-singers, starting with Kabīr (Lorenzen 1996), unlike the Brahman of the Brahmanic Advaita-Vedānta (which, however, also left caste-dependence in modern times to become more democratic and universal).

Perhaps the most incisive and interesting novelty was that the Nāda-Brahman migrated beyond the Hindu fold to become a truly transreligious and transnational category. It is known today in India as well as in the West and nurtures various forms of post-modern transreligious spirituality that cannot be reduced just to India or just to the West. Interestingly, already Śārṅgadeva's Nāda-Brahman was a transreligious category that synthesized all prominent sound and language discourses of the time. The Sants and Sikhs are transreligious as they incorporated Hindu and Muslim theologies, and even more transreligious are the Radhasoamis, who absorbed many different influences – from Hindu, Sant, Sikh, Muslim, Christian, Neoplatonic up to Theosophy and modern codes of science. In all – Śārṅgadeva, the Sants, the Sikhs, the Radhasoamis – we find profound sonic consciousness and practices to immerse oneself in sound, as different as the respective practices (and experiences) may be.

The Nāda-Brahman became transnational not least because of the Radhasoamis, but also other internationally acting global players like Swami Shivananda, according to Beck. If he is right, it was a yogified Nāda-Brahman which initiated a new sonic awareness in the West and found expression in new practices of *nāda-yoga*. Remarkably, precisely such practices of modern esotericism commonly refer to Berendt as authority (Wilke and Moebus 2011: 999, 1001). However, Berendt as we have seen introduced the musical Nāda-Brahma(n) to a Western audience, inspired by Indian musicians who taught him the formulas "sound is God" and "the world is sound". They created a new regime of attention in him about music-making, listening, and spirituality, and about the sonic character of the universe which he sees corresponding to both modal styles of music and modern sciences. Berendt even claims that a whole generation of young Western musicians gained a new sonic consciousness and spiritual attitude after discovering Indian modal music. What he calls "New Consciousness" refers to all these issues at once, which for him were indicative of a sea change towards a New Age world of greater harmony. Berendt did not exclude the field of *mantra* practices, resonances, and vibrations, but his major focus was undoubtedly audible music. It therefore remains an interesting fact that even contemporary spirituality that locates the Nāda-Brahman in vibrating sound, out-of-body experience, spiritual healing, and ecology refers to Berendt, whose own approach had obviously a synthesizing and integrative effect concerning the two native developments of the Nāda-Brahman: its emergence as a new innovative concept that denoted a new musical, audible Brahman and its later yogification with a focus on the inner *nāda* and sound vibrations. In this process, sonic consciousness transformed. There was a tendency to move away from Brahman ontology and the experience of Brahman bliss in favour of an expansion of sonic cosmology and altered states of consciousness.

It remains a fascinating task to investigate the diverse forms and interpretations of (sonic) consciousness. My approach has been one of the cultural and historical study of religion, which feeds on a contextualist-constructivist perspective, historical anthropology, the question of cultural hermeneutics, semiosis, cultural hierarchies of perception, and the aesthetics of religion. It is explicitly not an approach that asks about real and unreal, true and false, or better and worse forms of sonic consciousness. Such questions are insignificant, as of course people immersed in sonic consciousness will perceive their beliefs and experiences to be real. This reality even is an "augmented reality" (Johannsen and Kirsch 2020: 144, 149–152) and at the same time constructed, or in other words created, evoked, and generated by cultural regimes of expectation and attention, such as the habituation to the highly sonic life-world of India and its exceptional value for sound. Regimes of attention and augmented realities are produced, stimulated, and enhanced, as well as shaped and re-shaped by discourses, theologies, cosmologies, and performative displays, by narratives and narrative strategies (Johannsen and Kirsch 2020), by talent and training (Luhrmann 2020), and not least by sensory strategies (Mohr 2020) and sensational forms (Meyer 2008), such as hearing song and instrumental music, or sensing resonances and sound vibrations, both of which generate (and guarantee! if deep listening takes place) complete immersion in sound, and in consequence augmented realities, such as the experience of Brahman bliss, or of inner, immaterial, otherwise imperceptible sounds and higher celestial spheres. Both these experiential worlds are products of, or at least connected with, the powerful Nāda-Brahman concept and *nāda* practices.

References

Babb, Lawrence A. 1986. *Redemptive encounters: Three modern styles in the Hindu tradition. I: Spiritual recognition and the Radhasoami faith*. Berkeley: University of California Press.

Beck, Guy. 1993. *Sonic theology: Hinduism and sacred sound*. Columbia: University of South Carolina Press.

Berendt, Joachim-Ernst. 1983. *Nada Brahma: Die Welt ist Klang*. Frankfurt: Insel.

———. 1991. *The world is sound: Nada Brahma. Music and the landscape of consciousness*. Rochester: Destiny Books (First English Edition 1987).

Bergunder, Michael. 2016. "Religion" and "Science" within a global religious history. *Aries* 16: 86–141.

———. 2020. Umkämpfte Historisierung. Die Zwillingsgeburt von 'Religion' und 'Esoterik' in der zweiten Hälfte des 19. Jahrhunderts und das Programm einer globalen Religionsgeschichte. In *Wissen um Religion: Erkenntnis – Interesse*, ed. Klaus Hock, 47–131. Leipzig: Evangelische Verlagsanstalt.

Denosky, Jim. n.d. Paul Twitchell. *Founder of Eckankar: The Ancient science of soul travel*. http://om-guru.com/html/saints/twitchell01.html. Accessed 7 January 2022. n.p.

von Glasenapp, Helmuth. 1928. *Der Râdhâsvâmî-Satsang und die Mystik der Gottestöne. In: Helmut von Glasenapp. Religiöse Reformbewegungen im heutigen Indien* (Morgenland, Heft 17), 48–59. Leipzig: Hinrichs'sche Buchhandlung.http://www.om-guru.com/, see Denosky.

Jackson, William J. 1994. *The power of the sacred name. V. Raghavan's studies in Nāma-Siddhānta and Indian culture*. Delhi: Satguru Publications.

Johannsen, Dirk, and Anja Kirsch. 2020. Narrative strategies. In *The Bloomsbury handbook of the cultural and cognitive aesthetics of religion*, ed. Anne Koch and Katharina Wilkens, 143–153. London: Bloomsbury Academic.

Juergensmeyer, Mark. 1991. *Radhasoami reality. The logic of modern faith*. Princeton: Princeton University Press.

Katz, Jonathan. 2001. Mataṅga. In *The new Grove dictionary of music and musicians*, ed. Stanley Sadie and John Tyrrell. London: Macmillan.

Koch, Lars-Christian. 2008. Raga-Music in North India between religious meaning and courtly culture. In *Religion und music*, ed. Lydia Guzy, 61–74. Berlin: Weißensee Verlag.

Lorenzen, David N. 1996. *Praises to a formless God*. New York: SUNY.

Luhrmann, Tanja M. 2020. Absorption. In *The Bloomsbury handbook of the cultural and cognitive aesthetics of religion*, ed. Anne Koch and Katharina Wilkens, 85–95. London: Bloomsbury Academic.

Mataṅga. [1992] 1994. *Bṛhaddeśī of Śri Mataṅga Muni*. 3 vols. Prem Lata Sharma. Ed. New Delhi: Indira Gandhi Centre for the Arts.

Meyer, Birgit. 2008. Religious sensations: Why media, aesthetics and power matter in the study of contemporary religion. In *Religion: Beyond a concept*, ed. Hent de Vries, 704–723. New York: Fordham University Press.

Mohr, Hubert. 2020. Sensory strategies. In *The Bloomsbury handbook of the cultural and cognitive aesthetics of religion*, ed. Anne Koch and Katharina Wilkens, 129–141. London: Bloomsbury Academic.

Śārṅgadeva. [1943] 1953. *Saṅgītaratnākara of Śārṅgadeva. With Kalānidhi of Kallināṭha and Sudhākara of Simhabhūpāla*. 4 vols. S. Subrahmanya Sastri. Ed. Madras: Adyar.

Van der Veer, Peter. 2001. *Imperial encounters: Religion and modernity in India and Britain*. Princeton: Princeton University Press.

Wilke, Annette. 2017. Moving religion by sound: On the effectiveness of the Nāda-Brahman in India and modern Europe. In *Aesthetics of religion. A connective concept*, ed. Alexandra K. Grieser and Jay Johnston, 323–346. Berlin/Boston: de Gruyter.

Wilke, Annette, and Oliver Moebus. 2011. *Sound and communication. An aesthetic cultural history of Sanskrit Hinduism*. Berlin/New York: de Gruyter.

Improvisation and Consciousness: Some Recent Links

Ralph Yarrow

Abstract This presentation looks at links between improvisation and consciousness with reference to theatre and performance as reflected in recent work. It explores improvising and the improvisatory as a state, a process and a mode of production. 'Making things new' implies a condition allied to the origination of form, in tune with a generative capacity. What kind of consciousness (or perhaps Consciousness) might this be? Might it constitute a kind of 'group mind' and a reduction of 'self-consciousness'? What are its psycho-physiological and ecological parameters? Is it a kind of 'field awareness', and is its crucial impetus a key mode of will or decision? I will touch on work by Gary Peters, Adam Kahane, Samuel Beckett and others in order to signal some of these dimensions.

Keywords Consciousness · Improvisation · Drama · Mind · Psychology

'Improvisation', says Alessandro Bertinetto, citing Jankélévitch, 'is the practice of 'beginning'' (Bertinetto 2018: 80); it is also, he suggests, 'openness to emergent and new outcomes (in terms of values, meanings, and experiences)' (Bertinetto 2018: 77). The Italian all'improvviso identifies improvisation and improvising as spontaneity, a sudden or perhaps 'unconsidered' phenomenon, implying that it arises in a condition that is equivalent to or precedes what we normally think of as thought. Thus improvisation is of interest philosophically and aesthetically as a site at which to examine the nascence of form and being and pragmatically as the condition or state from which performance may emerge. This essay looks at links between

R. Yarrow (✉)
School of Literature, Drama and Creative Writing, University of East Anglia, Norwich, UK
e-mail: r.yarrow@uea.ac.uk

P. S. Satsangi et al. (eds.), *Consciousness Studies in Sciences and Humanities: Eastern and Western Perspectives*, Studies in Neuroscience, Consciousness and Spirituality 8, https://doi.org/10.1007/978-3-031-13920-8_14

consciousness and improvisation with reference to theatre and performance as reflected in recent research work. It considers improvising and the improvisatory as:

- a state
- a process
- production

In this context, my essay will specifically address studies that are concerned with modes and aspects, in which improvisation may change individual consciousness in the process of performing the self.

The following key studies concerning theatrical improvisation provide a fairly comprehensive overview of central aspects of recent research in this topic. Thus Gary Peters in his book *Improvising Improvisation* (2017) identifies improvising as commitment, choice and volition: the initial impulse or inhabiting of the decision to improvise or make different. Chloe Arros in *Balancing Acts: Aesthetics of Vulnerability in British Contemporary Improvised Theatre* (2021) examines the balance of the intuitive and the deliberate in the process of improvisation; Clayton Drinko's *Theatrical Improvisation, Consciousness and Cognition* (2013) discusses the functioning of individual and collective consciousness in improvisatory production; Anthony Frost and Ralph Yarrow in *Improvisation in Drama, Theatre and Performance* (2016) chart the history and practice of improvisation across the globe and point out both its effect as a nexus of aesthetic, psychological and political renewal, and the centrality of the condition of *disponibilité* (availability) in its genesis. They define theatrical improvisation as:

> the skill of using bodies, space, imagination, objects, all human resources, to generate or to reformulate a coherent physical expression of an idea, a situation, a character (even, perhaps, a text); to do this spontaneously, in response to the immediate stimuli of one's environment, without preconceptions. (2016: xiv)

The term 'improvisation' – specifically in drama, theatre and performance, but also more generally – refers to

- a history: lots of people have done it at different times in different contexts and in different ways, all of which relate to, comment on, challenge or redefine those contexts and the systems and structures which frame them
- a set of practices, similarly culturally embedded and implicitly transgressive
- an underpinning set of principles or theoretical positions

In this respect it covers both a diachronic and a synchronic dimension. It manifests itself continually across time and space, and it signals a set of understandings or claims about human performance and production; it is both disruptive and constructive. In summary, '[i]mprovisation is not just a style or an acting technique; it is a dynamic *principle* operating in many different spheres, an independent and transformative way of being, knowing and doing' (Frost and Yarrow 2016: 3).

Improvisation includes the following modes and procedures:

- individual production in many forms: verbal, tactile, tonal, rhythmic, somatic, etc.; and genres: drama/theatre, music/jazz, poetry/scripting, dance/movement, painting/sculpting etc.

- pair and group production as above
- processes of social, psychological, communal dialogue, interaction and negotiation – including co-creativity, therapy/therapeutic applications, entertainment and fun, information exchange, problem-solving, crisis intervention etc.
- accepting and using risk, failure, fragmentation, deconstruction, loss, unknowing, puzzle, vulnerability, chaos

All of these involve getting or being engaged in the generation of something, essentially as a process of restructuring or reformation from stimuli (verbal, sonic, visual, rhythmic, narratological, ideational etc.). We could ask: does it create, or recreate? Is it originary, or is it rearranging the deckchairs? (It's possible of course that the answer is 'both/and'.) On the one hand, the production of form necessarily involves both skills (a known and/or previously trained set of physical and intellectual acts) and materials (existing elements of matter, accessible to sense); and on the other hand, these ingredients are recombined in such a way as to generate something essentially 'different'. In this process, the identity of producer and product is foregrounded and interrogated. Not surprisingly, makers of performance throughout the ages have been drawn to explore the how and why of improvising.

The improviser is aiming for a 'new combination' of performance practice and performer training engendered in the moment, which is why these issues recur in different ways in the work of many well-known performance-trainers and theorists in the twentieth century. The performer is likewise beholden to both new and old, to a skills basis and a departure from or addition to that, to both familiarity and novelty, to what s/he has been and what s/he could be/do. Each (moment of) production is, at least potentially, an existential variable.

A symposium in Zurich in 2016, entitled *Impro Talks*, showcased the variety of dimensions, contexts, modalities and interfaces of improvisation, improve and the improvisatory – as practice, praxis, methodology, event structure, performance form, human capacity, cognitive functionality, aesthetics, politics, etc. The symposium identified, exemplified and demonstrated intellectually and practically that there are ways of grounding what improvisation is and does in psycho-physiological, interpersonal, communal and 'universal' terrains.

This conviction is also found in recent work on performance and ecology and on the psychophysiology and somatics of performance (Barba 1995; Schechner 2004; Zarrilli 2020). In this context it is useful to think about improvisation, particularly in terms of how it affects the relationships of self to self, self to other and self to surrounding contexts and environments, and how the kinds of bodily disposition that many practitioners believe it engenders signal changes in daily and extra-daily ways of thinking and acting, in both the narrow sense of theatrical and the wider sense of lifeworld performance.

Improvisation thus is about making things in a variety of forms and modes of being. But as Arros, Drinko and Peters suggest in different ways that making comes from somewhere; it occurs and takes place as a variety of psychophysical processes (2021, 2013, 2017). These aspects open up questions and links between consciousness and improvisation, both of which can be understood in a variety of ways, namely as a state or condition, as a process and as a mode of production. My essay in the following part will focus on one specific form of consciousness, namely, in what ways are improvisation and the process of improvising fundamental to the performance of 'self' and of the acts which locate and constitute it? How do I access what I am not yet? By incorporating some new phenomena and by forgetting the entity – or shifting process/set of events – which I know myself as? By knowing in a way which undoes and/or reconfigures the known? In this process, is a passage beyond 'the known' into the potentiality or impetus of 'knowing' useful or essential? Does this involve shifts, changes or refinements of physical and/or mental activity?

Richard Schechner proposes that the performer is 'not me' (the everyday persona) and also 'not not me' (Schechner 2013: 72); Eugenio Barba targets an 'extra-daily body' (Barba 1995: 15–16), Jacques Lecoq – following Jacques Copeau – seeks to access a 'neutral body' (Frost and Yarrow 2016: 68–76); Nicholas Nunez considers that theatre may be about embodying 'anthropocosmic' energies (Nunez 1996); Sanford Meisner attempts to work with the immediacy of interactive responsiveness, which bypasses calculation or 'thinking' (Meisner 1987). All these celebrated performance-trainers (and others like Peter Brook, Michael Chekhov, Jerzy Grotowski, Vsevolod Meyerhold and Ariane Mnouchkine) seek to extend or transform the 'self' of the performer and the range of its performative capacity 'beyond' the everyday, including kinds of energy, modes of operation and interactivity, sensitivity, openness to the unexpected or unfamiliar, honing of sensory awareness, physical flexibility, spontaneity, originality etc.; and they all employ techniques or practices that draw centrally on and explore improvisation in a variety of senses and applications.

In most cases, the process of individual change and theatrical production go hand in hand. For instance, there is a whole set of theatre-related practices in the USA which were first known as 'improv' or 'Chicago improv' and then extended to 'theatre games'/'improv games' or even 'improv olympics'. These involve real-time construction of scenes and scenarios of varying length by a team of actors, incorporating audience suggestions about character, incident and style. In the 'games' model, the audience is involved further in 'scoring' competing teams' performances against a variety of parameters including inventiveness, audacity, teamwork and so on. Thus improvisation here is both the process (working together, feeding off each other's suggestions and moves) and the product.

Other dimensions of improvisation involve dancers, musicians or poets responding to each other and creating new patterns or elements: Indian classical musicians' gauntlet-throwing dialogue *jugulbandi,* jazz musicians' solo or shared riffs, Bengali poets' spats, contact impro, rap interchanges, Commedia dell'Arte performers' extemporisation from the given circumstances of *lazzi*. All these display the recognition and redeployment of core elements or patterns (words, chords, steps/moves etc.), and all of them depend on the prior development of skillsets – physical, technical, psychological and aesthetic.

Many eastern performance forms work through extensive body/somatic training over many years to cultivate a base of performance flexibility and skill. Only after this thorough schooling is 'improvisation' possible: it presupposes 'mastery' and then operates on a basis of extended physical resource and interactive sensitivity to other performers – constituting a kind of extra-daily operational capacity. In this sense, performance theorist and renowned director Eugenio Barba locates extra-daily 'bios' or enhanced psychophysical disposition as an iconic performative capacity fundamental to expressiveness and communicative potentiality, which is rooted in a very precise command of individual skills and sensitivities (Barba 1995). Partly as a result of extensive intercultural engagement over the last five decades, much western practice now also incorporates similar training regimes, for ballet and music performances, obviously, but also in the now-standard curricula of drama schools and conservatoires, which frequently include training in improvisation.

In addition, it can be argued that such training manifests not only in the examples noted, but also in 'conscientious' improvisation work, which is displayed in the spontaneous response to a story featured by Playback companies as well as by the improv teams described above, many of whom undertake a lengthy, if not always so formally organised, process of developing individual response mechanisms, group interactivity, listening, accepting and extending, organising and recombining the elements of their practice. During these kinds of long-term performance training, it may be said that the mind and body not only develop capacities but also discover new kinds of *habitus*, that have significant mental and physical consequences.

Bertinetto holds that '[a]rtistic improvisational performances in which procedural knowledge (skills, habits and styles, shortly: artistic competence) is applied in specific situations are involved in the process of generating and transforming procedural knowledge, that is, in the evolution of artistic competence' (Bertinetto 2018: 83). He further points out that, in consequence:

> [h]abit and invention are not the two antipodes of human behavior, but they feed on each other. Hence, each performance not only expresses the artist's personality, but (re)shapes it as well. So, in an important sense, the self who is the subject of the action is generated through the action. [...] Briefly, improvisation illustrates that the shaping of habits is not the opposite of the exercise of creativity. (Bertinetto 2018: 84)

These features all exemplify and invite investigation as criteria of performance and their relation to the aesthetics of production and reception, and they have been central to major strands of performance theory, in particular in approaches establishing conjunctions between performance and anthropology (Schechner 1985; Barba and Savarese 2005); they also invite neurological, psychological and philosophical interrogation into the conditions and processes they present and the implications these raise. The following pages of my essay will focus on central issues of these approaches.

Improvisations are forms of interaction with the human performers/producers and with the elements, shapes, patterns and significances they produce. The resultant output varies in terms of originality, from a relatively stereotyped adoption of learned sections, styles or material to radical challenges to the 'rules' or conventions of syntax, sequence, format or function. Improvisation can thus be considered as a:

> tension between embedded habits and the unexpected emergence of the new, that performative tension between structured and creative practice that – in normative terms – generates

> meaning and makes possible its (always unsecured and fallible) understanding. [...] Improvisation illustrates how the conditions of human action and human practice are generated by exercising the action and the practice, at the same time building habits of behavior (from which the agents passively depend) and transforming creatively those habits. (Bertinetto 2018: 75 and 83)

So what are these 'habits' in relation to neurological activity and psychosomatic condition? Several kinds of consciousness (in the sense of perceptual interaction with environments and the 'performance' of modes of knowledge) are involved here, for example:

- heightened functioning of perception and performance, a sense of 'flow', or 'being in the zone'
- ability to be open to a range of stimuli, patterns and impulses
- group or inter-individual receptivity, field awareness
- ability to shift or restructure cognitive and motoric performativity

It is of course possible to investigate some of these aspects experimentally. Both Drinko and Arros draw on psychological terminology to frame and explicate specific kinds of mental activity occurring, for instance in the case of neurological investigations of jazz musicians. Thus Drinko reports that '[d]uring the fMRI brain scans of the improvising jazz musicians, Limb found that the lateral prefrontal cortex was inhibited and the medial prefrontal cortex increased in activity during improvisation but not during memorized scales or songs' (Drinko 2013: 95–96). Different zones of the brain thus 'fire' in particular ways, giving rise – or at least correlating – to shifts in perceptual and conceptual activity. These kinds of description help to chart what is taking place neurologically, although they do not necessarily fully 'explain' it, least of all from an aesthetic or philosophical point of view. But there are, without question, a number of interesting leads in this kind of research worth highlighting.

Arros's work picks up the discussions concerning the concept of 'expertise' by invoking Ryle's 'knowing what' and 'knowing how'. Expertise, derived from training of whatever kind, produces a sense of 'cognitive ease', i.e. you know *how* to do something, and you are confident in your skillset. That seems to give rise to an increased spontaneity of response and flexibility of production – namely choosing what. So in this understanding, 'cognitive ease', although derived from training, manifests as a ground awareness that 'I can' – an unconscious or preconscious level of confidence or coherence, which may then enable 'deliberative' choice. (Peters puts it slightly differently, identifying choice as the initial switch which sets the process in motion: that I am now in the improvisatory mode; in other words, I know how. But here too the *how* precedes the *what*.)

Drinko also asks: 'What happens to consciousness and cognition during improvisation, and what happens to the mind/ brain of skilled improvisers during exceptional performances?' He investigates 'improv consciousness' and focuses mainly on the structure of long-form improv – called 'the Harold' – from Chicago and the development of this format from Viola Spolin to Del Close, two of the internationally most influential theorists and teachers of improvisation theatre. Drinko focuses on aspects of individual and collective consciousness in improv work. These include 'flow', the sense of ease and heightened mental coherence, central to the work of Mihaly Csikszentmihalyi, and 'group mind', a concept referring to ways in which these attitudes and modes of consciousness bleed into collective functioning, rapid picking up of cues and anticipation of actions and words. For Drinko:

> [t]he key to understanding improvisation's effects on the mind is that it orders consciousness. By focusing outward on one's scene partners, the game of the scene, or the space, skilled improvisers can order consciousness in such a way that System 1 or intuitive, automatic thinking rises to the surface in the competition amongst different modes of thought. Whether it be Viola Spolin's focus (or point of concentration), long-form improvisation's perceptual shift toward patterns and themes, or Keith Johnstone's status, not thinking about one's own behavior leads to less self-conscious improv with better ensemble collaboration. (Drinko 2013: 93–94)

According to Drinko, such a moment 'connects cognitive studies with spontaneity experienced for the first time (not "as if" for the first time) in front of an audience' (2013: 9). In this way his analyses 'begin to explain what those unique structures and effects are that make improvisation so beneficial to its practitioners and the way their minds work' (Drinko 2013: 2–3).

Csikszentmihalyi's concept of 'flow' is linked to Spolin's notion of 'spontaneity', as both point to a state that includes a heightened awareness of fellow players and the audience ('instant feedback'), which is furthered through what Spolin calls 'mirroring'. Thus, Spolin's improvisation training (Spolin 1963) includes several mirror exercises in which participants mirror the gestures, expressions or bodily postures of their partners. In a similar vein, Drinko suggests links to the function of *mirror neurons* and suggests that this 'unconscious imitation [...] could be one of the reasons improvisers are able to be so in sync with their teammates' (Drinko 2013: 27). He refers to Ramachandran as an example, indicating that this ability 'could allow us to interpret other people's intentions and predict their actions' [...] and help people 'see the world from another person's *conceptual* vantage point' (cit. 27). In terms of individual consciousness, Csikszentmihalyi and Drinko both suggest a heightened mode of socialised or externalised functioning.

Drinko links his concepts of 'flow' and 'group mind' to Steven Johnson's notion of 'emergence', where the 'group or network is smarter than any one of its separate components' (Drinko 2013: 50) – like the social organisation of ants. An important premise of this convergence of individual minds is 'giv[ing] up control, not pre-planning' (Drinko 2013: 50), similar to many of the precepts outlined by iconic improvisation teacher Keith Johnstone for both performers and story-tellers (Johnstone 1981, 1999). Accepting this condition of not knowing leads to a further level of mental activity, in Drinko's words: '[W]hen people are not afraid of messing up and comfortable with not knowing what comes next, they are more able to also access this undermind processing' (Drinko 2013: 79). 'Undermind' here seems to refer to a kind of broad 'field' awareness, which does not edit out detail, but works instead with pattern recognition, so that the pattern generates – or stimulates – recall of detail. These notions recall Arros's enquiry into cognitive ease and – to change the genre – to novelist Russell Hoban's identification of the generative function of what he calls 'the moment under the moments' (Hoban 1992). In all of these cases and concepts, spontaneity, intuition, 'cognitive ease', 'flow' and interconnection, as well as a kind of extended or 'field' awareness, seem to co-exist both as a neurological state and as a cognitive condition that facilitates the production of new kinds of order and output.

Addressing the issue of vulnerability in such situations, Arros especially makes use of neuroscientific work on the functioning of memory, the development of expertise ('knowing how'), the concepts of cognitive ease and cognitive convergence, conscious and unconscious response and its relationship to fear and courage. With this conceptual toolkit, she carries out her analytical objectives, namely to

- examine the central place of vulnerability (what might be called 'cognitive unease') as an emotional state in the creative process of improvisational theatre and the expert strategies that improvisers develop to overcome it
- study the unique ways in which improvisers can perform vulnerability, use it as a strategy in order to achieve virtuosity and make it part of the aesthetics of the form

Vulnerability can emerge as a creative strategy for both performers and audiences – 'a singular way of artmaking' (Arros-Steen 2021: 163) – if it becomes part of a deliberately chosen aesthetic. Referring to a well-known British company, Arros says:

> [T]he members of Forced Entertainment turn failure and unpreparedness into a style and deliver a performance which fails the ideals of theatre but fulfils its own goals of artistic ambivalence. [They thus deliver] the delight [of] seeing how the mechanism works at the moment of breakdown. (Arros-Steen 2021: 200)

In this way, the negotiation of 'unease' becomes part of the process of acquiring 'ease', and this 'ease' may itself lead on to 'cognitive convergence' and 'cognitive consensus' with other players/performers/producers. So, both for performers and audiences, the situation of not knowing what will happen next (as word or action), instead of producing a paralysing stasis, becomes the springboard for a new departure or set of possibilities: 'success' is thus generated precisely from acknowledging and allowing oneself to work with 'failure'. Arros identifies similar strategies in the work of improvising groups and solo performers.

While Arros and Drinko have addressed neurological events and processes and in consequence opened up connections to underlying or more extensive conditions of being and acting, Peters shifts the argument further towards criteria for action/forming of all kinds on an existential scale, by concentrating on the renewal of thought and action, which is fundamental for any kind of improvising. For Peters:

> improvisation isn't so much a genre, idiom, style or technique – it's a predicament we are thrown into, one we find ourselves in. The predicament [...] is a complex entwinement of choice and decision. The performativity of choice during improvisation may happen "in the moment", but it is already determined by an a priori mode of decision. In this way, improvisation happens both within and around the actual moment, negotiating a simultaneous past, present and future. (Peters 2017: Front cover)

Essentially, he suggests that improvisation is not so much an act – though it is that also –, but an attitude, a commitment to operate in a way that Frost and Yarrow describe with reference to theatre and performance as *disponibilité*, the condition of being open and available to move in any direction via any modality, style or genre, at any given moment (Frost and Yarrow 2016: 71/2). This decision/obligation speaks of aesthetic purpose, even rigour. Improvisation in this sense – in contradistinction to improvisation as a matter of acts in moments, which Peters acknowledges also – is a quality of distance and judgement, and not at all solely under the sway of immediate sensory/sensible reaction. There is a drive towards the positionality of doing different in a radical sense, of always discovering new beginnings, or – as Peters explores with reference to Kant, Hegel, Heidegger, Deleuze and Beckett – it is about returning to or rather situating oneself at the point of origination of form. In Beckett's work, for example, grammatical constructions like multiple negatives in prose and dramatic structures of repetitive 'failure' (in *Waiting for Godot* 'nothing happens: twice') materialise this continual 'return to Zero' for both protagonists and spectators/readers, culminating in Molloy's recognition that:

> to know nothing is nothing, not to want to know anything likewise, but to be beyond knowing anything, to know you are beyond knowing anything, that is when peace enters in, to the soul of the incurious seeker. (Beckett 1954: 96; see also Yarrow 2001, 2020)

'Truth'/knowledge/virtue arise from the location of this condition. It is not just a question of how one knows objects (phenomena), but rather an equivalence with the 'beginning' of wisdom – being in/at/under-standing the dynamics of eventing/process/phenomenal existence from this ('still point') of *askesis*. For Peters, what 'improvisation really does offer' is 'the dramatization of the beginning' (2017: 9), and this 'phenomenological beginning' is *purpose* (2017: 11). In a similar vein, this state of being forms a central concern in Adam Kahane's discussion of 'will' in *Power and Love* (Kahane 2010) – i.e. the will always to be in the condition of beginning anew. In this sense Kahane maintains that love is an act of will, not a feeling. Referencing Tillich, he suggests that love as will denotes the drive of everything to realise – or in current performance terminology, to 'embody' – itself.

Bertinetto recalls Plato's and Aristotle's idea that philosophy begins in wonder (*thaumazein*) and that wonder is often associated with curiosity as impulse to intellectual exploration. Wonder is thus a generative attitude. He further establishes a conceptual analogy with Foucault's identification of 'curiosity as the will to knowledge', which can be 'understood as a critical practice of deconstructive autonomy: a critical practice that enables individuals to constitute themselves'. Curiosity in this sense is thus "a kind of openness to the unforeseen, that affords individuals to 'embrace the uncanniness present in experience' and that 'enables one to attend to what is happening now and disappearing and to begin to care for what is and what might be'" (Bertinetto 2018: 78).

And what are the psychophysical dimensions of these philosophical musings for improvisation and consciousness? Rehearsal and practice, perhaps, is Peters suggests; repetition and reduction, as Beckett exemplifies? Actors practise attending to what is happening now, so that the actions, expressions and words they have learned may manifest in new combinations each time. But to ensure that the journey to meaning starts from a place before mere repetition, they need to follow a path towards the unknown, however counter-intuitive that may seem. You have to let go of what you know and what you think you are if you want to have a chance to become (other). This would be the equivalent of practising a cultivation of a specific habit of body and mind, a disposition – putting oneself in the position from which one is able – to create and an ability to play a whole range of forms, because the technical mastery is already achieved and can be awakened/applied. In this situation or state then, the 'I' that is operating is not self-reflective but self-referential: it does not seek feedback or praise for its accomplishment or beauty; its intent is to maintain, 'strictly', the operational awareness of continual beginning, which is the quality of 'self' as coterminous with the productive force. Thus the 'training' here targets the skill of reducing or suspending the force of the current version of self, of form or practice to a 'moment' of neutrality, of consciousness as mind and body in its simplest form as 'pure' awareness or Merrell-Wolff's 'consciousness-without-an-object', perhaps (Merrell-Wolff 1973). This moment of an ultimate and existential beginning, elaborated by Arras, Drinko and Peters as the psychophysical and aesthetic crux of the improvisatory, might be captured in Indian mythology as the initial point of the dance of Shiva and the ever-renewed play of Krishna.

References

Arros-Steen, Chloé. 2021. *Balancing Acts: Aesthetics of Vulnerability in British Contemporary Improvisational Theatre*. Ph.D Dissertation, 228p. Royal Holloway, University of London.

Barba, Eugenio. 1995. *The Paper Canoe: A Guide to Theatre Anthropology*. Trans. Richard Fowler. London/New York: Routledge.

Barba, Eugenio, and Nicola Savarese. 2005. *A Dictionary of Theatre Anthropology: The Secret Art of the Performer*. London: Routledge.

Beckett, Samuel. 1954. *Molloy*. Paris: Minuit.

Bertinetto, Alessandro. 2018. Why is improvisation philosophically so appealing? Some modest and incomplete suggestions. In *Free Improvisation: History and Perspectives*, ed. Alessandro Spordoni and Antonio Rostagno, 73–100. Lucca: Libreria Musicale Italiana, Universita di Roma.

Drinko, Clayton D. 2013. *Theatrical Improvisation, Consciousness and Cognition*. New York: Palgrave.

Frost, Anthony, and Ralph Yarrow. 2016. *Improvisation in Drama, Theatre and Performance: History, Practice, Theory*. Basingstoke: Palgrave.

Hoban, Russell. 1992. *The Moment Under the Moment*. London: Picador.

Johnstone, Keith. 1981. *Improvisation and the Theatre*. London: Methuen.

———. 1999. *Impro for Storytellers: Theatresports and the Art of Making Things Happen*. London: Faber.

Kahane, Adam. 2010. *Power and Love: A Theory and Practice of Social Change*. San Francisco: Berrett Koehler.

Meisner, Sanford. 1987. *Sanford Meisner on Acting*. New York: Vintage.

Merrell-Wolff, Franklin. 1973. *The Philosophy of Consciousness Without an Object*. New York: Julian Press.

Nunez, Nicholas. 1996. *Anthropocosmic Theatre: Rite in the Dynamics of Theatre*. London: Routledge.
Peters, Gary. 2017. *Improvising Improvisation: From Out of Philosophy, Music, Dance and Literature*. Chicago & London: University of Chicago Press.
Schechner, Richard. 1985. *Between Theatre and Anthropology*. Philadelphia: University of Pennsylvania Press.
———. 2004. *Performance Theory*. London: Taylor and Francis.
———. 2013. *Performance Studies: An Introduction*. New York: Routledge.
Spolin, Viola. 1963. *Improvisation for the Theater*. Evanston: North Western University Press.
Yarrow, Ralph. 2001. Theatre degree zero. Studies in the literary imagination. 34: 2. In *Drama and Consciousness*, ed. Daniel Meyer-Dinkgräfe, 75–92.
———. 2020. Zero Plus One: Beckett and the Natyasastra. In *The Natyasastra and the Body in Performance*, ed. Sreenath Nair, 217–228. McFarland: Jefferson.
Zarrilli, Philip. 2020. *(Towards) a Phenomenology of Acting*. Abingdon/New York: Routledge.

Part VI
Consciousness and the Intellect

Beyond Consciousness in Early Christian Mysticism

Andreas Müller

Abstract Late antique Christian mysticism is a way of transcending consciousness. The culmination of the approaches of the time is represented by the writings of Gregory of Nyssa (approx. 335–390) and Pseudo-Dionys of the Areopagus (5th c.). According to them, God cannot be expressed in discursive language. Theology therefore serves to ultimately transcend consciousness and verbal language and to describe the unspeakable only in negation.

Keywords Mystical theology · Apophatism · Gregory of Nyssa · Dionysios Areopagites · Christian mysticism · Negative theology

Limits of consciousness and especially the limitedness of consciousness is a central theme in Christian mysticism, the roots of which are to be found in ancient and especially late ancient Christianity. Mysticism is centrally about a mental approach to the transcendent, the divine. In ecstasy, mystics transcend the boundaries of their sensually perceivable world. For some mystics, this meant an immersion in the divine primal, but for others it also meant a general transcendence of perception and consciousness. In doing so, they often followed approaches with a Platonic or Neoplatonic background consistently present in the Christian realm.

In the following, I would like to illustrate with two examples how mystical concepts strove to transcend consciousness in order to reach the deepest insights into the self and the world. I will focus on two fathers of the so-called negative theology, namely Gregory of Nyssa in the late fourth century and Dionys Pseudo-Areopagites in the fifth century. Gregory was part of a group of theologians who lived in Cappadocia in Asia minor. Dionysius probably is a pseudonym. Nothing is known

A. Müller (✉)
Faculty of Theology, Professor for the History of Church and Religion of the First Millenium, Kiel University, Kiel, Germany
e-mail: AMueller@kg.uni-kiel.de

P. S. Satsangi et al. (eds.), *Consciousness Studies in Sciences and Humanities: Eastern and Western Perspectives*, Studies in Neuroscience, Consciousness and Spirituality 8, https://doi.org/10.1007/978-3-031-13920-8_15

about his origin. Probably he lived in Syria in the region of Antioch (for details see Müller 2020, in German).

1 Gregory of Nyssa: Reaching Out to the Incomprehensible Divine

Gregory (c. 333–394 AD) can be considered the Church Father who for the first time elaborated a kind of mystical system. This is found primarily in two of his writings, one on the life of Moses (Gregory of Nyssa 1964), and the other in a very eloquent commentary on the Song of Songs (Gregory of Nyssa 1994).

A core idea of Gregory's mystical concept is the permanent unknowability of God (Ruhbach and Sudbrack 1989). Ultimately, God is therefore beyond all human reach. Nevertheless, Gregory sees possibilities for man to approach the divine. The starting point is the anthropology of the Cappadocian (see among others Volp 2006).

For him, the interiority of the human being is understood as a mirror that must always be cleaned. Already in his early writing *On Virginity,* the Cappadocian emphasised the image of God in the human being. In other passages of his writings, too, he emphasises that the "His-like" relationship between God and human beings is a prerequisite for the knowledge of God. In the inner-human mirror, God appears, but only as symbol, as an image, as it were, because human language and rationality in their conceptual form (Figura 1987) are incapable of grasping God. Symbolic language, on the other hand, in its pictorality, offers inexhaustible depths of meaning, "which every time the conceptual language has outlined something too clearly, call for a new transcending of the contents of understanding" (Canévet 1984). Man must constantly stretch himself deeper into God in order to come closer to him. The image of reaching out comes from Phil 3:13f., where Paul describes the pursuit of the heavenly treasure. The idea of reaching out for the divine, for the heavenly jewel in Paul's metaphor that even Paul has not yet grasped, in any case is central for Gregory. In Greek, this is called *epektasis – straining for.*[1] It is therefore a constant process of approaching the divine, which is probably not attainable in this world at all.

Already in Gregory's early writings, he developed – probably following Origen – a concept of gradually approaching the divine through concrete steps of asceticism and purification. However, a *unio mystica* is not to be found in his work (Figura 1987). Gregory only knows the steps on the way to the divine. On the path of assimilation to God, man progresses infinitely. Even a vision of God is not envisaged by him (Figura 1987).

Gregory of Nyssa represents a link between the Alexandrians of the third century and later Byzantine mysticism. Pseudo-Dionysius Areopagites, my second

[1] On epektasis in Gregory of Nyssa and the background in Plotinian philosophy, see Alexopoulos (2006); also more fundamentally on the concept Ramelli (2018).

example, had already received him extensively in the fifth century, as did Maximos the Confessor later (Canévet 1984). In particular, the doctrine of the infinity and incomprehensibility of God was newly introduced by Gregory of Nyssa into the stream of tradition (Figura 1987). In his *Vita of Moses,* he also speaks of the darkness of God in this context. Literally it says there:

> The more the spirit progresses, the more it attains to the knowledge of existence through ever greater and more perfect attention, and the closer it comes to perception, the more it sees that the divine nature is invisible. For when he has left all appearance, not only what sensual perception grasps, but also what the spirit believes to see, he penetrates ever deeper into the inner being until, with great spiritual effort, he reaches the invisible, the incomprehensible, and there then sees God. For therein lies the real knowledge of the sought, the seeing in the not-seeing, that the sought transcends all knowledge, closed on all sides as through darkness by its incomprehensibility. (Gregory of Nyssa 1994)[2]

2 Dionysios Areopagites: "Mystical Ecclesiality" in the Context of Neoplatonic Thought

The author, who worked in the fifth century and used the pseudonym of Dionysius of the Areopagus, further systematised the thoughts of Gregory of Nyssa, among others, and consistently integrated them into a large world view influenced by Neoplatonism. In doing so, Dionysius in part explicitly draws on the writings of the Neoplatonist Proclus, who died in 485 AD (Ritter 2018; Butorac and Layne 2017: 161–288).

The titles of his writings make clear that the author was by no means only concerned with spiritual practice, but with an entire cosmology and ecclesiology: The *Heavenly Hierarchy*, *The Ecclesiastical Hierarchy*, *The Divine Names,* and *Mystical Theology.*

---

[2] In the original Greek, the text reads as follows: Τί (sic!)δὲ βούλεται τὸ ἐντὸς γενέσθαι τοῦ γνόφου τὸν Μωϋσέα καὶ οὕτως έν αὐτῷ τὸν θεὸν ἰδεῖν; ἐναντίον μὲν γὰρ δοκεῖ πως εἶναι τῇ πρώτῃ θεοφανείᾳ τό νῦν ἱστορούμενον- τότε γὰρ ἐν φωτί, νῦν δὲ ἐν γνόφῳ τὸ θεῖον ὁρᾶται. μηδὲ τοῦτο τοῦ είρμοῦ τῶν κατὰ ἀναγωγὴν ἡμῖν θεωρηθέντων ἀπᾴδειν νομίσωμεν. διδάσκει γὰρ διὰ τούτων ὁ λόγος ὅτι τῆς εὐσέβειας ἡ γνῶσις φῶς γίνεται παρὰ τὴν πρώτην οἷς ἂν έγγένηται. διότι τὸ ἐξ ἐναντίου τῇ εὐσεβείᾳ νοούμενον σκότος ἐστίν-ἡ δὲ ἀποστροφὴ τοῦ σκότους τῇ μετουσίᾳ τοῦ φωτὸς γίνεται. προϊὼν δὲ ὁ νοῦς καὶ διὰ μείζονος ἀεὶ καὶ τελειοτέρας προσοχῆς έν περινοίᾳ γινόμενος τῆς τῶν ὄντων κατανοήσεως, ὅσον προσεγγίζει μᾶλλον τῇ θεωρίᾳ τοσοῦτον πλεῖον ὁρᾷ τὸ τῆς θείας φύσεως ἀθεώρητον. καταλιπὼν γὰρ πᾶν τὸ φαινόμενον, οὐ μόνον ὅσα καταλαμβάνει ἡ αἴσθησις ἀλλὰ και ὅσα ἡ διάνοια δοκεῖ βλέπειν, ἀεὶ πρὸς τὸ ἐνδότερον ἵεται ἕως ἂν διαδυῇ καὶ τῇ πολυπραγμοσύνῃ τῆς διανοίας πρὸς τὸ ἀθέατόν τε καὶ ἀκατάληπτον κἀκεῖ τὸν θεὸν ἴδῃ. ἐν τούτῳ γὰρ ἡ ἀληθής ἐστιν εἴδησις τοῦ ζητούμενου καὶ έν τούτῳ τὸ ἰδεῖν έν τῷ μὴ ἰδεῖν, ὅτι ὑπέρκειται πάσης εἰδήσεως τὸ ζητουμένον οἷόν τινι γνόφῳ τῇ ἀκαταληψίᾳ πανταχόθεν διειλημμένον. διό φησι καὶ ὁ ὑψηλὸς ωάννης, ὁ ἐν τῷ λαμπρῷ γνόφῳ τούτῳ γενόμενος, ὅτι Θεὸν οὐδεὶς ἑώρακε πώποτε, οὐ μόνον τοῖς ἀνθρώποις ἀλλὰ καὶ πάσῃ τῇ νοητῇ φύσει τῆς θείας οὐσίας τὴν γνῶσιν ἀνέφικτον εἶναι τῇ ἀποφάσῃ ταύτῃ διοριζόμενος.

The starting point of his approach is the Neoplatonic opposition of the world of diversity, including the diversity of human concepts, and the "One" from which this world is ultimately emanated (Leppin 2004). According to Dionysius, the conceptually differentiating and thus diversity-creating way of human language is always insufficient for the "primal" to be grasped (Leppin 2004), because this ground of all existence, the truth itself, eludes differentiating naming. The approach to the primal happens first through spiritual contemplation (Leppin 2004). Dionysius, like Gregory of Nyssa, can speak of an ascent. Similar to Gregory, this involves ascending beyond the light into a state that Dionysius calls the "superluminal darkness" (Pseudo-Dionysius 1991). This metaphoric oxymoron expresses the fact that the reality of God cannot be reached by differentiating human concepts. God cannot be grasped even with the basic distinction between light and darkness. Therefore, language must be used in a paradoxical way when it is applied to God. By using the opposing terms "superluminal darkness", the opposites are at least linguistically cancelled out, as it were. However, this does not provide a term for God. Accordingly, mysticism in Pseudo-Dionysius now really leads to the *mysterion*, to the conceptually incomprehensible mystery. Mysteria, and here especially the ecclesiastical sacraments, make possible precisely what language and conceptuality are ultimately incapable of achieving: they abolish diversity (Leppin 2004).

Dionysius' approach takes the personal God of the Bible as a starting point for orientation, but he does not stop at this God of revelation. Rather, in his thinking, this image of God is ultimately to be transcended – for the God confessed in the Bible points to a divinity that is above all personalities.

Dionysius ultimately emphasises the so-called *via eminentiae*, the way of transcending earthly concepts, when approaching the divine. God can best be grasped in such a way that one transcends all concepts of him and thus speaks of God as the super-good, the super-divine, etc. (Pseudo-Dionysius 1990). However, this succeeds less in the language of a theological treatise than, for example, in the form of a liturgical hymn. The treatises of Dionysius therefore also have a kind of prayer-like character (Leppin 2004).

However, a contemplative movement back to the primal in a cataphatic way – that is by positively describing the divine, in contrast to the apophatic strategy of stating what God is not – cannot be completely successful either. Neither the *via positiva,* nor the via *analogiae*, nor the *via eminentiae* are able to overcome the conceptual differentiation that separates the sign or the representation from the unity of God. On the contrary, God is essentially and indescribably the One (Leppin 2004). Accordingly, he cannot be grasped with human conceptuality. Rather, in the sense of a further path, we must ultimately speak of God or the divine, the so-called *via negativa,* which is characteristic of Pseudo-Dionysius. It is described in more detail in *Mystical Theology*:

> Rising still higher, we say of it (the All-Cause) that it is neither soul nor spirit; neither is imagination, opinion, reason, or thought to be ascribed to it, nor is it to be equated with reason and thought, nor is it predicated, nor thought. It is neither number nor order, neither size nor smallness, neither equality nor inequality, neither similarity nor dissimilarity. It has

> neither a firm stand, nor does it move, nor does it rest ... It eludes all (essential) determination, naming and knowledge. (Pseudo-Dionysius 1991)[3]

Accordingly, the only thing that can be said about God is what he is not. The *via negativa* is not content with describing God as dark instead of light. Rather, speaking of God as "superluminal light" or even "non-light" rejects the concept of light as such for God, thus ultimately renouncing any form of conceptuality. Accordingly, Dionysius challenges Paul's disciple Timotheos (Leppin 2004):

> [...] stretch yourself in a non-knowledgeable way, as far as it is at all possible, to unity with the One who transcends all being and knowledge. For only when you express yourself unconditionally and unrestrictedly of yourself as well as of all things, will you be carried up in purity to the over-eggifying ray of the divine dark, letting go of everything and detached from everything. (Pseudo-Dionysius 1991)[4]

It is precisely this complete being-out-of-self that Dionysius, like Plotinus before him, calls ecstasy or an ecstatic process (Pseudo-Dionysius 1990). In ecstasy, man and his consciousness are at least momentarily suspended in their primordial ground. In this state Dionysius sees a form of mystery that transcends consciousness, which is quite similar to the *mysteria*, the sacraments of the Church. In this respect, the Platonist Dionysius can also be considered an ecclesiastical theologian, who systematically introduced a way of transcending consciousness into Christianity and at the same time rationally justified it.

Comment by the Editor Rev. Prof. Prem Saran Satsangi
Perfect Theology embracing e.g. Topological Graph Theoretic Instrumentation Process of Across& True/Flow Measurements between each Pair of Vertices/Nodes of Infinitesimally Small ("Digitized/Discretized") Line Segments (representing Continuum).

References

Alexopoulos, Theodorus. 2006. Das unendliche Sichausstrecken (Epektasis) zum Guten bei Gregor von Nyssa und Plotin. Eine vergleichende Untersuchung. *Zeitschrift für Antikes Christentum* 10: 302–312.

Butorac, David D., and Dannielle A. Layne, eds. 2017. *Proclus and his legacy*. Berlin: de Gruyter.

Canévet, Mariette. 1984. Gregor von Nyssa. In *Große Mystiker. Leben und Wirken*, ed. Gerhard Ruhbach and Josef Subbrack, 17–35. München, Beck.

---

[3] The original Greek is: Αὖθις δὲ ἀνιόντες, ὡς οὔτε ψυχή ἐστιν οὔτε νοῦς, οὔτε φαντασίαν ἢ δόξαν ἢ λόγον ἢ νόησιν ἔχει- οὐδὲ λόγος ἐστὶν οὔτε νόησις, οὔτε λέγεται οὔτε νοεῖται- οὔτε ἀριθμός ἐστιν οὔτε τάξις, οὔτε μέγεθος οὔτε σμικρότης, οὔτε ἰσότης οὔτε ἀνισότης, οὔτε ὁμοιότης ἢ ἀνομοιότης-οὔτε ἕστηκεν οὔτε κινεῖται οὔτε ἡσυχίαν ἄγει- [...] οὔτε λόγος αὐτῆς ἐστιν οὔτε ὄνομα οὔτε γνῶσις-

[4] The original Greek is: καὶ πρὸς τὴν ἕνωσιν, ὡς ἐφικτόν, ἀγνώστως ἀνατάθητι τοῦ ὑπὲρ πᾶσαν οὐσίαν καὶ γνῶσιν- τῇ γὰρ ἑαυτοῦ καὶ πάντων ἀσχέτῳ καὶ ἀπολύτῳ καθαρῶς ἐκστάσει πρὸς τὸν ὑπερούσιον τοῦ θείου σκότους ἀκτῖνα, πάντα ἀφελὼν καὶ ἐκ πάντων ἀπολυθείς, ἀναχθήσῃ.

Figura, Michael. 1987. Mystische Gotteserkenntnis bei Gregor von Nyssa. In *Grundfragen christlicher Mystik: Wissenschaftliche Studientagung Theologia Mystica in Weingarten vom 7.-10.11.1985*, ed. Margot Schmidt, 25–38. Stuttgart-Bad Cannstatt: Frommann-Holzboog.

Gregory of Nyssa/Gregorii Nysseni. [4th Century CE] 1964 *De vita Moysis*... In *Gregorii Nysseni Opera. Vol. 7. Pars 1,* ed. Werner Jaeger, Hermann Langerbeck, and Herbert Musurillo, 11–87. Leiden: Brill.

Gregory of Nyssa/Gregorius Nyssenus. [4th Century CE] 1994. *In Canticum Canticorum Homiliae,* vol. 1–3, ed. Franz Dünzl. Freiburg im Breisgau i.a.: Herder.

Leppin, Volker. 2004. *Die Christliche Mystik*. München: Beck.

Müller, Andreas. 2020. Mystik in der Alten Kirche. In *Christliche und jüdische Mystik. Gemeinsamkeiten und Unterschiede*, ed. Christoph Markschies, 73–103. Leipzig: Evangelische Verlagsanstalt.

Pseudo-Dionysius. 1990. De divinis nominibus II,3. In *Corpus Dionysiacum I*, ed. Beate Regina Suchla. Berlin: de Gruyter.

———. 1991. Mystische Theologie I,1. In *Corpus Dionysiacum II. De coelesti Hierarchia i.a*, ed. Günter Heil and Adolf Martin Ritter. Berlin: De Gruyter.

Ramelli, Ilaria. 2018. Apokatastasis and Epektasis in Cant and Origen. In 2018. Gregory of Nyssa. In *Canticum Canticorum. Analytical and supporting studies*. Proceedings of the 13th International Colloquium on Gregory of Nyssa (Rome, 17–20 September 2014), ed. Giulio Maspero, Miguel Brugarolas, Ilaria Vigorelli, and Miguel Brugaroals Brufau, 312–339. Leiden/Boston: Brill.

Ritter, Adolf Martin. 2018. Neoplatonismus und Christentum am Beispiel des Verständnisses von Gebet und Gottesverehrung bei Proklos und Dionys. In *Dionys vom Areopag. Beiträge zu Werk und Wirkung eines philosophierenden Christen der Spätantike*, ed. Adolf Martin Ritter, 123–148. Tübingen: Mohr.

Ruhbach, Gerhard, and Josef Sudbrack, eds. 1989. *Christliche Mystik. Texte aus zwei Jahrtausenden*. München: Beck.

Volp, Ulrich. 2006. Die Würde des Menschen. A contribution to anthropology in the Early Church. *VigChr.* S 81, 172–182. Leiden/Boston: Brill.

Exploring Alternative Realities

Mark Juergensmeyer

Abstract The idea of alternative realties is fundamental to science, especially contemporary physics that suggests that reality is far different than the time-space world as we know it. Alternative realities are also the stuff of religion from ancient times to the present, countering our ordinary common-sense understanding of the world around us. Whether such views of alternative reality are threatening or ennobling depends in large part on the critical judgment that is brought to bear on them. As the current spate of covid misinformation shows us, pseudo-science can be as misleading as devious religion. Yet there is a promise of the possibility of a symbiotic relationship between progressive religion and critical science, rather than a competitive one, that in the future will continue to ennoble both.

Keywords Science · Quantum physics · Religion · Alternative reality · Robert Bellah · Dayalbagh

In the days prior to the pandemic I faced a long drive—almost four hours—through the concrete maze of Southern California freeways to get from my home near Palm Springs to Santa Barbara, where I taught three days a week. To make the voyage tolerable I would often listen to books on tape.

One title intrigued me: *Reality Is Not What It Seems* (Rovelli et al. 2017). It was a book about quantum gravity theory by the Italian physicist, Carlo Rovelli, and it turned out to be one of the most engrossing books that I have encountered. You might ask what a sociologist of religion like me would know about quantum gravity theory. The answer is "nothing." But that was the point. I wanted to find out what the new physics was all about.

Rovelli is an artful writer, but he admits at the outset that he is faced with a daunting task: trying to get someone like me, accustomed to our ordinary reality, to step

M. Juergensmeyer (✉)
Department of Global Studies, University of California, Santa Barbara, CA, USA
e-mail: juergens@global.ucsb.edu

P. S. Satsangi et al. (eds.), *Consciousness Studies in Sciences and Humanities: Eastern and Western Perspectives*, Studies in Neuroscience, Consciousness and Spirituality 8, https://doi.org/10.1007/978-3-031-13920-8_16

inside a different view of the world, one informed by the mathematics and theories of quantum physics. "If we try to put together what we have learned about the physical world in the course of the twentieth century," Rovelli writes, "the clues point toward something profoundly different from what we were taught at school." This elementary structure of the world that is emerging, according to Rovelli, is "generated by a swarm of quantum events, where time and space do not exist" (Rovelli et al. 2017: Introduction).

That was the point in the audio disk when I went back to listen again to what he said. Our ordinary world is so basically defined by time and space that it is hardly conceivable to imagine a world in which they are not fundamental constituents of the way we perceive reality. In Rovelli's world of quantum gravity, however, "reality is a network of granular events; the dynamic that connects them is probabilistic; between one event and another, space, time, matter, and energy melt into a cloud of probability" (Rovelli et al. 2017: Introduction).

Rovelli admitted that for most of us in our daily lives we are not burdened with this knowledge, and we live as if time and space actually existed. We are content to live in a world of Newtonian physics. But now that our eyes have been opened to the possibilities of different ways of seeing the world, it is hard to go back to ordinary reality without the disconcerting feeling that we are not always seeing the entire picture.

The character of Neo played by Keanu Reeves in the movie, *The Matrix,* has a similar feeling when he takes the pill of recognition that allows him to see that the daily reality in which he has been living is part of a computerized matrix in which humans have been exploited. The unsettling alternative reality that has been revealed to him is not easy to accept, but eventually he gives in to the notion that it is the more true and certain of the two realities between which he is trying to negotiate.

Long before the Matrix movie, there was ancient Greek thought, which trafficked on a similar theme. Rovelli in his book on quantum gravity cites Plato's analogy of the cave. Most people in the cave see shadows on the wall cast from the fireside and think that is reality. Only the most daring peek outside in the blinding sunlight and are aware of how limited is the cave dwellers' view of the world.

Plato, of course, was writing over two thousand years ago, and though the Matrix movie is more recent, these ways of thinking indicate that the notion of alternative reality is something that has been lurking in the shadows of human imagination for some time. In the sciences it is a reminder that scientific truth has always been a jarring discovery of alternative reality, something that upsets our ordinary view of the world.

Quantum gravity is simply the latest in a whole line of scientific observations that were initially thought implausible if not impossible. When in the sixteenth century Copernicus advanced his heliocentric theory that the earth revolved around the sun and not the other way around, the hierarchy of the Roman Catholic Church bitterly rejected it. Such a theory would suggest that humans played a lesser role in the cosmos that they were willing to admit. Today, of course, Copernicus's view is accepted as a matter of fact.

1 The Alternative Realities of Religion

One of the reasons why religious authorities were so upset with Copernicus is that it challenged an alternative reality of their own, the spiritual realm of religion. The notion of religion as alternative reality has been given currency in recent years by the writings of Robert Bellah, one of the most prominent scholars in my field, the comparative sociology of religion. At Harvard, Bellah was a student and later a colleague of Talcott Parsons, the leading sociologist of religion of his generation and an exponent of the classical sociological theories of Max Weber and Émile Durkheim. Bellah continued this tradition when he shifted from Harvard to Berkeley, where I was his colleague in the fifteen years I coordinated the religious studies program there. Bellah was the chair of my advisory committee, as well as a mentor to me in the social sciences. By that time Bellah had regarded himself as an interpretive sociologist, in order to distinguish himself from his more quantitatively driven colleagues.

Bellah entered the debate in recent years among scholars of religion who question the notion that religion is a thing that has an essential character aside from its cultural aspects. The Harvard historian of religion, Wilfred Cantwell Smith, wrote an engaging book on *The Meaning and End of Religion,* in which he advocated a scholarly moratorium on the use of the term, preferring to use "religious" as an adjective rather than as a noun (Smith 1962). Smith claimed that the term religion was a novelty in the English language and had no inherent analytical utility. More recently, Talal Assad has taken up Smith's charge but placed the creation of the idea of religion in historical and social contexts (Talal 2001).

Although Bellah respected both Smith and Asad, he asserted that it was in fact possible to talk about religion as something in and of itself. Bellah accepted the notion that religion was a product of the human imagination, but that was precisely what interested him. It was not just a matter of religious this or that, Bellah reasoned; the term signified a different way of looking at the world. It was an alternative view of reality.

To search for answers to the question of where and how this perception arose and what role has it played in humanity, Bellah embarked on a lengthy scholarly journey. He had retired from teaching at this point in his career, so he had the luxury of taking the whole sweep of history into consideration in trying to understand where religion had come from and how it had changed over time. I recall that when I met with Bellah during that time he was working on the book, he was enthralled with what knowledge he had acquired about ancient India and China and about astrophysical theories regarding early forms of life.

Bellah's project was published as *Religion in Human Evolution* in 2011 (Bellah 2011). It is a huge book, as impressive in its scope as it is rich in detail and insight. In it he takes the long view, beginning 13.8 billion years ago with the Big Bang and the creation of stars and planets, including our own, and then the emergence of living cells in the primal ooze, and the beginning of animate life forms. He ends the book at the Axial Age, the rise of new modes of conceptual activity in the sixth century BCE, a period when intellectualism was sprouting around the world, from Greek thought to philosophical developments at the end of India's Vedic period.

It is in this grand historical narrative that he addresses the idea of what religion is and relates it to the development of living species. Early life forms, Bellah suggests, are focused on material things, survival, and procreation. But later in the evolutionary process more evolved life forms have the leisure of spare time. Freed from the necessities of existence they can do whatever they want. And what they often do is unstructured and arbitrarily structured activity, doing things for no apparent purpose. They are like school children finally released from their boring classrooms for a few precious moments for recess. What they do during recess time is to run around and have fun and explore the world. It is something that we call "play."

Following the lead of the Dutch historian, Johan Huizinga, Bellah affirms that play is the beginning of all forms of culture, including religion (Huizinga [1938] 1949). It is the ability for humans to be creative, to roam and discover. Initially it is primarily an activity. This is true of religion as well. The early forms of religiosity—such as the rituals described in Leviticus and the rites detailed in the Vedas of ancient India—are focused on activity, on what priests do to interact with God or the gods. It is only later, in the Axial Age of the sixth century BCE, that religion becomes more introspective and cerebral, and this is when we can describe religion as a product not just of creative activity but of creative thought: the religious imagination.

One illustration that Bellah gives of the process by which activity related to religion becomes conceptualized is his description of the development of the Greek idea of *theoria*. Before Plato, this referred to a practice in which an emissary of one Greek state would go to another state to observe their religious festivals and come back and give a report on what they saw. *Theoria* was a report on a different kind of religion. Plato took this concept and related it to the intellectual adventure of going out to search, not just for religious festivals but for truth. The classic example that Plato gives is the analogy of the cave, in which most of us venturing into the cave see shadows cast on the wall and think that is what is truly real. If we turned around to see the source of those shadows, the bright light behind the objects that were casting the shadow would be so intense that most of us would flinch and turn back to the shadows. Only the bold would go where other people dare not go, in search of the real objects, the truth. These are the *theoria* that Plato admires, the searches for truth. Later Aristotle would refine this further in a way that we all know, in which the idea of theory is related to the identification of truthful concepts (Bellah 2011: 577).

Hence religion in Bellah's understanding is something, or rather some perception. It is an imagined world of being, "a general order of existence," as the anthropologist Clifford Geertz describes it. Bellah goes further in labeling it "religious reality," one of various multiple realities that "calls the world of daily life into question" (Bellah 2011: 5). Here Bellah is relying on a whole school of sociology associated with Austrian philosopher, Alfred Schütz, regarding the notion that reality is socially constructed, and before him the American philosopher, William James, who thought about cultural forms as constructions of the social imagination (Schütz 1967). According to this point of view, made popular by the book, *The Social*

Construction of Reality by Peter Berger and Thomas Luckmann, what we perceive as everyday reality is a social construction of what things are and what they mean (Berger and Luckmann 1966). A wooden table, to most humans, is a place to put books and plates of food, but to a termite it is an edible feast. It all depends on your point of view. What Bellah adds to this conversation—aided by the thinking of the pioneering French sociologist, Émile Durkheim—is the insistence that religious perceptions are one of these constructions of reality (Durkheim 1995). The table might be, for instance, an altar in a religious reality. These religious realities are among the various multiple realities that most people navigate among every day. These multiple realities are often overlapping views and sometimes contesting ones, but they can present levels of meaning and reality that are quite different from one another even though they relate to the same thing, just as we and termites see tables differently though the table remains the same.

The religious visions of reality are quite remarkable views of the world, and some—like the rapture theology of Evangelical Protestant Christianity—are quite apocalyptic. Others may not be as extreme. But the basic dichotomy between the ordinary world and the alternative reality of religion is maintained, even if the imagined religious reality is simply a subtle awareness of a deeper stratum of meaning in the world. It may just be the certainty that there is an ultimate reality, as the Protestant theologian Paul Tillich has described it. For many, however, and an increasingly large number of Evangelical Protestants in the United States, it is a vibrant and startling reality that could break into the normal world at any moment.

To accept the alternative reality of religion is to peer through the looking glass and see that there is a Wonderland behind the ordinary world, and it is only a few steps away. Like Wonderland it both mirrors ordinary reality and alters it. It provides an alternative source of authority and identity, and it can provide a sense of community, a kinship with the fellow faithful that can be more binding than any other form of social or political allegiance. The alternative reality of religion is a powerful construct, one that is the source of motivation and meaning that can alter people's lives.

2 Can Alternative Realities Be Trusted?

To those outside the realm of faith, the question that arises is whether it is true. To answer that question, however, requires an exploration of the idea of truth and what it means for any perception to be true.

One way of answering this is to respond that the criterion of truthfulness resides in the scientific sense of what is true. But therein lies a further dilemma, since the history of science is the creation of differing paradigms of truthfulness that are challenged and supplanted by other views.

In science, truth is an elusive term, and most scientists prefer to speak of verifiable theories. What makes a scientific theory valid (and in the eyes of many of us, "true") is that it covers everything within its purview, leaving no room for anomalies, and that it is verifiable by experience and experiments, at least as much as possible. I say that it is expected to be verifiable, but in many cases a theory that makes sense mathematically and that seems to cover all the bases cannot be verified immediately. Albert Einstein's General Theory of Relativity, for example, which was first proposed in 1915, received a welcomed verification in 2021 when his idea of how mass can curve the web of space-time was observed in mapping the gravitational field of a distant galaxy (Gohd 2018). Stephen Hawking's Area Theory, positing that the area of black holes in space does not decrease, was not verifiably observed until 2021 with evidence based on the gravitational waves created by the collision of two black holes in distance space (Chu 2021).

The interesting thing about such scientific theories—the closest thing we have to truth in science—is that they can change over time. The observable facts do not change, so perhaps they can be regarded as true, but we keep acquiring new facts, some of which do not seem to fit with the old facts, such as Einstein's evidence that time, space, and mass are convertible entities. What changes dramatically is how we interpret these facts and what we make of them. Einstein's scientific revolution made Newtonian physics passe, though it is still reliable for the proximate reality in which most of us live.

An interesting book that describes the history of these scientific revolutions is Thomas Kuhn's *Structure of Scientific Revolutions* (Kuhn 1962). According to Kuhn, scientific theories are paradigms of understanding that can change with newly found facts and with new ways of interpreting existing facts. Kuhn shows that often these new paradigms are rejected out of hand by the establishment, not just the clerical and political establishment but often in many cases by the scientific establishment as well. The old guard sees its life work belittled by the new frameworks of understanding and is threatened by them.

Kuhn's book is about the physical sciences, but when I was a graduate student in the social sciences at Berkeley in the 1960s when the book first appeared, it was required reading. Everyone, it seemed, was talking about it. We graduate students immediately made the connection between paradigms of scientific analysis in the physical sciences to the theories propounded by our elders in the social sciences. We were also yearning for new paradigms of analysis and were aware how bitterly our seniors would want to retain the theories that had made them famous within their fields.

People often approach religion the same way—skeptically, with an openness to religious approaches that seem to be comprehensive in scope and verifiable by their own experience. I don't mean this in a scientific sense, of course, since religion is not science. It is clouded in metaphorical language that often masks its role in providing a comprehensive worldview of morality and meaning that lies behind everyday life. It is not meant to be competitive with science, and it never is.

Still there are some interesting similarities between the alternatives realities as hypothesized by science and those proposed by religion. For one thing, there are competing theories in both science and religion about the right way—the most truthful way—of understanding the facts that are commonly accepted within each circle. Although most religious persons will adamantly assert that the fundamental elements of religion do not change over time, their way of thinking about and interpreting them certainly does.

Let us take Christianity, for example. The central fact of the religious tradition, for most of the faithful, is that Jesus Christ in some way bridges the divine realm from the everyday world. The way that bridging is done, however, is the subject of a great deal of differing interpretations.

These Christian differences are already apparent in the Bible and begin with the way the birth of Jesus is described in the four Gospel books of the New Testament. These four books were selected by a gathering of church elders, the Council of Carthage, in 387 CE, after considerable discussion. Some writings that were contenders for inclusion, such as the Gospel of Thomas and the Gospel of Judas, were considered too esoteric or heretical and ended up in the dustbin of history. The four that remained ranged from Mark, probably written about 30 years after the death of Jesus, to Mathew and Luke, written later in the first century, to John, which may have been written as late as 120 CE.

The popular nativity scene of the birth of Jesus, with Mary and Joseph and the baby in a manger with the wise men and shepherds looking on, appears in none of the gospels, at least not in just that way. The oldest gospel, Mark, does not include any birth story of Jesus but begins with his prophetic teachings. Mathew begins his gospel with an extensive patrilineal genealogy, beginning with Israel's great King David, and then his sons, and his sons' sons, and so on down to Jesus, thereby proving that he was born in that regal lineage. Mathew does describe Jesus as being born by virgin birth in a humble manger, but he pictures Jesus as being visited by kingly wise men from the East. Luke, whose gospel favors the poor and outcast, sees Jesus as forecasting a new community of equality, and in his nativity scene, Jesus's birth is visited not by regal wise men but by humble shepherds from nearby fields where they were keeping watch in the night. John, like Mark, has no birth narrative, but he does have an origin image. In the gospel of John it is the gnostic concept of the logos, the ethereal Word, that has existed before time and space that embodies the divine and zaps to earth in human form.

So which way of thinking about Jesus is correct? Was he a prophet, a divine king, a creator of a new community, or an ethereal spirit? By inclusion of all of the gospels the early Christian church affirmed that they were all correct. Outside the Christian community there were other views of Jesus, of course. Clearly the authorities at the time saw him as a zealot, a revolutionary who was trying to wage a Jewish insurrection against Roman rule; others likely saw him as a prophetic holy man of the type that were common during those years. The historian Josephus notes the existence of a wise man and popular teacher called Jesus, so there is independent verification of his existence. Everything else about him is speculation or, rather, subject to interpretation and one's perspective.

Behind most religious speculation is a conviction similar to that behind most scientific inquiry: that there is a truth, and it can be known. Perhaps it cannot be fully known, but one can gain a proximate understanding. Mohandas Gandhi, who appreciated all religious traditions, took the Christian phrase "God is love" and altered it, saying that "God is truth." He also inverted it, saying that in a similar vein,

"Truth is God." In making that statement he meant to include all forms of truth, whether they referred to the physical world, the social sphere, or to moral and spiritual characteristics. Gandhi regarded all forms of truth as ultimately absolute but also ultimately unknowable by human frames of reference. All we can hope for is the proximate truth that is gained by intelligence, experience, and experimentation. Gandhi titled his own autobiography as *The Story of My Experiments with Truth* (Gandhi 1927).

It is fair to conclude that the truth-searching in religion and science follows a similar pattern. Again, I am not saying that they are the same, and it would be a mistake to think of religion as a kind of science or of science as a kind of religion. All I am saying is that they approach the search for truth in a similar fashion. The criteria for measuring good science and good religion are similar: a comprehensive understanding verified by experience and open to critical assessment.

3 The Symbiosis Between the Alternative Realities of Science and Religion

Having noted that both religion and science point to alternative realities and that there are similarities in the way that both approach an awareness of these realities, the question arises whether there might be possibilities for symbiotic interaction between the two. The history of sour relations between scientists and religious authorities would seem to quickly give a negative response to that question. The Italian astronomer, Galileo, suffered from a church trial and ostracism for his defense of Copernicus's view that the earth revolved around the sun rather than the other way around. The pope at the time took the position that the earth, and by implication the humans that populated it, should be at the center of the world and not the periphery.

In more contemporary times, Evangelical Protestants have rejected the notion that the physical world and its animate population have evolved over hundreds of thousands of years. They prefer to take literally the description in Genesis, the first book of the Bible, that describes the creation as a feat by God in seven days. Though this position could be ignored as a bit of unscientific religious dogmatism, it has become a political football in the insistence of Evangelical Protestants that Biblical creation stories should be taught in schools alongside the scholarly accepted theories of evolution.

The COVID pandemic of recent years has been an arena for further religious rejection of science. While most of the world applauded the rapid development of effective vaccines against infection from the COVID-19 strain of SARS viral disease, an anti-vax movement gained momentum in the United States and elsewhere around the world. In addition to suspicions about whether the newly developed vaccines were reliable, there were religious hostilities to science that emerged in the rejection of the vaccines. "God will protect us" was the common refrain of anti-vax protestors who seemed to think that religious faith was a sufficient shield against the virus. At the same time, most established religious organizations embraced the vaccines and urged the faithful to protect themselves with them.

It seems that many religious figures are stymied regarding the claims of science, not knowing whether to regard them as threatening or to embrace them. In nineteenth-century America, Mary Baker Eddy proposed that religion could absorb and replace science, a position that she held in her book *Science and Health with Key to the Scriptures* ([1875] 2012), which launched a whole new Protestant Christian denomination, Christian Science. Adherents of the faith hold that sickness is not a physical malady but a spiritual one that can be resolved through the healing mediation of Christian Science Practitioners. Needless to say, most people outside this denominational fold do not concur and regard these beliefs to be a rejection of science, not an integration of it into the religious worldview.

So the arena of cooperation between science and religion seems to be fraught with tension. Still there are a great number of religious representatives who have tried to take science seriously as a complement to the religious searches for truth. They aver a great respect for scientific endeavors and claim that religion and science can be compatible if seen as simultaneous searches for truth, though on different paths.

Some years ago, a physicist, Frijof Capra, wrote a best-selling book, *The Tao of Physics,* which asserted that many of the theories of quantum physics verified the insights of ancient Buddhism (Capra 1975). The book gained a certain amount of fame in some quarters, notoriety in others, for its suggestion that religious insights were not only compatible with science but actually were confirmed by scientific discoveries. For people who were interested in Asian mysticism but were also firmly committed to modern science, the book seemed to buy into the idea that the two were related.

What Capra found intriguing was the mysterious properties of quantum physics that seemed to change with the way that it was viewed. As the physicist Erwin Schrodinger noted, once quantum systems are observed, the rules governing them no longer seem to apply. The way one looked them altered their reality, or at least seemed to do so. The Copenhagen interpretation developed by some atomic physicists noted that quantum mechanics was epistemic rather than ontological. That meant that it was more concerned with understanding how the world worked than determining a true perspective. This understanding seemed to open the door for classic notions of Buddhism that also aver that there is no objective reality but only our perceptions of it. Capra eagerly asserted that this meant that Buddhism and quantum physics were interrelated.

More recently another physicist, Adam Frank, who also happens to be a practicing Buddhist, revisited Capra's thesis and found it wanting (Frank 2021). As much as he was committed to science as an astrophysicist at the University of Rochester, and as much as he was a practicing Buddhist, he found that neither Buddhism nor physics really needed the other for its verification. Yet he thought that a conversation between the two was possible. In fact, he thought that a dialogue between religious philosophy (in his case Buddhist) and the natural sciences (in this case quantum mechanics) could be profitable for both sides.

A similar position has been taken by the Radhasoami religious community in India. One of the masters of the Beas branch of the tradition, Jagat Singh, titled his book of discourses *The Science of the Soul* (1972). To some extent, Jagat Singh's use of the term "science" was a kind of metaphor for the spiritual discipline required of the tradition and its verification through experience.

The Dayalbagh branch of the Radhasoami faith has taken the engagement with science to a whole new level. The community has been committed to scientific studies from the early years of the establishment of the Dayalbagh center near the city of Agra. The founding master, Anand Swarup, created an educational institute in 1917, and by 1973 the Dayalbagh Educational Institute had sufficiently developed in prestige and reputation to receive university status by the Government of India (the institute became a "deemed university"). Its curriculum aimed at providing for its students a well-rounded education, but central to it was a dedication to the sciences.

Several of the heads of the Institute have been scientists or engineers, including the current director, Prem Kumar Kalra, and the former head of the Institute, Dr. Prem Saran Satsangi, who later became anointed as the spiritual master of the community. Dr. Satsangi was trained at Michigan State and Waterloo University in Canada in electrical engineering and systems analysis and has devoted much of his efforts as leader of the community to exploring the relationship between spirituality and scientific endeavors.

Since 2012 Dr. Satsangi has supported the science of consciousness studies, relating spiritual states of consciousness to altered states understood through the approaches of quantum mechanics and other scientific analyses. In 2019 he launched the first of a series of conferences devoted to the scientific study of consciousness. In these forums, scientists and practitioners have been brought together to work on approaches to understanding the nature of consciousness and how it can be transformed. It is an interesting development in the most recent attempts to find arenas in which perspectives drawn from science and those from religion can relate and learn from one another.

As I tried to make clear in this article I think that science and religion can profit from engaged interaction. In saying this I do not mean to imply that they are the same kind of activity. I do not think that science can prove religion, nor religion prove science. But I also do not think that they are inherently competitive. This was the great mistake of many religious thinkers from the nineteenth century to the present, who tried to put religion on a scientific basis, and this was the problem with scientific hacks, who think that the very existence of the scientific method disproves the validity of religion. These comparisons do not work any more than if one were to compare music and medicine and assert that one disproves the other. They are different arenas of activity.

That being said, it seems to me that since both science and religion involve the projection of the mind into different realms of hidden reality, they have a symbiosis that makes it possible for them to learn from one another. Though they are different in subject and approach, religion and science both point to alternative realities that challenge our everyday view of the world. They can be mutually exclusive, though they need not be hostile partners; in some interesting ways they are capable of exploring areas of mutual interaction and understanding.

References

Bellah, Robert. 2011. *Religion in Human Evolution: From the Paleolithic to the Axial Age*. Cambridge: Belknap Press of Harvard University Press.

Berger, Peter L., and Thomas Luckmann. 1966. *The Social Construction of Reality. A Treatise in the Sociology of Knowledge*. New York: Anchor Books. A Division of Random House.

Capra, Fritjof. 1975. *The Tao of Physics*. Berkeley: Shambhala Publications.

Chu, Jennifer. 2021. Physicists observationally confirm Hawking's black hole theorem for the first time. *MIT News*. https://news.mit.edu/2021/hawkings-black-hole-theorem-confirm-0701. Accessed 15 June 2022.

Durkheim, Émile. 1995. *The Elementary Forms of Religious Life*. New York: Free Press.

Eddy, Mary Baker. [1875] 2012. *Science and Health with Key to the Scriptures*. SMK Books. (e-book).

Frank, Adam. 2021. Does quantum mechanics favor Buddhist philosophy? *Big Think*. https://bigthink.com/13-8/quantum-mechanics-buddhism/?utm_term=Autofeed&utm_medium=Social&utm_source=Facebook#Echobox=1640472492. Accessed 10 June 2022.

Gandhi, Mohandas K. 1927. *Satya Na Prayogo athva Atmakatha* [Experiments of Truth or Autobiography] in Gujarati. First English version published in two volumes, *An Autobiography: Experiments with Truth*. Ahmedabad: Navajivan Press, 1927 and 1929. Originally published in Gujarati as a series of articles in Gandhi's journal, *Navjivan*, in weekly installments from 1925 to 1929, with English versions appearing at the same time in his companion journal, *Young India*.

Ghod, Chelsea. 2018. Einstein was right! Scientists confirm general theory works with distant galaxy. *Space.com*. https://www.space.com/40958-einstein-general-relativity-test-distant-galaxy.html. Accessed 2 June 2022.

Huizinga, Johan. [1938] 1949. *Homo Ludens: A Study of the Play-Element in Culture*. London: Routledge. (English translation). First published in Dutch as *Homo Ludens: Proeve Ener Bepaling Van Het Spelelement Der Cultuur*. Groningen: Wolters-Noordhoff, 1938. German translation, *Homo Ludens: Versuch einter Bestimmung des spielelementes der Kultur*, was published in Basel, Switzerland by Pantheon Academisch Verlagsanstalt in 1944.

James, William. [1902] 1985. *The Varieties of Religious Experience*. London: Penguin Classics.

Kuhn, Thomas S. 1962. *The Structure of Scientific Revolutions*. Chicago: University of Chicago Press.

Rovelli, Carlo, Simon Carnell, and Erica Segre. 2017. *Reality Is Not What It Seems: The Journey to Quantum Gravity*. New York: Riverhead Books.

Schütz, Alfred. 1967. *Phenomenology of the Social World*. Evanston: Northwestern University Press.

Singh, Jagat. 1972. *The Science of the Soul: Consisting of Discourses and Excerpts from Letters*. Beas: Radhasoami Satsang Beas.

Smith, Wilfred Cantwell. 1962. *The Meaning and End of Religion: A New Approach*. New York: Charles Scribner's Sons.

Talal, Asad. 1993. *Genealogies of Religion: Discipline and Reasons of Power in Christianity and Islam*. Baltimore: Johns Hopkins University Press.

———. 2001. Reading a modern classic: Wilfred Cantwell Smith's "*The Meaning and End of Religion*". *History of Religions* 40 (3): 205–222.

Reflections on the Rôle of Time in (Astro-) Physics

Wolfgang J. Duschl

Abstract As inhabitants of the universe, understanding it is an integral part of our self-awareness and thus of our consciousness as parts of a greater whole. An important aspect in this context is the nature, the manifestation, and the experience of time. In this contribution, I will present some reflections on properties of time on scales of the universe. These properties go far beyond what we are used to in our everyday life. I will address how astrophysical processes define a unique arrow of time. And I will describe how astrophysics uses the immensity of the geometrical and temporal size of the universe, combined with a finite but universally identical maximum speed of communication to verify simulations of the evolution of cosmic objects.

Keywords Astrophysics · Stellar evolution · Newtonian time · Relativistic time · Quantum gravitation

1 Introduction: Consciousness and Time[1]

Manning et al. state that "the ability of travelling in mental time, is thought to be at the heart of consciousness and, by the same token, at the center of human cognition" (Manning et al. 2013: 232). In this contribution, I will present some reflections on properties of time, which go far beyond what we are used to in our everyday life. I will first describe the current understanding of time in Newtonian, relativistic, and quantum physics. Then, I will turn to astrophysical and cosmological aspects. I will

[1] This contribution is mainly addressed to non-physicists. If necessary, I renounce here and there the exactness in the interest of comprehensibility, but not so the correctness.

W. J. Duschl (✉)
Astrophysik Kiel, Kiel University, Kiel, Germany
e-mail: wjd@astrophysik.uni-kiel.de

P. S. Satsangi et al. (eds.), *Consciousness Studies in Sciences and Humanities: Eastern and Western Perspectives*, Studies in Neuroscience, Consciousness and Spirituality 8, https://doi.org/10.1007/978-3-031-13920-8_17

address how astrophysical processes define an arrow of time, but not at all a good clock. And finally, I will describe how astrophysics uses the immensity of the geometrical and temporal size of the universe, combined with a finite but universally identical maximum speed of communication to verify simulations of the evolution of cosmic objects.

2 The Nature of Time

The nature of time is a still ongoing debate. In different models of the physical world, or rather models of different aspects of it, the understanding of time differs fundamentally.

2.1 Time in Newtonian Physics

In many respects, Newtonian time is the time we are used to in everyday life. It defines a global and absolute sequence of events, and it is strictly ordered: Past, present, and future are the same everywhere. Newtonian time does not directly and/or explicitly couple to space. Such a universal time can, at least in principle, be measured by clocks that show the same time everywhere. It is a different question and beyond the scope of this contribution, what makes a good clock to measure Newtonian time.

2.2 Time in Relativistic Physics

One of Einstein's fundamental postulates is the universal and finite value of the speed of light, just a little below 300,000 km s^{-1}, and its limiting nature in the sense that no information can propagate at higher speeds. In this respect, it is also a maximum speed for any type of communication.

As for time, this requires to abandon the concept of an absolute and universal time. Events are no longer globally ordered. For every observer, there exists an absolute past and an absolute future, but this is different for different observers. The present becomes a local phenomenon. Space and time are no longer decoupled. It is, however, not true that time and space are indistinguishable. Time stays a separate quantity and quality but is no longer independent of space, and vice versa.

This plays an important role in particular when velocities not too different from that of light and/or strong gravity are involved. The first instance has led Einstein to his Special Theory of Relativity and the second to the General Theory of Relativity. Both theories are among the best-checked theoretical descriptions of what is going on in our world. The Newtonian description may be regarded as the limiting case for low velocities and/or weak gravitational fields.

2.3 Time in Quantum Physics

As far as time goes, the situation is rather more intricate in quantum physics. Lacking a consistent quantum gravity theory so far, there is no good concept of time in the quantum world (for a recent approach to this issue see, e.g. Dias 2021). So far, time is regarded as a parameter (Zeh 2009) rather than an observable. In that respect, time has the same meaning as in Newtonian or relativistic physics, respectively.

2.3.1 The Age and Size of the Universe

Since the late 1920s it is known that the universe expands (Hubble 1929); since the second half of the 1990s that this expansion is – contrary to all expectations – not slowing down but rather accelerating (e.g., Perlmutter et al. 1995; Riess et al. 1998); and since about the turn from the twentieth to the twenty-first century that the expansion motion started about 13 to 14 billion years ago (e.g., Freedman et al. 2001; Riess et al. 2019). The latter data are derived by determining the relation between the expansion velocity and the distance of an object, the so-called Hubble parameter.

The exact nature of the beginning of the universe is not yet understood, as at that time a combination of quantum and relativistic physics must have been in action, a theory of which we do not yet have at hand (see above). While we are not really able to describe the beginning of the universe in a physically consistent manner, the subsequent evolution all the way to the Universe of today is understood, at least conceptually.

To the best of our knowledge, the universe has a flat geometry and an infinite extension. The *visible* size of the Universe, however, is limited by the finite value of the speed of light and the finite age of the Universe.

The finite value of the speed of light, furthermore, leads to a coupling of geometrical and temporal distances. A look into a distance is always equivalent to a look into the past – the so-called lookback time. In the local universe, the ratio between the lookback time and the distance is linear. For large distance, the relation is more, intricate but well understood.

2.3.2 Stellar Evolution in a Nutshell[2]

Stars form out of interstellar material (ISM, for short). For most of its time, a star is a machine in which nuclear fusion transforms four hydrogen nuclei into one helium nucleus, releasing binding energy which we ultimately observe as stellar luminosity. In an environment with sufficiently large pressure and temperature, as is the case in stellar interiors, this is an exothermal process, i.e., it runs by itself. After this so-called helium-burning phase, heavier elements can be formed, if the respective star is massive enough. At the end of this evolution, part or all of the stellar material is returned to the ISM, again depending on the stellar mass. If part of the stellar material is returned to the ISM, this will lead to a White Dwarf star and a Planetary Nebula (this is a misleading historical term as it has nothing to do with planets); if all material is returned, this will lead to a supernova. The material returned, however, is material which has gone through stellar fusion processes and thus is chemically altered material in comparison with the material the star was originally formed from.

2.3.3 The Arrow of Time

Stellar evolution provides in a very simple way the possibility to define an arrow of time, despite the fact that already special relativity has forced us to abandon the simple universal concept of the sequence of events.

Stellar evolution irreversibly changes the chemical composition of stars and of the ISM. Thus locally one can discern material of different age. There is an important caveat, though: As the rate of change of the chemical composition depends on the local rate of star formation and the mass distribution of the newly forming stars, the "speed" of evolution may differ considerably from location to location. In other words, the chemical composition of interstellar material defines the direction of the arrow of time, but provides only a lousy clock.

Finally, it is important to note that this arrow of time definition does not depend on the respective theses concerning the supposed type of the expansion of the universe (ever faster expanding, ever slower expanding, re-collapsing, etc.), as the nuclear evolution of the ISM is independent of it and mostly irreversible.

[2] For a detailed description of stellar structure and evolution, see, for example, Kippenhahn et al. (2012).

2.3.4 Observing Stellar Evolution: Time Traveling … Sort of

Observing an event always means looking into a certain distance. Due to the finite velocity of light, this is tantamount to looking into the past – be it a very small distance, be it a very large one. All our senses like eyes or ears have a finite temporal resolution of tens to hundreds of milliseconds, depending on the individual sense. During that span of time, light travels thousands to tens of thousands of kilometers. This means that only for processes above such distances the finite nature of the velocity of light plays a role for our perception. In other words, in everyday processes of life, the lookback into the past does not play a role. In space and the universe, however, the situation is rather different and can lead to very tricky, but at the same time very useful situations.

Observing a cosmic object means observing it in *our* past. In what state of its evolution we see it depends on the distance and on when the object formed.

Let us, as an example, take the Sun. There are plenty of other stars around, which have the same mass and chemical composition as the Sun.[3] Our Sun formed about 4.5 billion years ago. Let us assume that we observe at the same distance of 50 million lightyears[4] three stars identical to the sun, which, however, were formed at different times. Let us assume that they were formed 4.6, 4.55, and 4.5 billion years ago. Then we see them when they were 4.55, 4.5, and 4.45 billion years old. In other words, we see one at today's age of our sun (4.5 billion years), one 50 million years younger than our Sun (4.45 billion years), and one 50 million years older (4.55 billion years). This means that we can test our models by looking back in time. But we can not only see Sun-like stars that then were younger than the Sun is today, but also ones which then were already older than our Sun is today. In that respect, one can observe the state of the Sun in its future. This, of course, does not only apply to the Sun, or stars in general, but to all sorts of cosmic objects.

I wish to finish this contribution with a well-known example, where one can literally see the situation just described, namely the constellation Orion (Fig. 1), which is particularly well visible at late-winter evenings. At the same time at (almost) the same location of the celestial sphere above observers on planet earth, we see stars between the earliest and latest phases of their evolution, and with different masses (Table 1):

[3] For formation times well before that of our Sun, one has to take into account the chemical evolution of the ISM. This is a well-understood effect, which, however, does not contribute to the topic here and thus is neglected.

[4] Already the first published images of the James Webb Space Telescope (JWST) demonstrated the possibility of observing individual stars at such distances.

Fig. 1 The constellation Orion with the three objects of Table 1 marked. The photo was taken on January 4, 2020, with a cell phone camera. It pretty much represents what one can see with the naked eye on a cloudless sky

Table 1 Characteristic quantities of two stars and the star-forming region Orion Nebula in the constellation Orion

Object	Distance	Mass	Age	Status
Betelgeuse	~550	~17	~8 million	Near supernova explosion
Orion nebula[a]	~1350	0.08 to ~100	Just forming	Star formation region
Bellatrix	~250	~8	25 million	Giant star

[a]The numbers for mass and age refer to the just forming stars in the Orion nebula
The distance is given in lightyears, the mass in solar masses, and the age in years. They are marked in Fig. 1

References

Dias, Eduardo O. 2021. Quantum formalism for events and how time can emerge from its foundations. *Physical Review A* 103: 16. https://doi.org/10.1103/PhysRevA.103.012219.

Freedman, Wendy L., Barry F. Madore, K. Brad, Laura Ferrarese Gibson, Daniel D. Kelson, Shoko Sakai, Jeremy R. Mould, Robert C. Kennicutt Jr, Holland C. Ford, John A. Graham, John P. Huchra, Shaun M.G. Hughes, Garth D. Illingworth, Lucas M. Macri, and Peter B. Stetson. 2001. Final results from the Hubble Space Telescope Key Project to measure the Hubble constant. *The Astrophysical Journal.* 553 (1): 47. https://doi.org/10.1086/320638.

Hubble, Edwin. 1929. A relation between distance and radial velocity among extra-galactic nebulae. *Proceedings of the National Academy of Sciences* 15 (3): 168–173. https://doi.org/10.1073/pnas.15.3.168.

Kippenhan, Rudolf, Achim Weigert, and Achim Weiss. 2012. *Stellar Structure and Evolution*, 2nd ed. Berlin: Springer.

Manning, Liliann, Daniel Cassel, and Jean-Christophe Cassel. 2013. St. Augustine's reflections on memory and time and the current concept of subjective time in mental time travel. *Behavioral Sciences* 3 (2): 232–243. https://doi.org/10.3390/bs3020232.

Perlmutter, Saul, Carl Pennypacker, Gerson Goldhaber, R.A. Ariel Goobar, H.J.M. Muller, J. Desai Newberg, Alex G. Kim, Mathew Y. Kim, Ivan A. Small, Brian J. Boyle, C.S. Crawford, R.G. McMahon, P.S. Bunclark, D. Carter, M.J. Irwin, R.J. Terlevich, R.S. Ellis, K. Glazebrook, W.J. Couch, J.R. Mould, T. Small, and R.G. Abraham. 1995. A supernova at z=0.458 and implications for measuring the cosmological deceleration. *Astrophysical Journal Letters* 440: L41–L44. https://doi.org/10.1086/187756.

Riess, Adam G., Alexei V. Filippenko, Peter Challis, Alejandro Clocchiatti, Alan Diercks, Peter M. Garnavich, Ron L. Gilliland, Craig J. Hogan, Saurabh Jha, and Robert P. Kirshner. 1998. Observational evidence from supernovae for an accelerating universe and a cosmological constant. *The Astronomical Journal* 116: 1009–1038. https://doi.org/10.1086/300499.

Riess, Adam G., Stefano Casertano, Wenlong Yuan, Lucas M. Macri, and Dan Scolnic. 2019. Large magellanic cloud cepheid standards provide a 1% foundation for the determination of the Hubble constant and stronger evidence for physics beyond LambdaCDM. *The Astrophysical Journal* 876: 85. https://doi.org/10.3847/1538-4357/ab1422.

Zeh, H. Dieter. 2009. Time in quantum theory. In *Compendium of Quantum Physics*, ed. D. Greenberger, K. Hentschel, and F. Weinert, 786–792. Berlin/Heidelberg: Springer. https://doi.org/10.1007/978-3-540-70626-7_221.

Cultural Encounters and Indo-German Consciousness: Prince Frederick August of Augustenburg in India

Martin Krieger

Abstract This chapter studies the scattered career of Prince Frederick August of Augustenburg (1830–1881). Caught between the German-Danish conflict within the Duchies of Schleswig-Holstein, the Prince escaped from parental pressure as well as a lack of perspectives by delving into the riches of the Indian past. As an amateur-Indologist, he reshaped his self-perception and finally wrote the first Western biography of the Mughal emperor Akbar. The study draws on the Prince's publications and his surviving handwritten documents and tries to highlight India's role in his process of gaining self-confidence and self-awareness.

Keywords Indo-German cultural encounters · Indology · Akbar · Identity

1 A Global Biography

In a historical perspective, the term "consciousness" can be perceived in a collective as well as in an individual way. This is especially true, if we concentrate on those aspects of consciousness that are comprised under the highly complex concept of "identity." Entire societies brought forward collective identities, while simultaneously the self-perception of individuals could be shaped by personal experience and biographical ruptures, but by collective memory as well. In the past, individual or collective identities were usually constructed against a local or a national background, and only in a few instances did transnational or even global aspects play a role. Quite an exception is the life of Prince Frederick August of Augustenburg from North Germany. Brought up in a German-Danish environment in the nineteenth century, India gained a leading role on a long way to give his scattered biography some kind of sense. This chapter endeavors to reconstruct Frederick August's

M. Krieger (✉)
History Department, Kiel University, Kiel, Germany
e-mail: krieger@email.uni-kiel.de

P. S. Satsangi et al. (eds.), *Consciousness Studies in Sciences and Humanities: Eastern and Western Perspectives*, Studies in Neuroscience, Consciousness and Spirituality 8, https://doi.org/10.1007/978-3-031-13920-8_18

changing attitudes toward life and the development of a very personal Indo-German consciousness that became manifest in the first Western biography of the Mughal emperor Akbar he wrote.

The two settings of this story could not be more different: The Duchies of Schleswig-Holstein constituted a small territory situated between the Baltic and the North Sea. It was an area of German-Danish cultural encounter but, in many instances, of conflict as well. Being part of the Danish monarchy since 1460, a powerful German-speaking aristocracy played a decisive role in the political debates. Even worse, the Duchies were further split among a number of competing side-lines of the Danish royal house. Finally, a number of tiny semi-autonomous states hardly comprising more than a handful of small towns existed within the small Duchies. Frederick August of Augustenburg was the offspring of one of them.[1]

While Schleswig-Holstein became divided into a number of petty-states from the sixteenth century onward, India set off to emerge as one of the world's leading powers under the reign of the Mughal emperors. During the period of Jahangir and Akbar, the Mughals confirmed their claim as a hegemonial power in the North of the subcontinent, while Aurangzeb turned toward the South but finally overstretched the Mughal's territorial ambitions (Kulke 2018: 112–138).

Since the seventeenth century, the Duchies of Schleswig-Holstein and the vast Indian subcontinent had maintained commercial and cultural contacts. As subjects of the Danish crown, merchants and civil servants from the Duchies served in the Danish trading settlements in India, in Tranquebar (in today's Tamil Nadu) and Serampore (near Kolkata). Christian missionaries from North Germany were sent to the subcontinent under Danish colors, while an abundance of trading commodities as well as objects of cultural and natural history and knowledge were transferred in the opposite direction. Indian cotton textiles shaped fashion from the eighteenth century onward, while saltpeter from Bihar ended up in Danish and North German cannons. (On Indo-Danish commercial activities during the latter half of the eighteenth century see Feldbæk (1969) and on cultural exchange see Krieger (2010).)

From the latter half of the eighteenth century onward, India increasingly gained importance as an object of academic research but also as a distant goal of imagination and phantasy in the German-Danish context. Notably the broad reception of "Shakuntala" shaped a very distinct image of the cultural riches of the subcontinent. During the nineteenth century, Indology finally emerged as a relevant and broadly represented subject among German and Danish universities. A huge amount of books and manuscripts in Indian languages entered the Danish monarchy, for example, in 1820, when the Danish Botanist Nathaniel Wallich, superintendent of the Botanic Garden at Calcutta, sent several boxes of costly texts to the Danish capital Copenhagen (Krieger 2022: 117–120). However, we do not know if young Frederick August ever heard of Shankunala during his childhood in the North German-Danish aristocratic environment of Schleswig-Holstein.

[1] No proper English survey of the history of Schleswig-Holstein exists. For a brief overview of the history of the Duchies Schleswig and Holstein in German see Lange (1996: 153–190, 281–287).

2 Growing Up During Wartime

Frederick August of Schleswig-Holstein-Augustenburg, later to be called the "Prince of Noer," was born on 16 November 1830 in the town Schleswig. While a younger brother died during his early childhood, Frederick August always maintained a very close relation with his sister Louise (1836–1866). During his early childhood, the parents purchased the manorial estate Noer, situated some 20 km north of the town Kiel. Here, Frederick August spent most of his time more or less secluded from the social and cultural life of the Duchies' towns. On long, lonely walks through fields, forests, and along the shores of the Baltic Sea, he gained a sense of greatness of nature, as he recollected much later: "... the grand loneliness of our forests, the closeness of the sea tended to deepen my thoughts and guided them into the vast world" (Noer and Augustenburg 1886: 2, transl M.K.).[2] Being the only surviving son and closely attached to his liberal mother, he later described himself as somewhat weak and sensitive. He blamed his teachers and notably his father Frederick of having been unable to educate his fragile mind with appropriate care and to supply him with proper knowledge – a deficiency, which he was never able to compensate during lifetime. Decades later, he still perceived his youth as a period of missed chances: "....thus I was the only son and as a consequence a rather spoilt boy, whose first instructions as a child by unsuitable teachers were conducted in a fashion so casual and insufficient that I have suffered from this lack my entire life and have never been able to make up for what I previously missed" (Noer and Augustenburg 1886: 1, transl M.K.).[3]

A closer look at his surviving extensive correspondence and other texts unveils the impression that Frederick August was dyslexic, which was not acknowledged as a mental deficiency at that time. Instead, it was identified as a lack of diligence and was countered on his father's orders by an even stricter upbringing characterized by physical hardships, time-pressure, and military discipline. At the age of 12, it was deemed that the young Prince was of a "very peculiar nature" – whatever this meant remains obscure.[4] "My father," as he later reported, "used to call me a dreamer" (Noer and Augustenburg 1886: 2). The young Prince's further school-education broadened the gap between his own expectations toward life and the educational goals of his father, for whom Latin, German, calculating, riding, shooting, and gymnastics were deemed to be more valuable than modern languages and literatures. And no time was to be wasted with painting or drawing (LASH 22. No. 1723).

---

[2] "... die grossartige Einsamkeit unserer Wälder, der stete Anblick des Meeres, vertieften meine Gedanken und führten sie in die Weite" (Noer and Augustenburg 1886: 2).

[3] "... so war ich der einzige Sohn und in Folge dessen ein etwas verzogener Knabe, dessen erster Jugendunterricht leider durch ungeeignete Lehrer so lässig und ungenügend betrieben wurde, dass ich an diesem Mangel mein ganzes Leben hindurch zu leiden hatte und nie im Stande war, das früher Versäumte völlig nachzuholen" (Noer and Augustenburg 1886: 1).

[4] LASH (=State Archives of Schleswig-Holstein) 22. 1723. Unterrichtsplan für den jungen Prinzen zu Schleswig-Holstein-Sonderburg-Augustenburg, 1843.

The mid-nineteenth century was subject to fundamental changes, and the Prince's narrow world soon fell apart. Frederick August's family initially had been closely related to the royal Danish house. His predecessors had held high ranks within the Danish monarchy, and one of his grandmothers, Louise Augusta, even was the daughter of the Danish king Christian VII. This did not prevent his father Frederick to feel and talk German rather than Danish. A striking rupture was apparent from the outset: while his mother, born Duchess of Danneskjold-Samsøe, originated from the heart of the Danish-speaking areas of the monarchy, his father Frederick was a member of the German-speaking aristocracy of the Duchies of Schleswig and Holstein.

In 1848, civil war broke out between the German-speaking population of the Duchies and the Danish government. The Prince's father Frederick became commander-in-chief of the rebellious German troops of the Duchies. With his troops, he fought against the soldiers of his own distant relative, the king of Denmark. It was out of question that young Frederick August was to join the rebellious forces of the Duchies on behalf of his father's desire. Even if he was not involved in the war at the forefront, he was very well confronted with death, injuries, and other hardships of mid-nineteenth-century warfare. Finally, Frederick August became sick. On the edges of the battlefields of Schleswig-Holstein, he caught a lung-disease, which would trouble him throughout his lifetime and which obviously was the cause of his early death at the age of 51.

In 1850, the German troops of Schleswig-Holstein were defeated by the Danish crown, which was supported by other European powers. The German-speaking Dukes of Augustenburg were exiled and had to leave their estate at Noer and other possessions in the Duchies. Any claims toward the Danish throne were declined; the political as well as professional perspectives of the young Prince of Noer now were desperate. He had not only lost his health but simultaneously any perspective for a future career in the Danish monarchy (Wulf 2006).

Sometime before the family had to leave the country, his parents sent their son on a long journey toward Adelaide for physical recovery and probably also to keep him away from the troubles at home. On his way back from Australia, where he had stayed for about half a year, the ship made a short stopover in Indian Madras. From the outset, the young Prince fell in love with India and its ancient cultures, as transpired from the vivid impressions of the colorful life along the beaches and inside the South Indian colonial town described in his diary (LASH 22. No. 1860). In contradiction to parental instructions, Frederick August decided to remain on the subcontinent for some time – a decision, which displays a hitherto unseen sense of self-confidence and upheaval against the father in the young prince. After having visited Madras, he left the South, deeply impressed by the social and cultural life of the Indian people. By ship, he proceeded to Calcutta to make a short trip along the river Hooghly northward through Bengal. Finally, he left India via Egypt and the Ottoman Empire back to Europe, heavily shaped by his impressions, obviously having in mind to return again later (Lohmeier 1987).

3 An Amateur-Indologist

Following his exiled parents, Frederick August stayed in England and France afterward. Obviously on his father's orders, he studied in Cambridge and in German Heidelberg. However, even here, his love for India could not be replaced by regular studies, and the formal lectures were subsequently supplemented and later more or less entirely replaced by the Prince's studies on India, rendering him a virtual amateur-Indologist. Theodor Goldstücker (1821–1872), an Indologist in London, did not only teach him Sanskrit but simultaneously made him familiar with the history and cultures of ancient India in a more academic manner. Even if Frederick August always would be hesitant to perceive himself as a professional Indologist, he acquired a very firm knowledge on ancient India during these years.

The Prince studied Sanskrit and, furthermore, prepared a book-manuscript from the notes of his travels through India, Egypt, and the Ottoman Empire – but not on Australia, where he had stayed the longest time. Without any doubt, the latter's colonial Western civilization hardly held any attraction for him, while he deeply delved into the "Orient" instead. This first publication was brought to light in 1859, however somewhat hesitantly under the pseudonym Onomander. At that time, Frederick August's compassion for India obviously was still not to be made known to the public (Onomander 1859).

Deeply taken aback by the demise of his beloved mother and his father's plans for a second marriage with an American lady, a certain Miss Lee, Frederick August left for India a second time in 1864, until he was called back to Europe by his father's sudden death the subsequent year. To Goldstücker he wrote on his plans for a second journey: "I want to study natural history, ethnology, science and art of India in its development over the millennia in the way only a personal encounter would allow" (Noer and Augustenburg 1886: 131, Frederick August to Goldstücker, May 1865, transl. M.K.).[5] Next to his academic endeavors, he actually expected his health to improve, as he wrote to Goldstücker on his way to India from Mauritius: "I probably may hope to regain some degree of my erstwhile freshness, which I had enjoyed during my first journey" (Noer and Augustenburg 1886: 113 f., transl. M.K.).[6]

On the other hand, he does not seem to have been very sure if he would return to living at all. Before his departure, on 29 May 1864, he penned down his last will with detailed instructions as to how his heritage was to be distributed. This document displays a very personal sense of identification with the Orient: "If I should happen to die in the East, let my remains be buried there" (LASH 22. No. 1877). Frederick August had the plan to cross India from the South to the North. In May

[5] "Ich möchte Natur- und Völkerkunde, Wissenschaft und Kunst in ihrer tausendjährigen Entwicklung studieren mit dem inneren Verständniss, welches nur die lebendige Anschauung zu geben vermag" (Noer and Augustenburg 1886: 131, Frederick August to Goldstücker, May 1865).

[6] "Vielleicht darf ich hoffen, wieder einen Theil der Frische zu erringen, die mich auf der ersten Reise so glücklich machte" (Noer and Augustenburg 1886: 113f.).

1865, during a longer stay in Ootacamund, he expresses a deeper historical interest in the Mughal empire for the first time.[7]

A diary with brief sketches of places visited by him survives among his papers (LASH 22. No. 1900). Frederick August's itinerary does not seem very systematic; it was obviously guided by a rather coarse knowledge of the architectural highlights, but to the same degree by random as well. To aggravate the situation, detailed travel-instructions promised to be sent by Goldstücker never reached him (Noer and Augustenburg 1886: 126, to Goldstücker 2nd April 1865). From Colombo, the Prince first traveled to Tuticorin and then further on to Tinnevelly, Ramnad, and Madurai. And yet, in Madurai, he did not display much interest in the town's outstanding temple-architecture. This lack of interest did not only result from his poor health condition at that time, but simultaneously from his passion for hunting. After having recovered health in Madurai, he immediately left for a three-day hunt without paying much attention even to Sri Meenakshi temple (Noer and Augustenburg 1886: 122, Frederick August to father, 28 January 1865).

After having passed some spots of minor historical interest, he reached Thiruchirapalli, to proceed toward Nagapattinam and Tranquebar. Tranquebar had been a Danish trading-settlement for more than two hundred years, but it had been sold to the British East India Company already in 1845. Frederick August did not pay much attention to the common Danish cultural heritage but at least felt delighted when meeting an old Indian from a village who was still able to converse in Danish. Via Cuddalore, and without even mentioning the magnificent temple of Chidambaram, he reached the ruined city of the former nayakas of Gingee (reminding him of ancient Ephesos while hunting two monkeys), Madras, and finally Ootacamund. In some instances, he noted down his impressions of temples and other cultural highlights, and in other instances, he just recorded peoples' gossip such as the story of a sadhu he came to know: "Still ten years ago, a man was to be met in Tinevelly, who had spent fifteen years to roll himself with his body from Benares to Cape Comorin. He was thirty five years old at that time" (LASH 22. No. No. 1900. 18th January 1865, transl M.K.).[8] Finally Frederick August spent about 3 months in the Nilgiris, cherishing his second passion in India: hunting. Then, he was called back to Europe.

Sometime after his father's estate had been settled, Frederick August decided to return to India again in 1867. This third journey is to be perceived as a resumption of the second one rather than another journey in its own right. Frederick August again displayed a sense of restlessness and decided to move around: "I have entirely lost my balance and felt to be completely restless" (Noer and Augustenburg 1886:

[7] "... ich habe die Reise begonnen als ein vom Schicksal verschlagener Mensch, fast willenlos, ziellos und doch getrieben von der unüberwindlichen Macht, von einer seelischen Gewalt, die keine Vernunft und keinen Widerspruch annimmt und mich vorwärts drängt ohne dass ich selbst weiss wohin und wozu" (Noer and Augustenburg 1886: 131f.).

[8] "Noch vor zehn Jahren sah man in Tinivelly einen Mann, der über 15 Jahre gebraucht hatte – um sich von Benares nach Comorin zu ‚rollen'! Er war damals 35 Jahre alt" (LASH 22. No. 1900. 18th January 1865).

147, transl. M.K.).[9] Furthermore, his beloved sister fell sick and finally died in September 1866, which likewise added to the impression that virtually nothing could keep him home. This attitude toward life and travel was not even altered by the fact that the situation in the Duchies of Schleswig-Holstein had improved for him. The Duchies had been conquered by the Prussians in 1864 in the course of the German-Danish war, and the Dukes of Augustenburg were permitted to return to their estates again. Noer's economic situation was, however, weak at that time and would render only a meager income. But the Prince's limited resources would enable a respectable life in India rather than in the Duchies, and he came to the conclusion: "I deem it to be better to live as an ordinary private person in a distant country until the situation at home becomes better" (Noer and Augustenburg 1886: 151, transl. M.K.).[10] However, facing his poor health condition, life in India still would not be as easy as actually hoped for.

Frederick August traveled to the Netherlands and then further on to France and England. In London, he intended to meet Goldstücker and to resume his Sanskrit studies. However, he increasingly encountered intellectual difficulties: "The more I deepened my studies in Sanskrit, the deeper I witnessed virtually insurmountable obstacles to comprehend that queen of languages" (Noer and Augustenburg 1886: 147, transl. M.K.).[11] It may be assumed that these difficulties confirmed his decision to shift to the history of the Mughals instead. With respect to sixteenth- and seventeenth-century Mughal history, quite a number of sources translated into European languages would be available, as he certainly must have known.

On 20 April 1867, Frederick August left Southampton toward Alexandria to proceed to Madras another time. Several months he spent at Ootacamund within the Nilgiri mountains, again rejoicing in the clear air and the cool climate. Now and then he left for hunting tours and spent his days in different camps somewhere in the wilderness. Finally, he turned toward the North and visited Mysore, Seringapatnam, as well as some other, less known places to study historical sites or just to go out for hunting. On Christmas Day 1867, the Prince reached Bellary, intending to visit the prominent ruined city of Vijayanagara. However, he never reached that place. His health was too weak to resume any sightseeing. "I thought, I was about to die," as he wrote by end of December (LASH 22 1860). After 2 weeks, he felt slightly better, but finally he decided to return to Madras without having visited Vijayanagara. By palankin and train, Frederick August reached Madras in an extremely poor condition. The European doctors strongly recommended to travel back to Europe as early as possible. But Frederick August declined, emphasizing that a life outside

[9] "Ich hatte mein Gleichgewicht ganz verloren und war völlig ruhelos" (Noer and Augustenburg 1886: 147).

[10] "Ich finde es daher besser, als einfacher Privatmann in der Fremde zu leben, bis sich die Verhältnisse günstiger gestalten" (Noer and Augustenburg 1886: 151).

[11] "Als ich die Sanskritstudien wieder aufnahm, war es mir wie ein Traum. Je mehr ich mich in das Studium des Sanskrit vertiefte, desto mehr lernte ich auch die unüberwindlichen Schwierigkeiten dieser Königin aller Sprachen begreifen, leider aber nicht überwinden" (Noer and Augustenburg 1886: 147).

India would be worthless to him in the meantime. He was permitted to travel to Calcutta, "after I have explained to them that life would possess any value only under certain preconditions" (Noer and Augustenburg 1886: 176, to Goldstücker 24. February 1868), as he wrote to Goldstücker. The Prince's intention now was to travel from Calcutta through the heartlands of the former Mughal empire to the cooler areas of Kashmir, where he could recover his health again.

4 Akbar's Biographer

In Calcutta, the Prince's decision to study the history of the Mughal empire and notably Akbar's life was further confirmed. He met the German philologist Heinrich Blochmann (1838–1878), who served as a professor and principal of the Calcutta Madrasah. Blochmann had already conducted historical research on that issue and had submitted translations of Mughal sources (such as Abul-Fazl's "A'in-i Akbari") when he met the Prince. Blochmann further encouraged Frederick August to study Mughal history.

Against this backdrop, there is no doubt that the Prince tried to stop at relevant monuments of Mughal history on his way to Kashmir. He visited Sekundra and Agra (though not Fatehpur Sikri), before he went on to Delhi. To Goldstücker, he wrote: "On my travels in the North of India, I encountered numerous traces of Akbar's life and the results of his impact – be it large monuments or people's tales or even the remnants of his wise rule, which rendered his country to flourish" (Noer and Augustenburg 1886: 183, to Goldstücker tenth April 1868, transl. M.K.).[12] During his stay in Delhi, he finally must have decided to devote his future life notably to Mughal emperor Akbar and to write a biography on his exceptional life. This study was not just meant to become a mere history book but, even more, a means to sharpen his self-confidence as he wrote to Goldstücker: "Will this project be able to drive chaos out of my mind and to render order and peace?" (Noer and Augustenburg 1886: 184, transl. M.K.).[13]

The Prince decided to leave the well-trodden paths of his European life and to enter a new intellectual sphere. Akbar's life was to emerge as a catalyst for his own biography and the scattered existence of his family, the Dukes of Augustenburg. He had found something peculiar in the Mughal emperor's life that his own family and he himself could never enjoy: personal, dynastic, and political continuity. Moreover,

[12] August, Frederick to Blochmann. 10 April 1868, …: "Dann, auf meiner Wanderung durch das nördliche Indien fand ich in aller Orten die Spuren von Akbar's Thätigkeit und die Folgen seines Wirkens, mochte es in den grossen Bauten sein, die er errichtet hat, oder in den Überlieferungen seiner gewaltigen Kriegsthaten, die noch heute in dem Munde des Volkes fortleben, oder in der weisen Anordnung der Staatsverhältnisse, durch welche er seine Länder reich und blühend machte."

[13] "Würde diese Idee nicht meine Gedanken aus dem Chaos lösen und meinem Geiste Ordnung und Ruhe bringen?" (Noer and Augustenburg 1886: 184).

Frederick August knew that Akbar was praised for his tolerance toward other cultures and religions – a tolerance, which the Duke himself had never experienced during his own life shaped by the German-Danish encounter.

Kashmir itself, although he stayed there for about 7 months and thus much longer than in the plains, did not play a similarly important role for Frederick August's self-perception as an upcoming specialist on Akbar, compared to his shorter travels through the plains. Even if he visited some historical sites related to the emperor in Kashmir as well, which had been conquered by Akbar in 1586/1587, his major attention was again devoted to hunting and the recreation of his health. In Kashmir, the Prince encountered poverty and negligence, which rendered a virtual counter-image in contrast with his positive image of the emperor: "Kashmir could be the most fertile part of India. It could be a paradise even today! In contrast, I never witnessed a similar degree of poverty, dirt and misery during all my travels" (LASH 22. No. 1900: 8 June 1868, transl. M.K.).[14] From Kashmir, the Prince finally shifted to Western Punjab, where he likewise visited relevant historical sites. Punjab again constituted a significant witness to Akbar's power and his sense for political unity (Hottinger 1998: 143). He notably visited the fort of Attock, which he compared with the Mughal architecture of Delhi. Finally, the Prince returned to Europe via Lahore, Amritsar, Benares, Calcutta, and Madras.

Back in Europe, Frederick August finally settled in Noer. Only now, after his extensive encounter with India, he found some kind of peace. He did not only marry Carmelita Eisenblat, the daughter of a merchant from Hamburg, but simultaneously started to work on his seminal biography on Akbar. For this purpose, he not only consulted books from his own ever-growing library, which comprised about 30,000 books but also contacted eminent Orientalists and Indologists from the university of nearby Kiel, and borrowed information from Blochmann and his translation of the "A'in-I Akbari." His study obviously led to content and patience, which he had always missed up to this point. In 1880, the first volume of the biography was brought to light. Shortly after, the Prince died at his estate Noer. A second volume came out only some years after his death (Noer 1880–1885).

Without any doubt, the life of Prince Frederick August of Augustenburg stands out compared to numerous other nineteenth-century noblemen from North Germany. Apart from hunting, he never rejoiced in the social preferences of the aristocracy and rather socialized himself with intellectuals, notably specialists on the "Orient." He did not marry a noble lady, which certainly would have been expected from him, but the daughter of a merchant from a trading city. As an amateur-Indologist, he pursued a quite unusual career, and he devoted his short life to a country actually none of his fellow-noblemen would have ever thought of: India. In the history of the subcontinent, Prince Frederick August found something that he was not able to find in his home country: a sense of unity among diversity, tolerance, and an immensely

[14] LASH 221900 August, Frederick. *Reisetagebuch*, 8 June 1868: "Kashmir ist der fruchtbarste Theil Indiens – und es war und könnte noch jetzt ein Paradies sein! Ich habe jedoch nirgends auf all meinen Streifzügen durch Indien so viel Armuth, Schmutz und Elend unter der Bevölkerung gesehen, wie hier."

rich and ancient culture. By devoting his life to India, he escaped the harsh political and military reality of the Duchies of Schleswig-Holstein and the diminishing political influence of his family and himself. His distinct image of India thus served as a means to reshape his own individual consciousness.

Addendum: Seeker of Truth and Ultimate Reality in Radhasoami Faith

Anand Srivastav:
In Part V of the present volume, Andreas Müller and Martin Krieger in their essays describe the search for personal or transcendental truth by churchfathers of ancient Christianity and by a German prince, respectively. In both cases, the individuals have to struggle with their embeddedness in cultural and institutional conditions that are archived in rules, traditions, and teachings that are accessible to the intellect.

A completely different way finding truth and ultimate reality is promised by the (Indian) Religion of Saints and the Radhasoami Faith through the act of formal initiation. Here the seeker is guided by the spiritual master, and through him the seeker is enabled to approach ultimate truth by reaching higher levels of spiritual consciousness. In Radhasoami Faith, the seeker is formally registered as such and called Jigyasu. Under the supervision of the spiritual master, which may encompass not only spiritual and meditational practices, but also aspects of everyday life, the Jigyasu can apply for the first initiation in order to become a member of Radhasoami Faith. The initiation is granted by the decision of the spiritual master only. The second and final initiation may follow thereafter, again exclusively granted by the spiritual master, the Sant Satguru of the time.

The following lines give the verbatim report of **Revered Prof. Prem Saran Satsangi**, the present guru of Radhasoami Faith Dayalbagh, of a very special case of initiation. However, the example illustrates the process of initiation by the guidance of the spiritual master as practiced in hundreds of cases each year in Dayalbagh. It also illustrates the close connection of spiritual concerns and scientific investigation in Dayalbagh, where the intellectual comprehension of spiritual processes is pursued via scientific methodology.

> Registered Seeker (of Truth/Ultimate Reality) Ms. P. Saranya (Deaf & Dumb), One-year Diploma in Computer Usage for Office Management & Accounting, D/O PB V. Paramanantham (Tamil Nadu) had applied for First Updesh/Initiation (i.e. Sumiran & Dhyan) in her Zonal Bhandara of August 2022.
>
> On her application form it was advised as follows: "May approve S & D Initiation, subject to her satisfactory completion of a minimum of 3 months program successfully under RAF- Mahila Sashaktikaran Initiative organised by I/C SFG (Retired) Col. S.P. Satsangi".
>
> She joined the aforesaid Mahila Sashaktikaran Initiative on 11.09.2022 and is accompanied by one of her parents everyday for the training sessions.
>
> Initially, she was introduced to Yoga and Lathi training. Her responses by observation were found to be very good. Now inducted to Mainstream- trainees.

It is heartening to note that Ms P. Saranya, is performing at par with the other participants of the Mahila Sashaktikaran Initiative as reported by I/C SFG (Retired) Col. S.P. Satsangi and she is able to follow the instructions and perform exercises based on her instincts and the visual instructions given to her. This has also been taken-up as a Case-study for research purposes.

16.09.2022

References

Feldbæk, Ole. 1969. *India Trade Under the Danish Flag 1772–1808. European Enterprise and Anglo-Indian Remittance and Trade*. Odense: Studentlitteratur.

Hottinger, Arnold. 1998. *Akbar der Grosse (1542–1605). Herrscher über Indien durch Versöhnung der Religionen*. München: Wilhelm Fink.

Krieger, Martin. 2010. Material culture, knowledge, and European society in colonial India Around 1800: Danish Tranquebar. In *Artistic and Cultural Exchanges Between Europe and Asia, 1400–1900*, ed. Michael North, 53–72. Farnham and Burlington: Ashgate.

———. 2022. *Nathaniel Wallich. Global Botany in Nineteenth Century India*. New Delhi: Routledge.

Kulke, Hermann. 2018. *History of Precolonial India. Issues and Debates*. New Delhi: Oxford University Press.

Lange, Ulrich, ed. 1996. *Geschichte Schleswig-Holsteins. Von den Anfängen bis zur Gegenwart*. Neumünster: Wachholtz.

LASH (Landesarchiv Schleswig-Holstein/ State Archives Schleswig-Holstein, Germany) 22. No. 1723. *Unterrichtsplan für den jungen Prinzen zu Schleswig-Holstein-Sonderburg-Augustenburg, 1843*.

——— 1860. *Tagebuch des Prinzen Friedrich August (von Noer) über seine Land- und Seereise, 1849–1850*.

———. 1877. *Frederick August's Last Will. London 29th May 1864*.

———. 1900. *Reisetagebuch*.

Lohmeier, Dieter. 1987. Friedrich Christian Carl August, Prinz von Schleswig-Holstein-Sonderburg-Augustenburg. In *Biographisches Lexikon für Schleswig-Holstein und Lübeck 8*, 120–122. Wachholtz: Neumünster.

North, Michael, ed. 2010. *Artistic and Cultural Exchanges Between Europe and Asia, 1400–1900*. Farnham and Burlington: Ashgate.

Onomander [Frederick August of Schleswig-Holstein-Augustenburg]. 1859. *Altes und Neues aus den Ländern des Ostens*, 2 vols. Hamburg: Perthes – Besser und Mauke.

von Noer, Friedrich August. 1880–1885. *Kaiser Akbar. Ein Versuch über die Geschichte Indiens im sechzehnten Jahrhundert*, 2 vols. Leiden: Brill.

von Noer, Carmen Gräfin, and Friedrich August Prinz von Schleswig-Holstein-Augustenburg Graf von Noer, eds. 1886. *Briefe und Aufzeichnungen aus seinem Nachlass*. Nördlingen: C.H. Beck.

Wulf, Peter. 2006. Stationen eines standesgemäßen Exils. Zur Biographie des Prinzen von Noer 1864/65. *Zeitschrift der Gesellschaft für Schleswig-Holsteinische Geschichte* 131: 51–66.

Part VII
Consciousness and Pragmatics

Complementarity and Quantum Cognition

Reinhard Blutner

Abstract The idea of complementarity is one of the key concepts of quantum mechanics. Yet, the idea was originally developed in William James' psychology of consciousness. Recently, it was re-applied to the humanities and forms one of the pillars of modern quantum cognition. I will explain two different concepts of complementarity: Niels Bohr's ontic conception and Werner Heisenberg's epistemic conception. Furthermore, I will give an independent motivation of the epistemic conception based on the so-called operational interpretation of quantum theory, which has powerfully been applied in the domain of quantum cognition. Finally, I will give examples illustrating the potency of complementarity in the domains of bounded rationality and survey research. Concerning the broad topic of consciousness, I will focus on the psychological aspects of awareness. This closes the circle spanning complementarity, quantum cognition, the operational interpretation, and consciousness.

Keywords Quantum cognition · Complementarity · Bounded rationality · Awareness

1 Introduction

The idea of complementarity goes back to William James (1842–1910), a famous psychologist. Niels Bohr applied the idea to physics. Even though he suggested applying the idea also to the social sciences as well, he did not make any important steps in this direction. Consequently, the development of psychological theories based on such insights was delayed by more than 100 years. Only recently the new

R. Blutner (✉)
Institute for Logic, Language and Computation, University of Amsterdam, Amsterdam, Netherlands
e-mail: blutner@contact.uva.nl

P. S. Satsangi et al. (eds.), *Consciousness Studies in Sciences and Humanities: Eastern and Western Perspectives*, Studies in Neuroscience, Consciousness and Spirituality 8, https://doi.org/10.1007/978-3-031-13920-8_19

field of quantum cognition has been developed by researchers such as Diederick Aerts (1982, 2009), Conte (1983), Conte et al. (2008), Harald Atmanspacher et al. (2002), Busemeyer and Bruza (2012), Peter beim Graben (2004), and many others.

Even though the protagonists of quantum cognition have always stressed the point that they do not intend to reduce cognitive psychology to physics in any way and that they only see fruitful applications of the mathematical ideas developed in physics to the domain of psychology, the development of quantum cognition has been hindered by many misunderstandings and confusions.[1]

The aim of this chapter is to argue that a particular conception of complementarity – Heisenberg's epistemic conception – is the key conception for understanding quantum cognition, both from the foundational and from the empirical perspective. This is especially valid if we adopt Görnitz' recent reconstruction of Heisenberg's conception (Görnitz 2011). This view paves the way for a novel view on quantum cognition – one that stresses the explanatory role of abstract qubits and at the same time is a starting point for a novel and scientific solution of the mind-body problem (Görnitz 2018; Görnitz and Görnitz 2016; Mann and Mann 2017).[2]

Here is a short outline of the present chapter. In Sect. 2, I will explain Bohr's and Heisenberg's different ideas of complementarity. Section 3 discusses the role of Heisenberg's epistemic conception of complementarity in the context of modern quantum cognition. I will give some examples illustrating the potency of complementarity in the domains of bounded rationality and survey research. Section 4 is devoted to a foundational issue. It will explain **why** we find non-commuting observables in those domains. We trace the problem to the operational interpretation of measurement (question answering) and the psychological idea of resource limitations in cognitive processing. This immediately relates to the role of consciousness as awareness (Sect. 5). This closes the circle spanning complementarity, quantum cognition, the operational interpretation, and consciousness.

2 Complementarity

Originally, the idea of complementarity came from the psychology of consciousness, in particular from the writings of William James:

> It must be admitted, therefore, that in certain persons, at least, the total possible consciousness may be split into parts which coexist but mutually ignore each other and share the objects of knowledge between them. More remarkable still, they are complementary. (James 1890: 206)

As noted by Max Jammer (1989), Niels Bohr was acquainted with the writings of James, and he borrowed that idea from him. Similarly, in a letter to Stapp, Werner

[1] An almost complete realization of such misunderstandings is accomplished in Hümmler (2017).
[2] See also the impressive overview provided by Walach (2019) concerning the topic of health.

Heisenberg mentions "that Niels Bohr was very interested in the ideas of William James" (Stapp 1972: 1112).

In turn, Bohr introduced the idea into physics (originally as complementarity of momentum and position), and he proposed its application beyond physics to human knowledge in general. However, his physical conception of complementarity is quite different from James', and his often-cited claim to apply it to human knowledge was put to practice by Bohr. In chapter VII of his book (James 1890) – on more than 10 pages – James describes several phenomena which illustrate the splitting of consciousness into parts that are not accessible from each other. For example, these phenomena concern the "unconsciousness in hysterics" (James 1890: 202), partial blindness under "post-hypnotic suggestion" (James 1890: 207), or the splitting of a person in several selves in "alcoholic delirium" (James 1890: 208). One example describes the common situation of partial anaesthesia:

> The mother who is asleep to every sound but the stirrings of her babe, evidently has the babe-portion of her auditory sensibility systematically awake. Relatively to that, the rest of her mind is in a state of systematized anaesthesia. That department, split off and disconnected from the sleeping part, can none the less wake the latter up in case of need. (James 1890: 213)

Another example refers to the famous subject "Lucie", who was in a state of "post-hypnotic suggestion" and who could, among all the cards covering her lap, only see those cards that were not a multiple of:

> She was particularly blind to numbers such as 9, 12, 15. Hence, the part consisting of the multiples of 3 was split off and disconnected from the part of numbers. However, under special conditions, when she was not asked to tell which cards she saw but to write them down, the other part of the numbers was accessible. (James 1890: 207)

Following Blutner and beim Graben (2016), it seems adequate to use the term 'auto-epistemic accessibility' to refer to these phenomena (beim Graben and Atmanspacher 2006). I will use the term 'autoepistemic' to refer to the epistemic states of human subjects who can reflect on their own epistemic states. If two different states are not simultaneously epistemically accessible to the subject under discussion, then they can be seen as complementary in James' sense.

By contrast, Bohr's concept of complementarity is clearly not a copy of James' epistemic conception.

Bohr's conception refers to the laws of nature rather than to the idea of (auto) epistemic accessibility as in James' writings. In other words, it is an ontic conception rather than an epistemic one.[3]

A concept closely related to Bohr's concept of complementarity is Heisenberg's famous uncertainty principle.[4] In his book *Die physikalischen Prinzipien der Quantentheorie*, Heisenberg (1944: 9) starts his introduction of the uncertainty relation with the idea that all facts of atomic physics that are describable in space and

---

[3] For more details, the reader is referred to Howard (2004) and to Blutner and beim Graben (2016).

[4] The term *uncertainty principle* is a translation of the German term *Unschärfeprinzip* or *Unbestimmtheitsprinzip*.

time have to be describable in the wave picture as well. In the simplest case, a particle can be described in the wave picture by a 'wave packet'. However, for a wave packet no precise location and no precise velocity can be defined since the wave packet tends to be dispersed over the whole space. According to the simple laws of optics the following uncertainty relation can be derived:

$$\Delta q \cdot \Delta p \geq h \tag{1}$$

Hereby, Δq and Δp denote the standard deviation (measuring the dispersion) of position and momentum, respectively. The constant h is given as Planck's quantum of action relating radiation energy (E) to frequency (f) in the eq. $E = h{\cdot}f$.

After explaining this optical picture of the uncertainty relation, Heisenberg makes clear that the uncertainty relations can be derived without reference to the wave picture by using the general schemes of quantum theory and its physical interpretation. Generally, it is the complementarity of certain observables (expressed by their non-commutativity or order-dependence of the relevant operators) that allows the derivation of uncertainty relations.

The following table gives a schematic illustration of the main differences of Bohr's and Heisenberg's view of complementarity demonstrating that the so-called Copenhagen interpretation of quantum theory is not as uniform as usually assumed (Table 1).

Let us return now to the interpretational problem for probabilities. In quantum theory, a deep problem concerns the nature of the state vector. Here, we will assume an epistemic interpretation of the state vector. This clearly is the view of Heisenberg's Copenhagen interpretation with a subjective interpretation of probabilities. As made clear recently, this view does not necessarily entail observer-induced wave packet collapse (Caves et al. 2002a, b). More importantly, this picture conforms to the predominant picture of the Bayesian interpretation in Artificial Intelligence and Cognitive Psychology. Hence, it is plausible to take this interpretation as the basic conception of probability in quantum cognition (for more details, see Blutner and beim Graben 2016).

In the following section, we will define the notion of (auto)epistemic accessibility based on the operational interpretation of quantum physics. The operational

Table 1 The Copenhagen interpretation

Niels Bohr	Werner Heisenberg
Following **Kant's** philosophy in assuming that human beings are endowed with the ability to think and imagine according to certain classical categories and schemas	Following **Einstein's** leadership in reinterpreting basic concepts of physics, such as time, position, momentum
Ontic interpretation of complementarity (referring to the laws of nature)	**Epistemic** interpretation of complementarity (referring to autoepistemic accessibility)
No wave packet collapse, no antirealism, no subjectivism	Measurement as wave packet collapse, subjective probabilities

See Howard (2004) for more details

setting suggests a particular algebraic structure for modelling propositions, one that is very different from the classical Boolean setting. The Boolean setting allows to model propositions as sets of possible worlds (with the operation of union, intersection, and complement for the basic propositional operations). By contrast, the operational setting motivates a non-Boolean algebraic structure that favours modelling propositions by subspaces of a Hilbert space or by projection operators of the Hilbert space and the corresponding lattice-theoretic operations. The algebra of propositions is defined by non-statistical axioms. Hence, the operational understanding does not require any notion of probability. The concept of probability will emerge by means of a measure function, and its subjective interpretation can be motivated by a (quantum) de Finetti representation theorem (Caves et al. 2002a, b). Hereby, probabilities are taken to be degrees of belief, which are justified by axioms of fair bedding behaviour.

3 Quantum Cognition

Quantum cognition is a research field that applies ideas from quantum physics and quantum information science to develop radically new models of a variety of cognitive phenomena ranging from human memory, information retrieval, and human language to decision making, social interaction, personality psychology, and philosophy of mind.

The initial motivation for this new research field is quite simple and rather unmysterious. It has to do with the assumed algebraic structure of the inner world of ideas, concepts, and propositions. Boole and other great logicians of the nineteenth century assumed that thinking is like doing regular algebra in following strict rules exhibiting associative, distributive, and commutative properties. These are the same rules we can observe when we consider the construction of sets by using union, intersection, and complementation (Boolean algebra).

However, modern cognitive psychology has challenged this view: natural concepts are based on prototypes. As such, natural concepts are geometrical concepts that best can be represented by convex sets (Gärdenfors 2000, 2014). In this way, a geometric understanding of the conceptual world was born. Now, it is not clear what exactly is the underlying algebra of convex sets. Obviously, the algebra underlying the operation with convex sets is different from the traditional Boolean algebra. Surprisingly, it comes close to the lattice underlying the closed subspaces of a complex vector space (so-called orthomodular lattice).

Based on the work of the great Hungarian mathematician and philosopher John von Neumann, it has become visible that the heart of quantum theory is a new kind of probability theory based on orthomodular lattices rather than Boolean lattices.[5] This theory is more general than the traditional (Boolean-based) probability theory.

[5] A classic observation is that the set of projections is naturally a complete orthomodular lattice.

Interestingly, this approach seems to be powerful enough to solve some hard puzzles known from standard approaches to rationality, logical thinking, and probabilistic reasoning. This opens new horizons for cognitive modelling and their rational foundation. Both for classical probabilities and for quantum probabilities, the probability functions of the events in a sample space *W* are additive measure functions, i.e.

$$P(X \cup Y) = P(X) + P(Y) \text{ } \textit{for disjoint } X, Y \subseteq W \tag{2}$$

In the classical case, we can derive the law of total probability (Eq. 3) immediately from the Boolean axiom of distributivity.

$$P(B) = P(A \cap B) + P(\bar{A} \cap B) \tag{3}$$

A simple consequence of this law is that

$$P(A \cap B) \leq P(B) \tag{4}$$

Obviously, the consequence (4) is valid because probabilities are always non-negative. Unfortunately, this result conflicts with common sense observations. A famous example is 'the conjunction fallacy' found by Tversky and Kahneman (1983). In one of their experiments, subjects are presented with a story such as the following one:

> Linda is 31 years old, single, outspoken and very bright. She majored in philosophy. As a student, she was deeply concerned with issues of discrimination and social justice, and also participated in anti- nuclear demonstrations. (Tversky and Kahneman 1983: 298)

After the presentation of the story the subjects are asked to assess the probabilities of several propositions on a numbered scale. We represent the critical propositions only (together with the averaged judgements of the probabilities):

(A)	Linda is active in the feminist movement	(0.61)
(B)	Linda is a bank teller	(0.38)
(A & then B)	Linda is a bank teller and is active in the feminist movement	(0.51)

Surprisingly, the probability for a conjunction of A and B is higher than that for the proposition B. Let us examine now how this situation can be handled using quantum probabilities. Fortunately, the example allows for the opportunity to be analysed using the simplest structure possible in quantum theory: a two-dimensional vector space. In this approach, one-dimensional subspaces (represented by unit vectors) realize the propositions *A* and *B* (and their orthogonal counterparts $\bar{A}$ and $\bar{B}$). Further, the relevant knowledge about the system (given by the description of Linda) can be represented by a simple vector *S*, called state vector. Basically, the lengths of the projections of the state vectors onto the vectors representing the

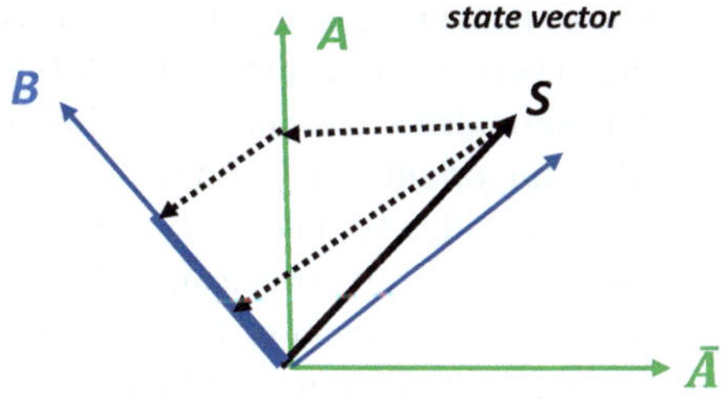

Fig. 1 A vector-based explanation of the conjunction fallacy. (Adapted from Pothos and Busemeyer 2011)

events under discussion are assumed to represent the (quantum) probabilities of the events (the so-called Born-rule).

The important comparison concerns the statements "Linda is a bank teller" (B) and "Linda is a feminist and a bank teller" (A & then B). The first statement corresponds with P(B) = 0.38 (see Fig. 1). The second statement is a conjunction of two statements. Our basic assumption for handling the conjunction is that first the state vector is projected to state A and second the resulting vector is projected to state B (Lüder's rule). The result of this operation is a vector of length 0.51 (see Fig. 1).[6]

Figure 1 illustrates that an ordinary two-dimensional vector space is sufficient to give a resolution of the conjunction puzzle. Complex numbers (as required for spanning a true Hilbert space) are not necessary in the present case. Hence, ordinary projections are sufficient to resolve the conjunction fallacy.

Another puzzle relates to the disjunction effect. As we will see immediately, this puzzle cannot simply be resolved with a real-valued vector spaces but requires a complex vector space (complex Hilbert space) in order to describe interferences. The disjunction effect occurs when conditioned decisions are considered. Obviously, we can rewrite Eq. (3) in the following form using conditional probabilities:

$$P(B) = P(B \mid A)P(A) + P(B \mid \bar{A})P(\bar{A}) \tag{5}$$

The disjunction effect is closely connected to violations of the 'sure-thing principle', one of the basic claims made by a (classically) rational theory of decision making. Let us assume that a decision maker prefers option *B* over option *B* when knowing that event *A* occurs (i.e. $P(B|A) > ½$) and likewise when knowing that event *A* does not occur (i.e. $P(B|\bar{A}) > ½$). Then the 'sure-thing principle' claims that the decision maker should prefer *B* over $\bar{B}$ when not knowing whether *A* occurs or not (i.e. $P(B) > ½$). If the decision maker refuses *B* (or prefers $\bar{B}$), we have a violation of this principle.

[6] Obviously, this operation does not correspond to the intersection $A \cap B$ of two vector spaces. We rename it by "A & then B".

In everyday reasoning, human behaviour is not always consistent with the 'sure-thing principle'. For example, Tversky and Kahneman (1983) reported that more students would purchase a non-refundable Hawaiian vacation if they were to know that they had passed or failed an important exam, compared to a situation where the exam outcome was unknown. Specifically, P(*B*|*A*) = 0.54, P(*B*| $\bar{A}$) = 0.57, and P(*B*) = 0.32, where *B* stands for the event of purchasing a Hawaiian vacation, *A* for the event of passing the exam, $\bar{A}$ for the event of not passing the exam, and P for the averaged judgements of probability. Disjunction fallacies are fairly common in behaviour (Busemeyer and Bruza 2012; Busemeyer et al. 2011).

Classical probability theory does not allow patterns such as $\{\mathrm{P}(B \mid A) > ½, \mathrm{P}((B \mid \bar{A})) > ½, \mathrm{P}(B) \leq ½\}$. Quantum probabilities allow a simple treatment of the puzzle, and a two-dimensional Hilbert space is sufficient for this analysis. As we have seen, in the quantum case, probabilities are calculated from state vectors by a squaring operation. For example, the probability P(B) can be calculated as follows if we introduce the corresponding projection operators ***A***, ***B***, $\bar{A}, \bar{B}$ projecting any state *S* into the subspace indicated by *A*, *B*, $\bar{A}, \bar{B}$:

$$P(B) = \left|BA(S) + B\bar{A}(S)\right|^2 \tag{6}$$

Obviously, we have the correspondences

$$P(A \,\&\, then\, B) = \left|BA(S)\right|^2 ; P(\bar{A} \,\&\, then\, B) = \left|B\bar{A}(S)\right| \tag{7}$$

With a bit of vector space arithmetic, we can rewrite Eq. (6) as follows (for the technical details, see Blutner and beim Graben, 2016).

$$\begin{aligned} P(B) &= P(A \,\&\, then\, B) + P(\bar{A} \,\&\, then\, B) \\ &+ \sqrt{P(A \,\&\, then\, B) \bullet P(\bar{A} \,\&\, then\, B)} \bullet \cos\Delta \end{aligned} \tag{8}$$

Hereby, the angle Δ is a phase angle describing a phase shift between the states ***BA***(*S*) and ***B*** $\bar{A}$ (*S*) making use of a complex vector space.[7]

Considering the numerical values of the Hawaiian vacation example, we get a value of −.23 for the interference term, i.e. the last term of the sum in Eq. (8). From this outcome we can fitting the phase shift parameter: $cos\Delta = -0.42$, i.e. $\Delta = 114°$.

A main topic in applied sociology is the investigation of questions and answers in attitude surveys. Survey researchers have demonstrated repeatedly that the same question often produces quite different answers, depending on the question context (Schuman and Presser 1981; Sudman and Bradburn 1982). To cite just one particularly well-documented example, a group of (North-American) subjects were asked whether "the United States should let Communist reporters come in here and send

[7] Note that in a real-valued vector space the states ***BA***(*S*) and ***B*** $\bar{A}$ (*S*) are both subspaces of B and cosΔ is either 0 or 1. Hence only by making use of Hilbert spaces the term can vary between −1 and + 1.

back to their papers the news as they see it?" The other group was asked whether "a Communist country like Russia should let American newspaper reporters come in and send back to their papers the news as they see it?" Support for free access for the Communist reporters varied sharply depending on whether that question preceded or followed the question on American reporters. The differences are quite dramatic: in a study of 1950, 36% accepted communist reporters when the communist question came first and 73% accepted them when the question came second.

Schumann and Presser (1981) described two kinds of ordering effects, which they called 'consistency' and 'contrast' effects. The example with the case of accepted communist reporters illustrates the *consistency effect*, where, in the context of the other question, the answer frequencies are assimilated. In the 'contrast' case, the differences of the answer frequencies are enlarged. In another article, Moore (2002) reports on the identification of two different types of question-order effects termed as 'additive' and 'subtractive'. All four types of question-order effects can effectively be handled by single qubits (for a detailed treatment, the reader is referred to Wang and Busemeyer 2013; Wang et al. 2014). Interestingly, it is only two parameters that are crucial to defining all four order effects: the angle between the two vectors representing the context and the target question and a single-phase parameter (for the technical details, see Blutner and beim Graben 2016).

From an empirical point of view, the framework of quantum states based on a two-dimensional Hilbert space is appropriate to account for an extended series of mental phenomena. Even though many researchers are satisfied with this situation, there are people who have asked for an independent motivation of this vector framework. What are the final reasons for accepting vector spaces, quantum probabilities, and the idea of complementarity? And what are the ultimate instruments for bringing together order dependencies and uncertainty relations for handling mental entities? A tentative answer is given in the subsequent section.

4 The Operational Interpretation of Quantum Theory

According to Birkhoff and von Neumann (1936), an operational proposition describing a physical entity (i.e. a, propositions being testable by yes/no experiments) can be represented by an orthogonal projection operator (or by the corresponding closed subspace of the Hilbert space). Hence the Hilbert space stands at the beginning of the theory. This fact poses a deep foundational problem: why using a Hilbert space and not any other geometrical structure?

Interestingly, Mackey (1963) started a distinct project for founding quantum mechanics. He considered the set L of all operational propositions, which was restricted in an axiomatic way. In fact, he introduced five axioms on L, and he proved that L is isomorphic to the set of closed subspaces of a generalized Hilbert space. This kind of rational reconstruction of quantum mechanics in terms of the actual operational meaning of the fundamental quantum mechanical concepts was further developed by Gleason (1957), the Genova school (Jauch 1968; Piron 1976), Foulis and Randall (1972), Solér (1995), and many others.

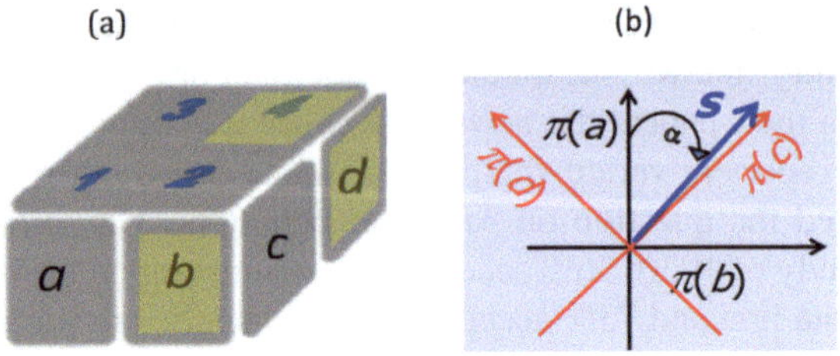

Fig. 2 (**a**) Illustration of the firefly box. If we look from the front perspective, we can find the firefly in the left part (proposition a) or in the right part (proposition b). If we look from the side perspective, we can see the firefly in the left part (proposition c) or in the right part (proposition d). (**b**) Vector representation of the same situation. Hereby, the function $\pi(x)$ assigns vectors to the corresponding propositions (based on Piron's law)

In these approaches, the set of operational propositions is structured by an orthomodular lattice instead of an ordinary Boolean algebra.[8] For orthomodular lattices, two main theorems can be proven, which I represent here in an extremely simplified way:

- Piron's theorem (Piron 1976): Under very general conditions, an orthomodular lattice can be represented by considering the subspaces of a given vector space – realized by map π.
- Gleason's theorem (Gleason 1957): Probabilities are the squares of the lengths of the projections of a state vector into a given vector space (or the convex hull of such projections).

Foulis and colleagues (Foulis 1999; Foulis and Randall 1972) give a handy illustration of the basic ideas. It defines the firefly box and its event logic. Assume that there is a *firefly* erratically moving inside the box depicted in Fig. 2 (left-hand side). The box has two translucent (but not transparent) windows, one at the front and another one at the right. All other sides of the box are opaque. In principle, the firefly can be situated in one of the four quadrants {1,2,3,4}.[9]

For testing whether the firefly is flashing and where it is, the external observer can take one of two perspectives: (i) looking at the front windows, the flash can be seen on the left-hand side (outcome a) or at the right-hand side (outcome b); (ii) looking at the side part, the flash can be seen on the left-hand side (outcome c) or on

[8] Mathematically, an orthomodular lattice has to satisfy the following axioms (the complement operation is indicated by′, conjunction by $\wedge$, and disjunction by $\vee$): (i) $x'' = x$; (ii) if $x \leq y$ then $y' \leq x'$; (iii) $x \wedge x' = 0$; (iv) if $x \leq y$ then $y = x \vee (x' \wedge y)$ (orthomodular law). The main difference between an orthomodular lattice and a Boolean lattice is that for the latter the law of distributivity is valid but not for the former. Hence, the law of total probability can be derived for Boolean lattices only.

[9] In the original example, the firefly can be flashing or not (the latter is indicated by being in world 5). We simplify a bit and ignore the world number 5. For a more detailed discussion, cf. Blutner and beim Graben (2016).

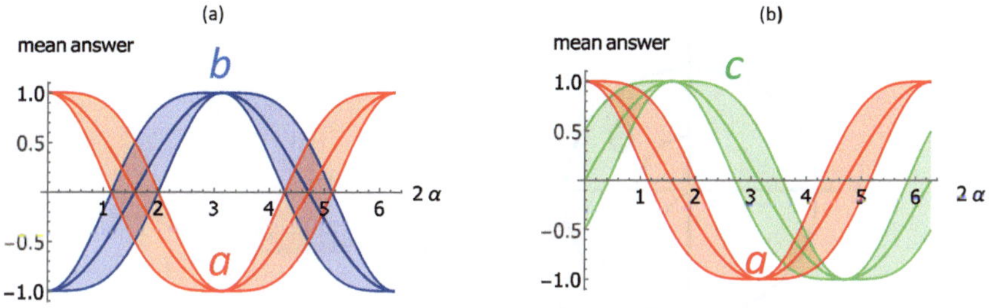

Fig. 3 (**a**) Mean answers for the verification of the opponent propositions a and b (yes = +1, no = −1); (**b**) mean answers for the verification of the complementary propositions a and c. The parameter α is the angle between the state vector s and the vector $\pi(a)$

the right-hand side (outcome d). Technically, the two perspectives are given by two partitions of the domain $W = \{1,2,3,4\}$: (i) front part = $\{a,b\}$; (ii) side part = $\{c,d\}$ (with $a = \{1,3\}$, $b = \{2,4\}$, $c = \{1,2\}$, $d = \{3,4\}$). These two partitions correspond to two Boolean blocks. However, the *union* of these two blocks no longer represents a Boolean lattice. It is weaker and realizes an orthomodular lattice (violating distributivity).

Assume now that the firefly box would have a third window at the top. Then a particular partition would result: top = $\{\{1\}, \{2\}, \{3\}, \{4\}\}$. It relates to an atomic Boolean lattice, the most informative lattice structure that is possible for the given domain W. It allows for exactly asking where the firefly is, in segment 1, 2, 3, or 4. Of course, it is possible to define an operation of 'integration'[10] that would integrate the front perspective with the side perspective into a perspective equivalent to the top perspective.

However, *integration* is a very resource-demanding operation quite different from the operation *union* of the lattice structure.[11]

In Fig. 2b, the state vector s is shown. It allows the calculation of concrete probabilities. The angle between the complementary propositions, represented by the vectors $\pi(a)$ and $\pi(c)$, is assumed being $\pi/4$. In Fig. 3, the calculated probabilities are used for presenting the expected mean answers for the verification of the opponent propositions a and b (yes = +1, no = −1) and for the verification of the complementary propositions a and c. The parameter α is the angle between the state vector s and the vector π/a. Further, the picture shows the standard deviations. In case of the opposite propositions, we get definite answers at angle $\alpha = \pi/2$. In the case of complementary proposition, we do not find an angle with two definite answers for a and c. As expected from complementary proposition, a kind of uncertainty principle is valid.

In the present examples, we have illustrated the role of opponent and complementary propositions with real numbers only. However, we should avoid the impression that the introduction of complex numbers is not essential for the treatment of cognitive phenomena. This is visible already in the Hawaiian vacation example and,

[10]This operation is also called 'refinement' and builds a product partition (beim Graben and Atmanspacher 2006, 2009).

[11]At least, this is true if a theory of resources is assumed as proposed by Halford et al. (1998).

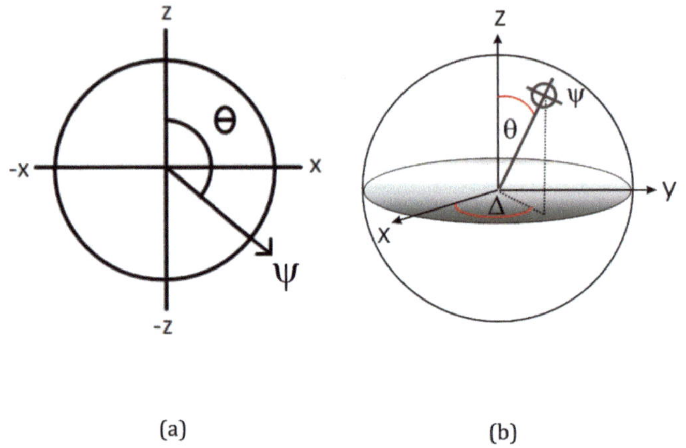

Fig. 4 Qubits: (**a**) in the real vector space (Bloch circle); (**b**) in the complex vector space (Bloch sphere)

further, in the mentioned examples of question-order effects. Figure 4 shows a parameterization of the state vector ψ in the two-dimensional real vector space and complex vector space. The Bloch circle (left-hand side) characterizes the vector ψ with the azimuthal angle θ; the Bloch sphere (right-hand side) characterizes the vector ψ with the azimuthal angle θ and the phase angle Δ.

$$y = \cos\frac{q}{\mathbf{2}}\bullet\begin{pmatrix}\mathbf{1}\\ \mathbf{0}\end{pmatrix} + \sin\frac{q}{\mathbf{2}}\bullet\begin{pmatrix}\mathbf{0}\\ \mathbf{1}\end{pmatrix} \qquad y = \cos\frac{q}{\mathbf{2}}\bullet\begin{pmatrix}\mathbf{1}\\ \mathbf{0}\end{pmatrix} + \sin\frac{q}{\mathbf{2}}\bullet e^{i\Delta}\bullet\begin{pmatrix}\mathbf{0}\\ \mathbf{1}\end{pmatrix}$$

The next illustration (Fig. 5) shows two applications of the qubit model.

The left-hand side, Fig. 5a, illustrates the Bloch circle with one axis representing the rational opposites **T**hinking and **F**eeling and a complementary axis representing the irrational opposites (i**N**tuition and **S**ensation) of Jung's personality theory (Jung 1921). The treatment in terms of qubits was first proposed by Blutner and Hochnadel (2010). The right-hand side, Fig. 5b, illustrates how we can use the Bloch circle for representing the circle of fifths in music theory. As pointed out firstly in Blutner (2017), this way of representation has an empirical impact and provides a new way to represent the attraction potential of chords and scales, which were carefully investigated in cognitive music theory (Krumhansl 1990).

In both cases, the use of complex vector spaces with phase parameters can considerably improve the agreement with the available empirical data, as shown by Blutner and Hochnadel (2010) and Blutner (2017). It is not only the observation of complementarity that is essential for quantum theory. Görnitz and Schomäcker (2018) stress the point that the idea of using complex numbers and analytic

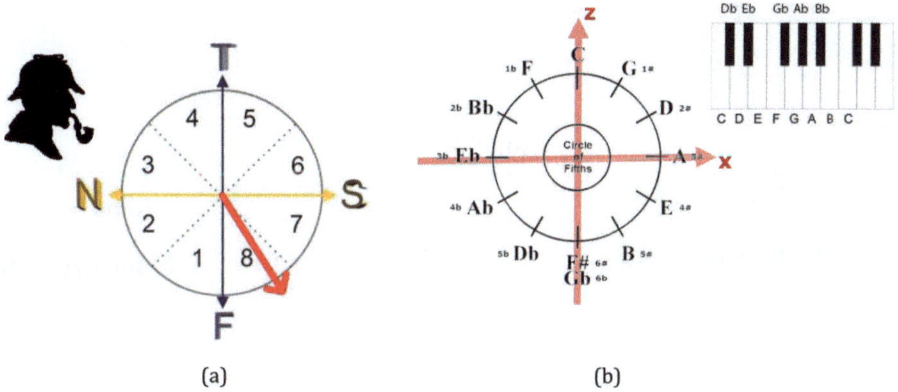

Fig. 5 Two applications. (**a**) A qubit realizing C.G. Jung's personality theory; (**b**) a qubit realization of the circle of fifths in music theory

functions is essential for quantum theory. It is especially the "requirement of complex differentiability analytic behavior" that ensures the holistic character of quantum theory.[12]

5 Consciousness

A problem for the scientific analysis of consciousness is that there is no unique notion of consciousness but a whole field of different but related notions.

In *Philosophical Investigations*, Wittgenstein (1953) introduces the notion of a 'family resemblance' to deal with certain problems of concept formation. Famously, he explained it by the concept family related to the word 'game'. The very same idea can straightforwardly be applied to the term 'consciousness'. Here are some concepts relating to the corresponding family, cf. Chalmers (1996: 25 ff):

- Consciousness as attentiveness including the ability to differentiate whether the object of cognition is internal or external. Access Consciousness
- Consciousness as raw experience realizing our Qualia. Phenomenal Consciousness
- Consciousness as mechanism for constructing our Self. Self-Consciousness
- Spiritual Consciousness, connoting the relationship between the mind and God and including the experience of meditation
- Stream of consciousness, altered states of consciousness, animal consciousness, etc. For Chalmers (1996) the family can be divided into two parts:

[12] In case of analysing the phenomenon of 'tonal attraction', a number of different empirical observations can be introduced in terms of a musical gauge field based on the internal symmetry group SU(2) (beim Graben and Blutner 2019).

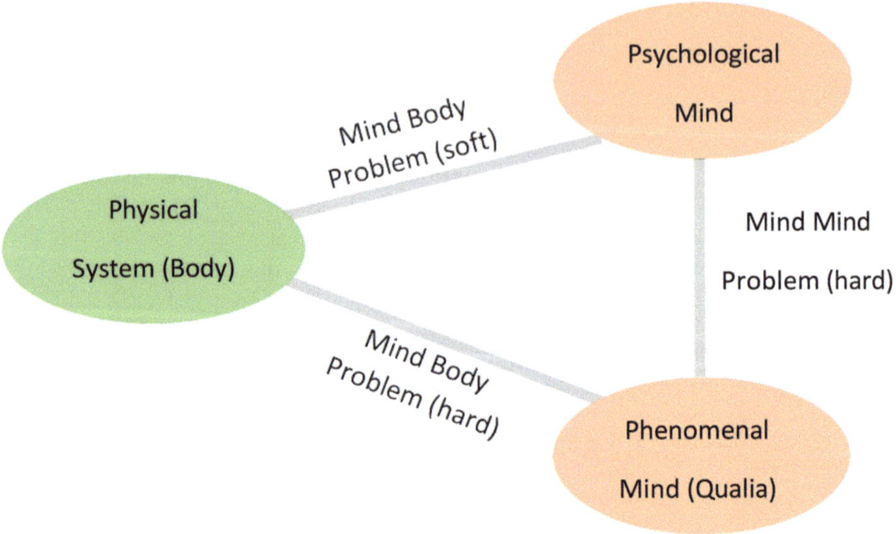

Fig. 6 Mind-body relations. (Following Chalmers 1996)

For now, all that counts is the conceptual distinction between the two notions: what it means for a state to be phenomenal is for it to feel a certain way, and what it means for a state to be psychological is for it to play an appropriate causal role (in explaining behaviour, RB). … At a first approximation, phenomenal concepts deal with the first-person aspects of mind and psychological concepts deal with the third-person aspects (Chalmers 1996: 12–16).

Taking the psychological and phenomenal conception of mind into account and adding a material body (physical system), we get the following picture of mind-body relations (Fig. 6):

We consider attentiveness and awareness as related to psychological consciousness. It is the most relevant concept for the present discussion. Ignoring the phenomenal mind for the moment, this allows to see the mind-body problem as a "soft" problem (in the sense of Chalmers 1996). Possibly, it can be cracked along the lines pursued by Price and Barrell (2012).[13,14]

[13] Also, the notion of spiritual consciousness including the experience of meditation deserves attention. However, even in this case, I see the notion of psychological consciousness in the foreground, and the topic of investigation is its correlation with certain physiological parameters. In the sense of Planck (1947), I consider the mind-mind problem and the hard mind-body problem as Schein problems of science. However, this does not have any visible consequences because the "real" scientific problems can be solved based on the psychological conception of mind.

[14] Recently, several models with network-like abilities have been proposed for the modelling of conscious and subconscious processes (Anderson 1990; Blutner 2004; Grossberg 2021). In contrast with Görnitz (2018), I cannot see that these ideas give an explication of consciousness in terms of quantum theory. I know a handful of papers only that directly connect neural networks with quantum effects (e.g. Acacio de Barros and Suppes 2009). These papers, however, do not refer

Awareness is a graded conception. We can be aware of something to different degrees. The simplified distinction between automatic and controlled processing (Schneider and Shiffrin 1977) makes a binary distinction instead of the finer classification by degrees. Awareness is related to cognitive resources. Automatic processing demands less resources than controlled processing. Authors such as Logan (1988) have pointed out that also the allocation of resources is a graded matter.

The foundational approaches to quantum theory discussed in the previous section exploit ideas that do not acknowledge the Hilbert space as a given conceptual framework but rather try to motivate it.

Following the research by Foulis, Randall, and colleagues, such a foundational framework is motivated by assuming partial Boolean algebras that describe the perspectives of a cognitive agent. Then there are two possibilities: (i) the perspectives are unified resulting in an orthomodular lattice – a structure that can violate distributivity; (ii) using Boolean refinement, the perspectives are integrated into a new perspective which is still Boolean. The latter kind of processing demands much more resources than the former. It is at this point that one important foundational aspect enters the theoretical scenery of quantum cognition: resource limitation. Resource limitation has the effect that a cognitive agent cannot simultaneously maintain all possible perspectives and integrate them into a new and more refined structure.

Hence, the whole quantum framework becomes dependent on the psychological concept of awareness and the allocation of (limited) cognitive resources (in the sense of Halford et al. 1998). If sufficient cognitive resources are not available, there is simply no way to combine the different Boolean blocks in a conjunctive way. Hence, it is not possible to generate a more refined Boolean algebra that allows for rational decisions. Instead, we must accept an orthomodular lattice and the weaker decision structures that are based on quantum probabilities (according to Gleason's theorem).

This argument properly relates to Damasio's theory of consciousness (Damasio 1994, 1999), in particular to his idea to distinguish a "high-reason" view of decision making from a more emotional but less rational view of decisions. The quantum mode with very fast decisions comes into play if not enough cognitive resources are allocated. By contrast, the more rational classical Boolean mode matters if sufficient resources are available (conscious, controlled processing). In Sect. 3 we have seen that the idea of qubits forms a powerful instrument for analysing and solving different puzzles and problems of cognitive psychology. For the last 30 years Thomas Görnitz and colleagues have been exploring if a qubit-based physics is possible (Görnitz 2011, 2014, 2018; Görnitz and Görnitz 2016; Görnitz and Schomäcker 2018). This millennium project is based on ideas of Carl Friedrich von Weizsäcker of seeing matter as information. According to Görnitz, qubits "provide a pre-structure for all entities in natural sciences. They are the basic entities, whereof the physical nature of the brain, on the one hand, and the mental nature of

to awareness or consciousness. Hence, the present ideas do not contain a novel (quantum) mechanism for handling consciousness. Rather, they provide some constraints for the route to this goal.

consciousness, on the other hand, were formed during the cosmological and the following biological evolution" (Görnitz 2018: 475). In other words, on the basis of qubits, massless and massive quantum particles can be constructed as well as grammatically formed thoughts (Aerts and Beltran 2021), as well as musically shaped ideas and mental structures (beim Graben and Blutner 2019). It goes without saying that handling the physical part of the problem is much more demanding than handling the psychological part.

From a philosophical perspective, the idea that qubits form the base for all quantum structures can help to achieve a deeper understanding of the mind-body problem. Qubits are the "substance" for constructing both mental structures and physical ones. Hence, mind and body are different aspects of the same elementary basic structures made of information (qubits). Of course, this picture is completely different from Descartes' dualist view of the mind-body problem. Concerning the many different proposals for clarifying the mind-body problem, I think the present idea comes closest to Spinoza's double-aspect view and the Indian philosophy in the Vedanta tradition. God is the ultimate for Spinoza, Brahman that for the Hindus, and abstract information that for von Weizsäcker, Görnitz, and their followers.[15] Concerning the future development of the idea of complementarity, I think the establishment of a vital connection between the efforts of Thomas Görnitz and mainstream quantum cognition can be beneficial.

Acknowledgement My thanks go to Anand Srivastav for inviting me to contribute to this issue and for directing my research in a certain direction – trying to connect the field of quantum cognition with the work of Thomas Görnitz. Further, I would like to thank Peter beim Graben for his vital and critical comments and Stefan Blutner-Montaño for feedback on an earlier draft of this chapter.

References

Aerts, Diederik. 1982. Example of a macroscopical classical situation that violates Bell inequalities. *Lettere Al Nuovo Cimento* 34 (4): 107–111.

———. 2009. Quantum structure in cognition. *Journal of Mathematical Psychology* 53: 314–348.

Aerts, Diederik, and Lester Beltran. 2021. Are words the quanta of human language? Extending the domain of quantum cognition. *Entropy* 24: 27.

Anderson, John R. 1990. *The Adaptive Character of Thought*. Hillsdale: Erlbaum.

Atmanspacher, Harald, Hartmann Römer, and Harald Walach. 2002. Weak quantum theory: Complementarity and entanglement in physics and beyond. *Foundations of Physics* 32 (3): 379–406.

beim Graben, Peter. 2004. Incompatible implementations of physical symbol systems. *Mind and Matter* 2 (2): 29–51.

beim Graben, Peter, and Harald Atmanspacher. 2006. Complementarity in classical dynamical systems. *Foundations of Physics* 36 (2): 291–306.

[15] In the field of music, Mannone (2018) refers to another aspect of the mind-body problem referring to musical gestures that connect the cognitive-symbolic layer of music to the physical layer of sound.

———. 2009. Extending the philosophical significance of the idea of complementarity. In *Recasting Reality. Wolfgang Pauli's Philosophical Ideas and Contemporary Science*, ed. H. Atmanspacher and H. Primas, 99–113. Berlin: Springer.
beim Graben, Peter, and Reinhard Blutner. 2019. Quantum approaches to music cognition. *Journal of Mathematical Psychology* 91: 38–50.
Birkhoff, Garret, and John von Neumann. 1936. The logic of quantum mechanics. *Annals of Mathematics* 37 (4): 823–843.
Blutner, Reinhard. 2004. *Neural Networks, Penalty Logic and Optimality Theory*. Amsterdam: ILLC.
———. 2017. Modelling tonal attraction: Tonal hierarchies, interval cycles, and quantum probabilities. In *Logic, Music and Quantum Information*, ed. M.L. Dalla Chiara, R. Giuntini, E. Negri, and S. Smets. Berlin: Springer. https://doi.org/10.1007/s00500-015-1801-7.
Blutner, Reinhard, and Peter beim Graben. 2016. Quantum cognition and bounded rationality. *Synthese* 193 (10): 3239–3291.
Blutner, Reinhard, and Elena Hochnadel. 2010. Two qubits for C.G. Jung's theory of personality. *Cognitive Systems Research* 11 (3): 243–259. https://doi.org/10.1016/j.cogsys.2009.12.002.
Busemeyer, Jerome R., and Peter D. Bruza. 2012. *Quantum Cognition and Decision*. Cambridge: Cambridge University Press.
Busemeyer, Jerome R., Emmanuel M. Pothos, Riccardo Franco, and Jennifer S. Trueblood. 2011. A quantum theoretical explanation for probability judgment errors. *Psychological Review* 118 (2): 193–218.
Caves, C.M., C.A. Fuchs, and R. Schack. 2002a. Quantum probabilities as Bayesian probabilities. *Physical Review A* 65 (2): 022305.
———. 2002b. Unknown quantum states: the quantum de Finetti representation. *Journal of Mathematical Physics* 43: 4537–4559.
Chalmers, David J. 1996. *The Conscious Mind. In Search of a Fundamental Theory*. Oxford: Oxford University Press.
Conte, Elio 1983. Exploration of biological function by quantum mechanics. In: *Proceedings of the 10th International Congress on Cybernetics, Namur-Belgique*, 16–24.
Conte, Elio, Aandrei Y. Khrennikov, Orlando Todarello, Roberta De Robertis, Antonio Federici, and Joseph P. Zbilut. 2008. A preliminary experimental verification on the possibility of Bell inequality violation in mental states. *NeuroQuantology* 6 (3): 214–221.
Damasio, Antonio R. 1994. *Descartes' Error: Emotion, Reason, and the Human Brain*. New York: Penguin Putnam.
———. 1999. *The Feeling of What Happens – Body and Emotion in the Making of Consciousness*. New York: Harcourt Brace.
De Barros, J.A., and P. Suppes. 2009. Quantum mechanics, interference, and the brain. *Journal of Mathematical Psychology* 53 (5): 306–313.
Foulis, David J. 1999. A half century of quantum logic – What have we learned? In *Quantum Structures and the Nature of Reality: The Indigo Book of Einstein Meets Magritte*, ed. D. Aerts and J. Pykacz, 36. Dordrecht: Kluwer. https://doi.org/10.1007/978-94-017-2834-8_1.
Foulis, David J., and C.H. Randall. 1972. Operational statistics I: Basic concepts. *Journal of Mathematical Physics* 13: 1667–1675.
Gärdenfors, Peter. 2000. *Conceptual Spaces: The Geometry of Thought*. Cambridge: The MIT Press.
———. 2014. *The Geometry of Meaning: Semantics Based on Conceptual Spaces*. Cambridge: The MIT Press.
Gleason, Andrew M. 1957. Measures on the closed subspaces of a Hilbert space. *Journal of Mathematics and Mechanics* 6: 885–894.
Görnitz, Thomas. 2011. The meaning of quantum theory: Reinterpreting the Copenhagen interpretation. *Advanced Science Letters* 4: 3727–3734.
Görnitz, Tomas. 2014. The basis for an understanding of matter and mind. *Foundations of Science* 19: 257–262.
Görnitz, Thomas. 2018. Quantum theory and the nature of consciousness. *Foundations of Science* 23: 475–510.

Görnitz, Thomas, and Brigitte Görnitz. 2016. *Von der Quantenphysik zum Bewusstsein: Kosmos, Geist und Materie*. Berlin/Heidelberg: Springer.
Görnitz, Thomas, and Uwe Schomäcker. 2018. The structures of interactions: How to explain the gauge groups U(1), SU(2) and SU(3). *Foundations of Science* 23 (1): 51–73.
Grossberg, Stephen. 2021. *Conscious Mind, Resonant Brain: How Each Brain Makes a Mind*. Oxford: Oxford University Press.
Halford, Graeme S., William H. Wilson, and Steven Phillips. 1998. Processing capacity defined by relational complexity: Implications for comparative, developmental, and cognitive psychology. *Behavioral and Brain Sciences* 21 (6): 803–831.
Heisenberg, Werner. 1944. *Die physikalischen Prinzipien der Quantentheory*. Leipzig: S. Hirzel.
Howard, Don. 2004. Who invented the "Copenhagen Interpretation"? A study in mythology. *Philosophy of Science* 71 (5): 669–682.
Hümmler, Holm Gero. 2017. *Relativer Quantenquark*. Berlin/Heidelberg: Springer.
James, William. 1890. *The Principles of Psychology*. New York/London: Holt and Macmillan.
Jauch, Josef Maria. 1968. *Foundations of Quantum Mechanics*. Reading: Addison-Wesley.
Jung, Carl Gustav. 1921. *Psychologische Typen*. Zürich: Rascher.
Krumhansl, Carol L. 1990. *Cognitive Foundations of Musical Pitch*. New York: Oxford University Press.
Logan, Gordon D. 1988. Toward an instance theory of automatization. *Psychological Review* 95: 492–527.
Mackey, George W. 1963. *Mathematical Foundations of Quantum Mechanics*. Reading: Benjamin.
Mann, Frido, and Christine Mann. 2017. *Es werde Licht. Die Einheit von Geist und Materie in der Quantenphysik*. Frankfurt: S. Fischer.
Mannone, Maria. 2018. Knots, music, and DNA. *Journal of Creative Music Systems* 2 (2): 1–23.
Moore, David W. 2002. Measuring new types of question-order effects: Additive and subtractive. *The Public Opinion Quarterly* 66 (1): 80–91.
Piron, Constantin. 1976. *Foundations of Quantum Physics*. Reading: W. A. Benjamin, Inc.
Planck, Max. 1947. *Scheinprobleme der Wissenschaft von Max Planck. Vortag gehalten in Göttingen am 17. Juni 1946*. Leipzig: J. A. Barth-Verlag.
Pothos, Emmanuel M., and Jerome R. Busemeyer. 2011. Formalizing heuristics in decision-making: a quantum probability perspective. *Frontiers in Psychology* 2: 3.
Price, Donald D., and James J. Barrell. 2012. *Inner Experience and Neuroscience: Merging Both Perspectives*. Cambridge: The MIT Press.
Schneider, Walter, and Richard M. Shiffrin. 1977. Controlled and automatic human information processing. *Psychological Review* 84 (1): 1–66.
Schuman, Howard, and Stanley Presser. 1981. *Questions and Answers in Attitude Surveys: Experiments in Question Form, Wording, and Context*. New York: Academic.
Solér, M.P. 1995. Characterization of Hilbert spaces by orthomodular spaces. *Communications in Algebra* 23 (1): 219–243.
Sudman, Seymour, and Norman M. Bradburn. 1982. *Asking Questions*. San Francisco: Jossey-Bass Inc. Pub.
Tversky, Amos, and Daniel Kahneman. 1983. Extension versus intuitive reasoning: The conjunction fallacy in probability judgment. *Psychological Review* 90 (4): 293–315.
Walach, Harald. 2019. *Beyond Materialist Worldview. Towards an Expanded Science*. London: The Scientific and Medical Network.
Wang, Zheng, and Jerome R. Busemeyer. 2013. A quantum question order model supported by empirical tests of an a priori and precise prediction. *Topics in Cognitive Science* 5 (4): 689–710.
Wang, Zheng, Tyler Solloway, Richard M. Shiffrin, and Jerome R. Busemeyer. 2014. Context effects produced by question orders reveal quantum nature of human judgments. *Proceedings of the National Academy of Sciences* 111 (26): 9431–9436.
Wittgenstein, Ludwig. 1953. *Philosophical Investigations*. New York: Macmillan.

Talking About Consciousness in the Sciences and the Arts

Anna Margaretha Horatschek

Abstract The format and mediality of how consciousness is represented – in mathematical equations, neurological images, graphs, scientific essays, novels, or poems – on the one hand depend on the premises of what consciousness is and the ensuing methodology of investigation, and on the other hand communicate (only) specific aspects of the phenomenon we experience as consciousness. My essay will address the differences in concept and representation of consciousness in a secular-materialistic and spiritual worldview, with the examples of the science fiction neuro-novel *Blindsight* (2006) by Canadian author Peter Watts, and a selection of poems by the Indian mystic Kabir (1440–1518). On the basis of these literary negotiations about the potential and limits of verbal communication, the essay explores the problems of first-person reports in Consciousness Studies.

Keywords Consciousness and representation · First-person reports · Kabir · Mysticism · Consciousness Studies and the neuro-novel · Peter Watts *Blindsight* (2006) · Consciousness Studies and literature

1 What's the Problem? A Philosophical Prequel

"We are unlikely to arrive at an adequate understanding of consciousness in the absence of a transdisciplinary perspective" (Zelazo et al. 2007: 2). However, the central problem of transdisciplinary studies is to find – or produce – a shared conceptual language. Thus, the term 'consciousness' means different concepts for different researchers, suggesting shared knowledge where there is only linguistic

A. M. Horatschek (✉)
Department of English and American Literatures, Cultures, and Media, Kiel University, Kiel, Germany
e-mail: horatschek@anglistik.uni-kiel.de

P. S. Satsangi et al. (eds.), *Consciousness Studies in Sciences and Humanities: Eastern and Western Perspectives*, Studies in Neuroscience, Consciousness and Spirituality 8, https://doi.org/10.1007/978-3-031-13920-8_20

homonymy.[1] For one group consciousness consists in nothing but brain activities and therefore can be studied and explained comprehensively in models and concepts of the natural sciences; a second group sees consciousness as qualitatively different from or ontologically prior to materiality. Here we find dualism, secular Panpsychism, and speculative realism (Shaviro 2014), but also spiritual and religious models, where consciousness may be understood as a trans-individual, dynamic, and all-pervading divine presence (substance dualism). For Western adherents of this second group, talking about consciousness presupposes the representation of the phenomenon 'consciousness' in verbal language (only recently did the idea of grasping metaphysical phenomena and concepts in mathematical formulas enter the academic discourse, e.g. in experimental or computational metaphysics). Philosophers like Steven Katz (1978) draw attention to the epistemological aspects of this process of translation. They hold that all representations of a phenomenon in verbally explicated concepts – including the concept of 'consciousness' – are bound to historically and culturally specific premises, media, and forms of representation that are archived, manifest in, and transferred via language. This means that what consciousness *is* cannot be captured comprehensively in any one model, no matter how elaborate, but that each model will open up new facets of approaching a phenomenon that involves

1. Physiology in terms of sensual body awareness and body memory,
2. Intellect and mentality in terms of historically and culturally specific discourses on knowledge, emotion, and their place in a specific overall image of world and cosmos,
3. Spirituality grounded in or emanating from experiences of transcendence.

All three aspects merge in the exclusive manifestation of and access to consciousness, namely in individual experience, yet for scientific analysis they have to be differentiated heuristically and studied with culturally and historically specific disciplinary methodologies, depending on which aspect is the focus of investigation.

For Eastern – more specifically Indian – proponents of a substance-dualistic model, the question of representing the pre-linguistic phenomenon – or experience – of consciousness may look very different. As Annette Wilke in her contribution to this anthology points out with reference to language and 'sonic consciousness':

> [E]ven beyond Sanskrit Hinduism, texts in India are invariably not only discursive bodies, but sound events and performative scripts as well. They turn into highly aesthetic, sensory-affective media, being vocalized, recited, sung, danced, and dramatically staged, affecting not only the performers but also the listeners. Having been composed for hearing, for memorizing, for embodied perception, they create emotional and body knowledge beyond intellectual cognition. (Wilke, Sonic Consciousness)

[1] For systematic distinctions of various meanings of the term 'consciousness' in the horizon of Western epistemology cf. Kriegel 2007; for a historical perspective on the development of concepts of consciousness cf. Seager 2007, both in Zelazo et al. 2007; for a representative history of concepts of consciousness in British literature cf. Horatschek 2016.

Ralph Yarrow in his essay on theatrical improvisation makes a similar point, namely that in Indian aesthetics representation does not necessarily imply distancing from the ontologically prior object of representation, but through elaborate strategies of performativity and ritual, semiotic representation produces an experience in the observer that gives presence to the represented object.[2]

My essay will address the differences in concept and representation of consciousness in a materialistic and a spiritual worldview as illustrated and negotiated in the 2006 neuro-novel *Blindsight* by Canadian author Peter Watts, a hard science fiction emphasising scientific accuracy, and a selection of poems by the fifteenth-century Indian mystic Kabir. Taking my lead from these literary texts, I shall then assess the role of language in first-person reports in the frame of Consciousness Studies. The theoretical premise behind this methodological approach via literature is the claim that literature is the laboratory of each culture, in which the books function like test tubes, enacting specific world models and knowledge formations in human life-worlds. In my examples, they offer highly self-reflexive experiments concerning the potential and limits to capture individual experiences of consciousness in verbal language.

The literary texts by Watts and Kabir negotiate radically different models of consciousness: The fictional society of Watts' *Blindsight* holds the conviction that consciousness is matter, namely brain mass that can be surgically modelled – or 'enhanced' in their diction – in order to produce an attitude in tune with the demands of imperial superpower dreams. The mystical poet Kabir on the other hand in his lines evokes the presence of a divine power of which he says: "The conscious and the unconscious, both are His footstools" (Kabir 2015: 52–53).

Already the literary genres of the respective texts, namely novel and poetry, illustrate in exemplary fashion different forms of representation as outlined above: The novel, a distinctly 'Western' literary genre that emerged in Europe in the eighteenth century, presents the (fictional) reality discursively as a narrative for an educated reading public; poetry, a genre originated in oral cultures all over the world, aims to re-produce the depicted experience in the listener by complex strategies of language, tone, style, and performance.

The first part of my chapter shall present the different concepts of consciousness in Watts' novel and in Kabir's poetry. In a second step I shall assess the status of first-person reports in Consciousness Studies in light of these consciousness models and the concomitant forms of representation. My focus will lie on the qualitative transformation of first-person experience in the process of translating it into the third-person report of scientific data in order to align the format of representation with the normative demands of empirical research. The conclusion will highlight the particular significance of diverse representations of consciousness for Consciousness Studies.

[2] For the epistemological implications of this model see Annette Wilke, "Sonic Consciousness in Hindu India" in this volume.

2 Peter Watts' *Blindsight*: Consciousness Detached

The first-person narrator of *Blindsight*, Siri Keeton, has had a "*radical* hemispherectomy: half the brain thrown out" (17).[3] It has been replaced by hardware enhancements, which plug him into an algorithmic knowledge reservoir geared at 'objective' observation. By way of self-description, Siri says: "Imagine that your purpose is not to replicate, or even to survive, but to gather information" (40). His objectivity depends on him not getting entangled with the objects of his observation. Entanglement is produced by emotions, because – in the words of Alfred North Whitehead – 'feeling' is "the basic generic operation of passing from the objectivity of the data to the subjectivity of the actual entity in question [...] effecting a transition into subjectivity. [Feelings] replace the 'neutral stuff' [...]" (Whitehead 1978 [1929]: 40f.). Accordingly, the brain surgery has eliminated Siri's feelings as far as possible (cf. 232). As a consequence, Siri incarnates an artificial intelligence with outstanding cognitive abilities for the price of not knowing what it feels like to be a person. However, because he still remembers his human condition before the lobotomy, he has become part of a crew on an imperial mission in order to translate the vast knowledge of his largely artificial intelligence fed by global networks of algorithms into a comprehensive verbal narrative for 'unenhanced' human beings in the spaceship and on earth. The unenhanced human beings on earth – on a different narratological level – of course include the reader of the book. The text itself thus testifies to Frank's observation that "narrative is central to human cognition in a way that database is not, that 'narrative models show how the mind thinks and how the world works'" (179) (Frank 2017: 232f.; quoting Hayles 2012).

With the severance of "cognition from consciousness" in his "post-human mind [...]" (Galant 2017: 27), Siri is a pure medium. Since on "February 13, 2082" aliens that look like a cosmic network of fireflies (37) surrounded the earth and then vanished again, the narrator together with four more "alters" or post-humans is sent on an interstellar mission to find out where the aliens came from. The constellation of the crew according to Galant "offers a vivisection of consciousness" (Galant 2017: 28). My analysis shall concentrate on the consciousness of Siri Keeton.

The novel starts with the sentence: "It didn't start out here" (28). The oxymoron 'out here' refers to Siri's location – 'out' in the interstellar world from a terrestrial perspective – and 'here' in the spaceship. It also refers to his loss of what French philosopher Paul Ricoeur calls *ipse* identity (Ricoeur 1991) and the consequential split of competence from comprehension, in Dennett's terms (Dennett 2017: 97–101). The *ipse identity* for Ricoeur is the awareness that it is me who has an experience, while what he calls *idem identity* refers to the surmised continuity of a self as it is constructed in biographies. Damasio calls analogous aspects the "*core* consciousness" and the "*extended* or *autobiographical* consciousness" (Damasio 2010: 130), or the self-as-subject and the-self-as-object (Damasio 2010: 8), and

[3] Page references in brackets following quotes without further specifications in this part of the essay refer to the edition Peter Watts. *Blindsight*. New York: Tor Books, 2006.

organises them in an ontological hierarchy: "The self-as-knower is grounded on the self-as-object" (Damasio 2010: 18).

For Siri, people have transformed into semiotic networks: "[I]f you […] ignored meat and studied *topology* [, e]very facial tick was a data point [, e]very twitch of the phenotype cried aloud to anyone who knew the language" (24). To fit into society despite his near-total absence of feelings, Siri after his lobotomy "observed, recorded, derived the algorithms and mimicked appropriate behaviours. Not much of it was ... *heartfelt* I guess the word is" (17).

With feelings lobotomised, in Damasio's model of consciousness there can be no ethics because "emotion returns us to the matter of […] value. [...] Social emotions […] form a natural grounding for ethical systems" (Damasio 2010: 100). Siri is lacking such a grounding. In his own words: "It's not so much about imagining how the other guy feels. It's more about imagining how *you'd* feel in the same place, right? […] So what if you don't *know* how you'd feel?" (233). His inability to infer the state of the other's subjectivity from his own sense of self manifests in his physical brutality against others, as he perceives them as 'thing' and 'it', devoid of a shared humanity and thus exempt from any responsibility on his part (cf. 13–15).

The novel's title *Blindsight* comments on such a mode of cognition without consciousness in various ways. Blindsight is a psychological condition discovered in the early 1970s by British psychologist Lawrence Weiskrantz (1926–2018), when blind persons discriminated visually presented objects rather accurately. The title in this context firstly highlights Siri's abilities to translate data into (narrative) information, although he is blind for their meaning, understood as "an organization of mind contents centered on the organism that produces and motivates those contents" (Damasio 2010: 18). Secondly, the title refers to the functional organisation of the aliens, as their actions are orchestrated like a huge network by some consciousness stuff or intentionality outside their own bodies. They function without knowing the meaning of their actions. And thirdly, Blindsight refers to what Siri's unenhanced former girlfriend calls 'a gut feeling' that feeds her conviction that life-like robots "aren't *people*, even if it can't put its finger on *how* it knows. They just don't *feel* real" (39).

But Siri is different from a machine robot in that he still has remnants of feelings. These feelings are activated in a situation when he fears for his life: Being assaulted by the leader of the team, the Vampire Sarasti, Siri regains his feeling of an individual self and thus his *ipse identity*, because his panic reduces his perception to his very specific situatedness and thus reinstalls his phenomenal consciousness when – in Chalmers' terms – "there is something it is like to be that system, from the first-person point of view" (Chalmers 2018: 6). With his *ipseity* regained, Siri "[a]t long last […] can empathize with Sarasti […]" (362).

Although this retrieval of sentience in terms of the book makes Siri human again, he does not perceive it as a gain, because from the purely functional point of view predominant in the fictional society he lives in, which has been 'biologised' and embodied in him with his lobotomy, empathy is a hindrance in the attempt to conquer the universe. Additionally, by having retrieved his sentience, Siri experiences a phenomenon he cannot reify as an object of investigation without eliminating the 'actual occasion' or 'actual entity' – in Whitehead's diction – of processual experience. "An actual entity is a process, and is not describable in terms of the morphology of a 'stuff' [...] these actual entities are drops of experience, complex and interdependent" (Whitehead 1978 [1929]: 40f. and 18).

Siri is fully aware that his core consciousness cannot be captured as an object, a thing, or a 'stuff' in any semiotic system, be it verbal or numerical, because its processual subjectivity jars with scientific norms of representation geared at objectification and externalisation. At the end of the novel, he asserts the impossibility of proving the existence of his *ipseity* and refers the reader to their own *ipse* identity as a reference point for 'grounding' what he's talking about: "You'll just have to imagine you're Siri Keeton" (362).

By narrating his retransformation from a medium of observation into a conscious human being, Siri creates a new 'narrative identity' (Ricoeur) for himself, comprising his regained *ipse identity* with his biotechnologically enhanced artificial intelligence still intact. Additionally, as the perfect interface between the more spatially conceptualised data bases of his algorithmic intelligence and the more temporal technologies of narratives, his text "mak[es] comprehensibility of incomprehensibility" (Frank 2017: 232), albeit only in the realm of mental and intellectual representation, and thus at an ontological distance from the emotional involvement and experience of the recipients on earth, including the reader of the book.

3 Kabir's Poems: Consciousness Immersed

Kabir was a late fifteenth-century mystic poet and a weaver, born in Banares (today Varanasi) and raised in a Muslim family. Very few of Kabir's poems can be attributed to him with certainty, which for my purposes is no drawback, as they thus exhibit several spiritual traditions of India (Bhakti, Sikhs, Vaishnavis, and Sant Mat). For the English versions I am relying on the 1915 translations by Rabindranath Tagore.[4]

Kabir belonged to the devotional Bhakti movement, "whose focus is on inward love for the One Deity, in opposition to religious orthodoxies and social hierarchies" (Mehrotha 2011). Accordingly, he explicitly rejects all religious or spiritual authorities, books, and institutions. Assuming the voice of his God, he says: "I am neither in temple nor in mosque: I am neither in Kaaba nor in Kailash: Neither am I in rites and ceremonies, nor in Yoga and renunciation" (43). "It is needless to ask of a saint the caste to which he belongs; [...] Hindus and Moslems alike have achieved that End, where remains no mark of distinction" (45–46). "Yoga and the telling of beads, virtue and vice – These are naught to Him" (71). The only authority Kabir acknowledges is his immersion in a sentient presence variously called "God" (45), "the

[4] All page numbers without additional bibliographical reference in brackets after quotes in this section refer to the edition *Songs of Kabir*. 2015. Trans. Rabindranath Tagore. Assisted by Evelyn Underhill. Mansfield Centre (USA): Martino Publishing.

Creator" (52), "my Lover" (54), "Lord Increate" (56), "the Supreme Brahma" (57), "the Self-poised One" (63), "The Supreme Soul" (51), or "the Supreme Spirit, the great Master, your Beloved" (68).

Repeatedly, the voice reflects on the limits of language and other symbolic forms to comprehend and represent "the Unattainable One", "that Hidden One" (62), "that formless God" (75):

> Oh how may I ever express that secret word?
> Oh how can I say He is not like this, and he is like that?
> [...] The conscious and the unconscious, both are His footstools. (52–53)

As there is no way to adequately describe this realm discursively, representation is replaced by performativity: the feelings of joy, love, and longing are not described but evoked in the reader by artful language, adopting Hindu-, Sufi-, Sikh-, and Sant Mat formal traditions like frequent addresses of his deity, exclamations, paradoxes, passionate love poetry, metaphors of sensual perception, gender crossing, and changing subject positions subverting caste and gender hierarchies.

Such a performative function of artistic forms is best theorised in the Indian rasa aesthetics, set down in the *Natyashastra*, a poetics of the arts from between 200 BCE and 200 CE. Sriramamurti expounds this strategy of knowledge production through experience with reference to the central term 'Camatkara': "[...] 'camat' is the present participle of the verbal root 'cam' meaning taste, eat and hence enjoy. Thus, 'camattvam' means getting immersed in tasting or enjoyment of something, particularly an aesthetic or mystical kind" (Sriramamurti 2016: 391). The grammatical form of the present participle indicates that something is going on right now and thus points towards an actually present experience in the moment of hearing – or reading.

In contrast to the world of *Blindsight*, the medium of knowledge transfer and production here is not the intellect, but the recipients' sensual body in their culturally specific semiotic encodedness to respond to specific tastes, sounds, gestures, colours, and smells. The result is not discursive knowledge, but a pre- or transdiscursive experience. In the words of Sriramamurti: "Camatkara [...] is not of a discursive order" (Sriramamurti 2016: 391). Accordingly, Kabir explicitly marks his knowledge and its representation as non-discursive and compares it to the bird's song, who "says not a word of that which it means" (79), because "the wise man is speechless; for this truth may never be found in Vadas or in books" (63). Only "the Ultimate Word" (64), where semantics do not arise from structural difference, but from sensually experienced sound – *shabda* – can performatively evoke his experience of consciousness. In Kabir's words: "The unstruck drum of Eternity is within me [...]" (49). "[...] the sound of the unseen bells is heard" (60); "the Unstruck Music is sounded" (63).

4 Comparative Synthesis

In Watts' *Blindsight* and Kabir's poetry, the term consciousness has a fundamentally different meaning: in Siri's world, consciousness in a chain of upward causation (cf. Fenwick 2004: 111) is a dispensable attribute of intellectual knowledge production, founded in physiology, that adds self-awareness and allows for empathy and ethics, and which therefore has been lobotomised in order to not interfere with the imperial strategies of his society. In this model, consciousness figures as – in Dennet's terms – a "decoration, an ideological amplification" (Dennett 2017: 100). However, as the novel makes clear, Siri experiences the world in a different way after he has regained his consciousness in the sense of self-awareness: He is aware that he no longer only reads the other people as semiotic objects in order to calculate the adequate behaviour, but that he can identify with them, 'taking the role of the other', and thus feel sympathy even with somebody he perceives as enemy. If one would want to stick with Dennet's model, one might say that Siri has regained the capability to gather the data input he receives in a meta-perspective that includes not only emotions, but also what Ricoeur would call an *ipse identity*. This does not preclude that it might – only? – be a quality of mind, but this quality colours any perception by adding the assured knowledge that it is an 'I' comparable to the I-ness I assume in other human beings – and perhaps in some animals as well. In short, Siri has regained the qualia of 'what-it-is-like-to-be-me', in Chalmers' diction.

The case is entirely different with Kabir. In Kabir's poetry, consciousness in a world of downward causation is a holistic state of being, grounded in intimate individual experience and embedded in a cosmology of transcendence. Under these premises, the sensual corpo-reality becomes a site of knowledge production that resists integration into the systems of rational knowledge: "The Purana and the Koran are mere words; Kabir gives utterance to the words of experience; [...] all other things are untrue" (90). "[...] stand fast in that which you are" (70). This consciousness transcends 'consciousness' understood as attribute or trait of psychological processes and – spread out in time – oscillates between the sayable and the unsayable, language and being, intellect and divine vision: "Between the poles of the conscious and the unconscious, there has the mind made a swing: Thereon hang all beings and all worlds, [...] and the Lord Himself taking form [...]" (59).

5 First-Person Reports: Consciousness Transformed

Zelazo, Moscovitch, and Thompson, the editors of *The Cambridge Handbook of Consciousness*, suggest that "it will not be possible to make serious headway in understanding consciousness without confronting the issue of how to acquire more precise descriptive first-person reports about subjective experience (Anthony and Roepstorff 2003, 2004)" (Zelazo, Moscovitch, and Thompson 2007: 2, quoting Anthony and Roepstorff 2003/2004). The inclusion of first-person reports into their empirical investigations is prominently represented by research done in Dayalbagh (Satsangi 2013), in research done by Schmidt and Walach (Schmidt and Walach 2014), and in the experiments of Price and Barrell (Price and Barrell 2012), praised by Max Velmans as "a paradigm shift in psychological research" (Velmans 2013: 213).

Thomas R. Smith holds that literary texts can be taken as first-person reports (Smith 2007: 412). I do not agree with his assessment, because any literary text is framed by far more conditioning parameters than even a first-person report, i.e. conventions of genre, literary traditions, mediality etc. But I shall start with his suggestion nevertheless and ask: What can Watts' novel *Blindsight* and Kabir's poetry teach us about the codified representation of first-person experiential consciousness in the natural sciences, as is the case in first-person reports? Both text genres deal with the verbal representation of individual knowledge gained by experience. How do they assess the potential and limits of language – the central medium of first-person reports – to represent and/or express what they perceive as consciousness?

In Watts' *Blindsight*, strategies of knowledge production are illustrated with reference to gathering information about the home base of the aliens, which they themselves call Rorschach. The name is a message: The Rorschach test is a diagnostic strategy in psychology, asking the patient to associate meaningful figures with selected ink blotches. This implies that knowledge production starts with the hermeneutic assumption of the observer, which is projected onto the phenomenon. The reference to Rorschach in this context alerts the reader to the fact that a phenomenon like the Rorschach formation of the aliens does not mean anything, unless the vast data reservoirs the crew commands are transformed into meaningful information for the human home base on earth to decide on any kind of action to be taken. In order to transform data into meaning, they have to be translated into language, and this is the function of Siri, the first-person narrator of the novel. However, his narration illustrates how deeply any translation is an interpretation, conditioned by the state of the natural, computer, and technical sciences and by dominant value discourses on earth, because in the words of American philosopher Hilary Putnam, "[...] factual judgments, even in physics, depend on and presuppose epistemic values" (Putnam 2012: 291). Despite these framing conditions, Siri successfully 'translates' his knowledge into the language of unenhanced humans, but significantly, as soon as he retrieves his ego-consciousness, he reaches the limits of translating the qualia of what-it-is-like-to-be-me into verbal language. Instead, he performatively refers the readers of his 'report' to their own me-ness and advises them to deduce not intellectually, but by empathy, what he, Siri, is going through.

First-person reports face the same problems as Siri when it comes to rendering in verbal language individual experiences in such a way that they yield scientific 'facts' in accordance with the normative requirements of replicability, generalisation, and empirical validity (Mausfeld 2002).

In an exemplary fashion, the highly self-reflexive methodology of Price and Barrell, inspired by the challenge of "developing a human science based on experience and meaning" (Price and Barrell 2012: 13), explicates the many steps that are necessary to transform the sensation of pain into a scientific object: First, the probands are primed to isolate the event of feeling pain from the experiential flow, then there is a "translation of meanings into words" according to specific rules (not too long, not too complex, no vagueness, no metaphors), and the resulting reports are processed by strategies of consensus-building. Finally, these data form the basis for a tentative model of conscious experience. What has been eliminated to a considerable extent from the verbalisation of experience by way of abstraction from individual contexts is the hermeneutic embeddedness of the individual experience (what remains of course are basic premises implied in the grammar and semantics of the English language, Western assumptions about what it means to be human etc.). But by eliminating central traits of individuality, what is also eliminated is that instance of comprehension called consciousness that marks any experience as 'my' experience and as 'experience', because it sorts out something as an experience from the constant flow of mental and physical impressions called life in the first place. The first-person reports, purified from significant traces of individual hermeneutics by abstraction in order to yield generalisable third person perspectives as scientific facts, will find out a lot about various mental competences – in Dennet's sense – to isolate the feeling of pain, but they cannot say anything about the consciousness of the person who comprehends that he or she feels that pain. This does not detract any value from the respective experiments, but it makes clear that the methodology itself makes sure that the result can only be informative about a reductive version of consciousness, where the comprehending or core consciousness (Damasio) has been eliminated for the sake of scientific normativity.

Kabir speaks from entirely different premises than any scientist, because he is not in search of knowledge about an object of investigation, but he already knows what he is talking about – and he knows that it cannot be captured in language. In contrast with the epistemological premises of the scientific methodology outlined above, Kabir's consciousness does not claim – and does not strive for – any subjective agency ascribed to the observing subject in the sciences. Ganesh Devy expounds:

> The *Kena Upanishad* comments on the need for the dissolution of the consciousness 'trying-to-know': – 'By who willed and directed does the mind light on its objects? [...]' In answer to this question, it proposes, 'That which is not thought by the mind but by which, they say, the mind is thought'. (Radhakrishnan, 582–3) (Devy 2020: 40)

Accordingly, Kabir's validity claims are grounded not in methodological accuracy, but in intensely experienced feelings. As a consequence, Kabir's strategy of knowledge production and knowledge transfer rests on the identificatory experience of his readers/listeners with "'a sense of knowing' [where] there is no difference between the experience and knowing that you have it'" (Strawson 2019: 8). In Kabir's words: "Where is the need of words, when love has made drunken the heart?" (81). In light of these observations, the differences between empirical models and Kabir's claims concerning the nature of consciousness form a *différend* in the sense of Lyotard, which cannot be resolved because there is no rule applicable to all parties involved (Lyotard 1983: 9). The appeal to let the experiment decide about the incompatible validity claims is no solution, as it requires the counterparty to accept empiricism as the benchmark of valid knowledge production in the first place.

Kabir's absolute emphasis on experience corresponds with what Strawson calls 'real realism', namely "to take [...] experience [...] to be exactly what one took it to be [...] before one did any philosophy. [...] It's [...] a fundamental kind of knowing" (Strawson 2019: 8). For Whitehead, this subjective experiential knowledge acquisition is consciousness: "Consciousness is the subjective form involved in feeling the contrast between the 'theory' which *may* be erroneous and the fact which is 'given'. Thus consciousness involves [...] the contrast between [...] the words 'any' and 'just that'" (Whitehead 1978: 162).

However, a frequent strategy to legitimate validity claims of the natural sciences in consciousness studies is the denigration of experience. Dennett for example denounces pre-theoretical experience as "a user-illusion brilliantly designed by evolution to fit the needs of its users" (Dennett 2017: 222), claiming that the truth beyond that illusion can only be discovered by the natural sciences (Burkemann 2015). At the basis of this denigration lies what Strawson calls physics-alism in contrast to physicalism. Physics-alism means to take "the information expressible in the terms of [...] the physical sciences" (Strawson 2019: 12) as the sole basis to construct legitimate world models. From this perspective, Kabir's consciousness is "folk theory" (Dennett 2017: 222) at best, and only what the methodology of the natural sciences can get into view counts as consciousness. Yet according to philosopher of science Paul Feyerabend: "*First-world science is one science among many*; by claiming to be more it ceases to be an instrument of research and turns into a (political) pressure group" (Feyerabend 2010: XXI).

For Consciousness Studies this means that the methodology of the natural sciences is a regionally, historically, and culturally specific representation and manifestation of consciousness that addresses specific aspects and produces specific concepts of consciousness while excluding others. It follows firstly, that we need a critical explication of the cultural coding and the potential and limits of distinct disciplinary and trans-disciplinary approaches, and secondly, that we need a variety of methodologies and forms of representation to capture the complexity of what is called 'consciousness' in different contexts. Talking about consciousness in such a highly self-reflexive way would indeed turn Consciousness Studies into a transdisciplinary field of research, where – in the words of Bakhtin and Medvedev – "[n]ot only the meaning of the utterance but also the very fact of its performance is of [...] significance" (Bakhtin and Medvedev 1978: 120).

References

Primary Sources

Songs of Kabir. 2015. Trans. Rabindranath Tagore. Assisted by Evelyn Underhill. Mansfield Centre (USA): Martino Publishing.

Watts, Peter. 2006. *Blindsight*. New York: Tor Books.

Secondary Sources

Anthony, Jack and Andreas Roepstorff, eds. 2003/2004. *Trusting the subject? 2 Vol.* Exeter: Imprint Academic.

Bakhtin, M.M., and P.N. Medvedev. 1978. *The formal method in literary scholarship: A critical introduction to sociological poetics*. Baltimore/London: Johns Hopkins University Press.

Burkeman, Oliver. 2015. Why can't the world's greatest minds solve the mystery of consciousness. *The Guardian Long Read* 21.1.2015. Accessed 22 March 2019.

Chalmers, David J. 2018. The meta-problem of consciousness. *Journal of Consciousness Studies* 25 (9–10): 6–61.

Damasio, Antonio. 2010. *Self comes to mind: Constructing the conscious brain*. New York: Pantheon Books.

Dennett, Daniel C. 2017. *From bacteria to Bach and back: The evolution of minds*. New York, London: W. W. Norton & Company.

Devy, Ganesh. 2020. Competing knowledges: An Indian perspective. In *Competing knowledges/ Wissen im Widerstreit*, ed. Anna M. Horatschek, 37–50. De Gruyter.

Fenwick, Peter. 2004. Neurophysiology, consciousness and ultimate reality. In *Science, Consciousness and Ultimate Reality*, ed. David Lorimer, 109–128. Exeter: Imprint Academic.

Feyerabend, Paul. 2010. *Against method.* 4th edn. Introduced by Ian Hacking. London/New York: Verso.

Frank, Nathan D. 2017. The mind of *Then we came to the end*: A transmental approach to contemporary metafiction. In *Explorations of consciousness in contemporary fiction*, ed. Grzegorz Maziarczyk and Joanna Klara Teske, 225–244. Leiden: Koninklijke Brill NV.

Galant, Justyna. 2017. Creations of the Posthuman mind: Consciousness in Peter Watts's Blindsight. In *Explorations of consciousness in contemporary fiction,* ed. Grzegorz Maziarczyk and Joanna Klara Teske, 27–37. Leiden: Koninklijke Brill NV.

Hayles, N. Katherine. 2012. *How we think: Digital media and contemporary technogenesis*. Chicago: University of Chicago Press.

Horatschek, Anna Margaretha. 2016. 'Inhabiting landscapes of the known and the unknown alike': Representation and consciousness in British Literature. In *Consciousness: Integrating Eastern and Western perspectives,* ed. Prem Saran Satsangi, Stuart Hameroff, and Vishal Sahni, 155–187. Delhi: New Age Books.

Katz, Steven T. 1978. Language, epistemology, and mysticism. In *Mysticism and philosophical analysis*, ed. T. Katz, 22–74. New York: Oxford University Press.

Kriegel, Uriah. 2007. Philosophical theories of consciousness: Contemporary Western perspectives. In *The Cambridge handbook of consciousness*, ed. Philip David Zelazo, Morris Moscovitch, and Evan Thompson, 35–66. Cambridge: Cambridge University Press.

Lyotard, Jean-François. 1983. *Le Différend*. Paris: Les Éditions de Minuit/Lyotard, Jean-François. 2017. *Der Widerstreit*. Übersetzt von Joseph Vogl. München: Wilhelm Fink.

Mausfeld, Rainer. 2002. The physicalistic trap in perception theory. In *Perception and the physical world*, ed. D. Heyer and R. Mausfeld. Chichester: Wiley.

Mehrotra, Arvind Krishna. 2011. Translator's notes: 'Chewing slowly' by Kabir. *Poetry Foundation.* https://www.poetryfoundation.org/poetrymagazine/articles/145883/translator39s-notes-chewing-slowly-by-kabir. Accessed 06 June 2019.

Price, Donald D., and James J. Barrell. 2012. *Inner experience and neuroscience. Merging both perspectives*. Cambridge/London: MIT Press.

Putnam, Hilary. 2012. *Philosophy in an age of science. Physics, mathematics, and scepticism*. Cambridge/London: Harvard University Press.

Ricœur, Paul. 1991. Narrative identity. *Philosophy Today* 35 (1).

Satsangi, Prem Saran. 2013. Vision Talk TSC 2013. https://www.dayalbagh.org.in/specialTalks/ index.htm. Accessed 14 August 2022.

Schmidt, Stefan, and Harald Walach, eds. 2014. *Meditation – Neuroscientific approaches and philosophical implications*. Heidelberg: Springer International Publishing.

Seager, William. 2007. A brief history of the philosophical problem of consciousness. In *The Cambridge handbook of consciousness*, ed. Philip David Zelazo, Morris Moscovitch, and Evan Thompson, 9–33. Cambridge: Cambridge University Press.

Shaviro, Steven. 2014. *The Universe of things. On speculative realism*. Minneapolis: University of Minneapolis Press.

Smith, Thomas R. 2007. Literature and consciousness. In *The Cambridge handbook of consciousness*, ed. Philip David Zelazo, Morris Moscovitch, and Eva Thompson, 412–414. Cambridge: Cambridge University Press.

Sriramamurti, P. 2016. Perspectives on consciousness: Indian philosophical traditions. In *Consciousness: Integrating Eastern and Western perspectives*, ed. Prem Saran Satsangi and Stuart Hameroff, 377–410. Delhi: New Age Books.

Strawson, Galen. 2019. The Mary-go-round. In *The knowledge argument: Then and now*, ed. Sam Coleman. Cambridge: Cambridge University Press. https://www.academia.edu/31517753/Strawson_The_Mary_Go_Round. Accessed 7 September 2019.

Velmans, Max. 2013. A review of Donald D. Price and James J. Barrell (2012) *Inner experience and neuroscience: Merging both perspectives*. *Journal of Consciousness Studies* 20: 208–214.

Whitehead, Alfred North. 1978 [1929]. *Process and Reality: An Essay in Cosmology*. David Ray Griffin and Donald W. Sherburne. Eds. New York, London: Collier Macmillan Publishers.

Zelazo, Philip David, Morris Moscovitch, and Evan Thompson. 2007. Consciousness: An introduction. In *The Cambridge Handbook of Consciousness*, ed. Philip David Zelazo, Morris Moscovitch, and Eva Thompson, 1–3. Cambridge: Cambridge University Press.

Religion, Leadership, and Consciousness: Interactions and Opportunities

Andrew Davies

Abstract Effective leadership is not about profit, goals, and vision alone – it is also about the curation of shared values and their implementation in the life and culture of an organisation. This chapter argues that values play an important, but often massively underestimated, role in the life of any organisation, and that the role religion plays in the shaping of those values needs to be recognised and respected. The chapter concludes with some brief observations as to the practical application of the model in addressing great global challenges such as climate change.

Keywords Leadership · Faith · Values · Ethics · Organisations

At many different levels and in a variety of contexts, the global pandemic afforded us a variety of opportunities for seeing the very best of leadership in operation. We saw so often the remarkable impact of inspirational, empathetic, and supportive leaders who steered their organisations, corporations, and nations through incredible challenges, often at considerable personal, emotional, and physical cost, and yet inspired hope and offered comfort to those in need. We also, sadly, observed the dreadful consequences of ineffectual, self-serving leadership – leaders who were unresponsive, uncaring, or unthinking, and who I am sure will inevitably suffer the electoral, commercial, or professional consequences of their failures, if they have not done so already. For good or ill, however, in this time of crisis it seems to me that it was not only the *actions* of leaders that made a difference to people's lives, but also their *attitudes*. We saw again that, crucial as it is, competence alone is not enough to make a leader; we needed also to see evidence of their values impacting their decisions and demonstrated in their interventions. We needed to know that they were doing their best, yes, but also that they were doing so because it really mattered to them and they really cared about the outcome, if we were to be confident

A. Davies (✉)
Department of Theology and Religion, University of Birmingham, Birmingham, UK
e-mail: a.davies.4@bham.ac.uk

P. S. Satsangi et al. (eds.), *Consciousness Studies in Sciences and Humanities: Eastern and Western Perspectives*, Studies in Neuroscience, Consciousness and Spirituality 8, https://doi.org/10.1007/978-3-031-13920-8_21

and secure in their leadership. The crisis reminded us, I believe, that all leadership, in one capacity or another, needs to be anchored in the values of the leader and the organisation they lead, if their leadership is really to be trusted by those who look to it.

However, I also believe that there is an important corollary to this assertion. Leaders need to act out of their values ... but also, there is a significant sense in which acting out of one's personal or organisational values is in itself an exercise of leadership. Leadership is not about position, status, and recognition. Being a leader is not a special class or category, an elite status only for those who were always destined for power. Rather, for me, leadership is all about choosing and committing to take responsibility for delivering positive change and improved outcomes in your sphere of influence. This means that fundamentally all are invited to exercise that responsibility in some capacity or another, and that leadership is for everybody. And although such an assertion might seem unduly simplistic, in essence, the best way to help leaders emerge is therefore to encourage talented and gifted individuals to have confidence in their values, to be prepared to live by them, and enable them to take them into the public square, so that the benefit of their leadership, insight, and capacity can be more widely shared. Since religion is so frequently a factor in the development and shaping of people's values and attitudes, it seems to me that an exploration of the ways in which religious values can promote effective leadership might prove valuable. The model that has become known as 'conscious leadership' might be a useful starting point for us to explore the potential value that faith-based leadership can add for us.

1 Conscious Leadership

What does it mean, therefore, to lead consciously? A consciousness-focused approach to leadership encourages the development of a thoughtful, self-aware approach to leadership, which invites leaders to engage with their internal experiences and emotions as well as their external environment (Marinčič and Marič 2018). The benefit that such a style of leadership might bring across a variety of professional contexts and disciplines has been widely recognised. Writing from the context of nursing, for example, Suzanne Ward suggests:

> Consciousness is defined as a state of awareness; when expanded, it is often referred to as mindfulness. Conscious leadership [...] combines the science of leadership theories [...] with the art of understanding the human experience. (Ward 2016: 383)

That 'experience' for Ward is the experience of the leader as well as those for whom she is responsible. Conscious leadership brings humanity in all its fullness to the decision-making process. It is not, therefore, something that is measurable purely by empirical scientific method. Its outcomes are not always directly transferable or reproducible. Following identical approaches in identical contexts will not always lead to identical actions, because it is neither the approach nor the context that makes the difference, but the leader's engagement with their humanity in the process of leading. Whilst some of the more mechanistic approaches to leadership in the past have tried to train leaders to be disengaged and distant from their colleagues and as objective, cold, and clinical as possible to teach this out of would-be leaders, there is inevitably a measure of human judgement and subjectivity in leaders. And one of the things I love about conscious leadership models is that they actually cherish that subjectivity and foreground it, so that we can more fully appreciate how leadership is not just about following processes, but about understanding our people, our environment, and indeed ourselves, and reflecting (not in a critically-distanced manner, but in a deeply personally engaged one) upon how we work together.

For Gina Hayden (2016), conscious leadership depends upon the intersection of four elements: 'self mastery' (developing an awareness of one's values, purpose, and priorities and learning to manage and engage with these in the process of leadership); 'conscious relating' (choosing to listen intently to and engage deeply with those we lead; being prepared to recognise their gifts and abilities and empower them to take personal responsibility); 'systems insight' (an appreciation of the complexity and interconnectedness of every element of life and one's leadership context); and 'collective responsibility', the "urge or calling to contribute positively to the areas they believe need attention and reformation" (Hayden 2016: n.p.). For Hayden, then, the role of consciousness is to transform the leader's perspective and remodel the criteria against which leaders should measure their success. The consciousness 'bottom line' evaluates how effectively a leader is true to themselves and empowers their team, and not just the traditional economic and productivity measures. Conscious leadership is more focused on connections than outputs, believing this will deliver the best in terms of outcomes for an organisation. This means that conscious leaders need to be concerned for both the short-term and long-term implications of their interventions, and appreciate the significance of their contributions in the broadest and most holistic context possible. It is, of course, impossible to hold all that data in one's mind and navigate all its complexity on an ongoing basis; but it is possible to be aware of wider connections and the potential opportunities and challenges they pose. Conscious leadership and a systems approach to thinking, to a very large extent, go hand in hand. It is so important for us to train our minds to be appreciative of who we are and how we contribute to broader society. Especially in the West, but maybe this is a global challenge as well, we are so quick to cling to our own values and our own priorities, the things that we want to do, the things that will benefit us. Instead, we need to learn to appreciate just how much we are part of a bigger whole, and that our success feeds into the success of others. A consciousness of how we connect to our world, to our community, and to those that we are leading will help us appreciate what we have to offer to the world rather than merely attending to whatever benefit we can bring to ourselves.

Conscious leadership, then, is about an internal motivation that is not entirely commercial and financial. It is not about where the *profit* is. It is about where the *benefit* is, and who the benefit is for. So, where does the motivation and drive that motivates and promotes such leaders come from? Is there perhaps an internal consciousness that guides them into the values that they promote? Does religion make a contribution here, perhaps?

2 Religion and Values

Values are, of course, not owned by any one culture or tradition, nor are they owned by religion. It is entirely possible for the most staunchly non-religious to be intensely committed to their values. But the narrative that 'we live in an increasingly secular world' is very much a Eurocentric one. Eighty-four per cent of the world's population claims some kind of religious affiliation (Pew 2012). Across most of the world, religion is still a hugely important factor across cultural, gender, and racial divides. Religious values and priorities shape the responses of many individuals to the big policy challenges that we face as societies, and also to the smaller practicalities of everyday life. If we are to be genuinely inclusive and listen to the full variety of global voices, then the positive role of religious belief in shaping character and values must be recognised and affirmed. That is not to say we should prioritise religious ideologies or agendas, but we surely have to acknowledge their existence and ask for them to be taken seriously as one factor among the many others that contribute to the development of effective policy. Surely faith communities should never dominate any policy debate or determine its outcome, but that does not negate the importance of religious voices being actively present at the decision-making table. Religious values still matter. Let me suggest, furthermore, that those values also provide an important and often underestimated opportunity to build cohesive relationships across faith groups.

2.1 Values and Interreligious Relations

There is an approach to interfaith work that suggests that all religions are fundamentally the same, and that they share the same ultimate aspirations – that they are different roads to the same goal. I have never shared this perspective, and nor I think do most religiously-conservative communities worldwide. To my mind it is vital that we emphasise the diversity of religious beliefs and traditions, and stress how all the world's great religions have their own insights, ideas, and perspectives on human and divine existence. They do not all hold to the same core principles – they are incredibly diverse in their ideas. And this means they will never come to universal philosophical and theological agreement, because, by definition, it is those ideological perspectives which contribute to their uniqueness and to the essence of who they are as people of faith – by definition, a Christian is not a Hindu and is not a Muslim. There will always be many theological ideas that they do not have in common. To my mind, it is important that we cherish that diversity and do not try to minimise it. To underestimate it is to do a severe injustice to the distinctive heritage and history of these great faiths. We must, of course, ensure that diversity of thought and tradition is never used as an opportunity for 'othering' or sidelining those who are not like us. We can respect other perspectives even as we cling to our own beliefs. But we should not pretend that there is universal agreement on our understanding of the divine, the nature of human experience, and our future destiny – these fundamental theological and ideological differences are important elements of the various faiths.

However, there is much more commonality in terms of value systems. The 'golden rule', fundamentally the assertion that we must treat people as we would like to be treated, is common to almost every principal religious tradition. Most faiths, if not all, share a love for values such as justice, beauty, love, mercy, hope, kindness, and grace. Many religious systems require their adherents to show ongoing practical help for the poor and disadvantaged. It seems to me, therefore, that, rather than trying to identify theological commonalities that are not really there, we should recognise the shared commitment to building a better world that most faiths promote, and use this agreement as an incentive to help us promote and seize opportunities to work together for societal benefit. Despite our important differences, our religions should motivate us to work together with those who are going in the same direction. Interreligious relations, for me, should be centred around shared practices, commitments, and obligations rather than shared philosophies, and this emphasis on common values provides a more coherent and inclusive opportunity for those values to be welcomed in the workplace or in public life.

3 Values and the Secular

There is a further important observation to be made, and whilst this is of course very much a generalisation, I propose that it is generally true. The secular West is often incredibly uncomfortable with any sort of recognition of religion in public life; we so often fail to appreciate how religion shapes people's lives and attitudes and motivations. The East, widely, and India in particular, has a much better track record in that regard, and a more holistic and integrative approach to spirituality and everyday life. The West, especially Europe, has often wanted to think of (spiritual) consciousness as (psychological) mindfulness, and sought to distance mindful practices from their meditative roots, to rationalise them to make them appear a little more scientific, secular, and 'businesslike'. It is important for us, on the contrary, to recognise the religious element behind such mindful practices, and not try to pretend they are entirely secular phenomena. But, furthermore, we also need to develop in Europe an appreciation that the sacred–secular divide is infinitely more porous and indeed collapsible than we usually admit. The two are not eternally separate spheres that will never be joined. Indeed, trying to exclude faith from public life is, I believe, actively harmful, and damages individuals as well as societal structures, because religious traditions are much more important to people's everyday lives than secular perspectives acknowledge or appreciate. I happen to be a Christian, for example, and this is very important to me. I am a Christian not only on Sunday morning, when I am in church, but I am also a Christian on Wednesday afternoon, when I am in a committee meeting, and on Friday afternoon, when I am in the lecture theatre. I cannot pretend to stop being a Christian in those cases, or disengage my religious perspectives from my actions, because to me, my life and faith are inseparable. My Christianity shapes my attitudes. Sometimes it should shape them rather more, I readily admit. But I cannot separate who I am from the religious values that motivate me.

It seems to me that this is a critically important principle. People of faith cannot leave their religious obligations and traditions behind when they interact with the wider world. But even more fundamentally, I would argue that there is a hugely positive contribution that they can make when they allow their faith values to inspire, encourage, and motivate civic engagement and public leadership. The importance of a value-conscious approach to leadership cannot be underestimated, nor can the contribution that religion makes here. Perhaps we need to try and inculcate within people of faith, yes, a commitment to their religious tradition, but more broadly an appreciation that their tradition can and should shape their engagement with society – that they should always seek to live by their religious values and not be frightened to do so. This does not mean, of course, that they should ever attempt to impose those religious values on other people, and again I do not mean to imply that only religions have values. Whilst many non-religious ideologies and philosophies have values as well, a commitment to specified values and an obligation to live by those values is, to my mind, a fundamental part of what it means to be religious. So, what might that mean practically?

4 Value-Conscious Leadership

I view value-conscious or value-based leadership as representing a leader's quest or commitment to be guided, inspired, or motivated by their values in their leadership beliefs and behaviours. These values will principally and most frequently be the leader's own, but they can in some contexts be a set of values imposed externally (e.g. organisational or cultural values), which may or may not perfectly align with the leader's own personal values, but which nevertheless are the core framework for their leadership within that organisation.

These values can be informally held as private opinions, shared by the leader in staff meetings and communications, or formally expressed in an organisational values statement. They can be religiously motivated or entirely secular; they can comprise a detailed and comprehensive code of practice for life, or can be a single, simple assertion that has relevance across a variety of contexts (e.g. the 'Golden Rule').

What makes a leader value-based, in my estimation, is their conscious attempt to view each leadership decision and engage each action deliberately through the lens of their value system, so that their values motivate and shape their choices each step of the way. This is not to say, however, that values are the only or even primary factor in every decision – other compulsions or obligations may come into play, too. But a value-based leader will implement even calls that seem intuitively counter to their values, if they have to, in the light of those values and without compromising them, and perhaps this implies there is a further underpinning characteristic that pervades their leadership: the choice to take responsibility. This is evident in many practical ways in the leader's everyday decision-making.

First, for example, adopting a value-conscious approach would mean that leaders choose to do the right thing, even when it might not be convenient for them. For value-conscious leaders, doing the right thing becomes a greater priority than having an easy life. These leaders aspire to do the right thing, no matter what the consequences of their actions are. Now of course, factoring in those consequences is an important part of deciding what the right thing to do is (at least in many ethical models). But a brief reflection on the history of business ethics rapidly teaches us that organisations and institutions that put profit before conduct and try to protect their reputation and market position rather than confronting failures head on ultimately suffer for this approach.

Second, value-conscious leaders do not fall back on the easy way out of a challenge; rather they have the courage to do the right thing simply because it is the right thing to do, even when that makes life harder and comes at a personal or organisational cost. Value-conscious leaders take a stand on matters of principle and not just on policy. Of course, policy and process are both important, but it's the work of management, not leadership, to put policies and procedures and codes of practice in place. Helping people understand the values that motivate and guide and lead us invites us to do the right thing, not just follow correct procedure. And so this is one area where some facility in ethical thinking is essential for today's leadership.

Third, the value-conscious approach means that we need to allow ourselves to be held to account on our values. We must be actively accountable, especially on our ethical standards and on our decision-making processes. Now, all leaders need to have the right to lead – I am not saying we should question their decisions all the time and undermine their authority. But leadership decisions cannot be made in isolation and must arise out of corporate culture. There is no need for any leader to feel threatened by having their leadership examined and decisions evaluated. In fact, we should create a leadership culture that positively encourages constructive critique, and encourages an environment where we are accountable to one another. It seems to me that religion can speak a lot into that in terms of building the security and the confidence of our teams and ourselves as leaders.

5 People-Conscious Leadership

Perhaps I can narrow my focus a little more, now, to think specifically about one particular value. Religion in all its many forms has much to say about the importance of valuing people and respecting individuals. Drawing on those traditions, I think that anyone seeking to be value-based or value-conscious in their leadership needs to develop a people-conscious approach to their work, which facilitates community and cherishes distinctiveness.

Diversity is a blessing, as long as it is never used as a tool for isolation and exclusion. It is in confronting that diversity and engaging across that difference that we find real fulfillment, and begin to better understand what our own purpose is. A person-conscious approach would embrace and applaud that distinctiveness and seek to lead people with their unique perspectives and best interests in mind. It would of course also need to retain a commitment to corporate values and culture ... these are entrusted to our care as well as the people we lead, and we cannot permit them to be compromised in any way. I have seen firsthand that leaders who are 'people people', who put the preferences and expectations of their teams first in every regard, sometimes feel unable to execute what they know is the right decision organisationally, because they want to sustain their relational commitments to work colleagues – they place people above organisational culture and values. I think there is an important balance to be found here, which conscious and holistic reflection can help us navigate: We need to lead people out of and into our values. In other words, our values need to inspire and guide the way we lead people, and that means respecting them, caring for them, acknowledging their needs as well as their contributions; but we cannot permit anyone to undermine or flout our espoused values, no matter how much we respect and appreciate this person or need their input. We need to lead them back to our shared value system and encourage them to operate within its framework. That does not require empty compliance, nor does it demand homogeneity. When those we lead work out of the value system we share, and when we lead them from those values, their individuality will flourish, but their commitment to the team will be unimpeachable.

Organisational culture, I believe, needs to be distinctive, systemic, and comprehensively embedded in the organisation. However, sometimes because we are conscious of our values and our people, we need to accept and promote diversity of practice within the organisation – what sometimes is called 'harmonious difference'. We can work together for the same ends and goals, but we do not have to be the same. It is organisational culture, fundamentally, that distinguishes one organisation from another, but how that culture is outworked in the practical running of any individual team should vary, because the leader needs to take account of the needs, skills, talents, and best working practices of their team members. Team members need to be treated with equity, not always with equality. Conscious approaches to leadership can help us balance the needs of individuals and the 'greater good' of the organisation in a people-conscious manner.

This holistic approach is important for the leader to adopt in their practice and promote among their teams. Those we lead also need to develop their personal and professional consciousness. We need to encourage our colleagues to bring their whole selves to work. They should not have to leave their values at the door. They should not have to overlook their core values, personal priorities, and other aspects of their life to be accommodated or tolerated. They should be able to be a whole person in the workplace without any fear of rejection or exclusion. In fact, I think it is really important that we encourage colleagues to draw on their religious and spiritual insights in their everyday decision-making. Their values need to feed into who they are as a leader, because they are fundamental to their personhood.

A person-conscious approach to leadership can also contribute a long-term mindset. Conscious leadership sees beyond just one generation. It views our current challenges as just one step in an eternal journey. We have to engage with the challenges and opportunities that confront us at this moment, but this moment is only passing. Life will go on, and we need to lead into the future as well as just lead the present. That kind of attitude will, I believe, build resilience and build strength within us, and also encourages what I might call 'slowness' in leadership. We have made leadership about speed and hurry and momentum, when often it really should be about care and consideration and reflection. Slow and steady often really does win the day. Sometimes we're too quick to leap in and make a call and make a decision, because we think that leadership means activity, that leaders have to 'say things and do stuff'.

I suggest, however, that sometimes what we need is reflective, responsive, and steady leadership that does not just leap in and make a pronouncement, but pauses to take a breath, waits for the right moment, ensures as far as possible that it has all the evidence, and then makes a careful and consultative decision clearly. We need to create an atmosphere that encourages leaders to be reflective and not just to be immediately responsive for the sake of it. Agility is a crucial quality in today's world, but it must not be confused with speed. There is a beauty and an elegance to slow-motion agility, too. 'Reflective' should never be thought of as the opposite of 'decisive' – the two qualities can absolutely work hand-in-hand, and need to do so, if our organisations are to be sustainable and effective, and our teams are to feel included and valued. Whenever we identify a problem, of course, we all want it addressed there and then, and this is entirely understandable. But perhaps business schools could profitably start to teach 'slow leadership', reflective leadership, which creates atmospheres in our businesses and in our institutions where we can pause for thought and reflect, rather than simply and cynically leaping into action because of the pressure to do something or appear leader-like.

6 Implications and Conclusions

To conclude, then, let me offer a few final observations on the practical contribution of this kind of approach to leadership. What would a religiously-informed, value-conscious approach to leadership, which puts people first, mean in terms of its implementation and delivery?

First, I think, it helps us build an appreciation of the spiritual needs of those we lead. It helps us develop an acknowledgement, not just an accommodation, of religion in the workplace; we don't just put up with or tolerate religion in our places of work or in our societies, but we actually seek to take advantage of the strengths that it brings. I can see why people are nervous about this if they don't come from a faith perspective themselves (or even if they do come from a particular tradition and are cautious of alternative perspectives). That is understandable in many ways. I can see that people think, well, actually we don't want to bring religious conflict into the workplace. Nevertheless, it seems to me that this is not what a religiously-conscious approach to leadership looks like. Indeed I would argue that the conflict comes from telling people who are deeply religiously committed that they cannot think as they do, that they have to pretend to be someone they are not when they are in the workplace. We need to proactively dismantle some of these barriers between sacred and secular in our thinking.

Second, I believe that value-conscious leadership encourages the proactive inclusion of diverse voices and that deep sense of personal responsibility and accountability. Taking our responsibility for our values and our people as well as our organisations seriously reinforces the sense of purpose that we have that underpins our actions and ensures that our organisations are living for their values. We, as leaders operating out of our values, interacting with people at the level of their values, do not just treat them as resources that bring economic benefit, but as individuals to whom we are very deeply committed and with whom we have something significant to contribute to the development of our organisations and mutual flourishing of our society.

Our world faces many grave challenges – none more serious than the climate crisis. This, it seems to me, is an important forum in which we should consider the impact that religion can make in leadership. If 84% of the world's population has some kind of religious obligation, that amounts to some 5.8 billion people spread out across the globe, from diverse cultures, ethnic backgrounds, and lived realities, who, despite their differences that must not be underestimated, all find themselves facing specific intersectional challenges, ranging from the educational to the environmental, from gender equality to globalised economies, from public health to private wealth. Though such challenges all have local impact and demand local interventions, they also require global collaboration. And we can enlist people of faith as agents of change by engaging with the values and practices embedded within their rich and emotive religious traditions. It is entirely possible to advocate a theory of change for religious engagement, one that harnesses the power of religious people towards making a positive impact, driven by the teachings of their respective faith-based traditions. By working within – rather than against – their worldview and cultural frameworks, recognising the breadth of the tradition and attending to its historical development, listening to underheard and underrepresented voices on the ground as well as the religious hierarchies, faith communities can identify opportunities to leverage their traditions for transformation. I could point to work being undertaken by Pasifika Christian theologians in Fiji to draw on their traditional spirituality as well as their Christian heritage and values to develop a model for activism which reflects and reinterprets their historic and contemporary cultures equitably and ethically through their 'reweaving the ecological mat' agenda, for example (Bird et al. 2020). Or I could highlight the example of the 'female forest guardians' of Indonesia (Gulf News 2020), who use Islamic teachings to promote their climate resilience endeavour of protecting their local environment by challenging deforestation. Both offer excellent examples of how religiously-shaped values inform leadership that inspires action, which is led by people most affected by the challenges of the climate catastrophe, but which is also holistic in that it merges both the secular and the sacred in resolving social problems. Possibly, just possibly, adopting and implementing a value-conscious approach to leadership more broadly might just make it easier to leverage faith to deliver the genuinely transformative change our world needs.

References

Bird, Cliff, Arnie Saiki, and Meretui Ratunabuabua. 2020. *Reweaving the Ecological Mat Framework: Toward an Ecological Framework for Development*. Suva: Pacific Theological College.

Gulf News. 2020. *Look: Indonesia's Female Forest Guardians Fight for the Environment*. https://gulfnews.com/photos/news/look-indonesias-female-forest-guardians-fight-for-the-environment-1.1608115834385. Accessed 1 Aug 2022.

Hayden, Gina. 2016. *Coaching Questions for Conscious Leaders*. http://consciousleadershipconsultancy.com/coaching-questions-for-conscious-leaders/. Accessed 1 Aug 2022.

Marinčič, Dejan, and Miha Marič. 2018. Conceptualisation of conscious leadership. *Škola Biznisa* 8 (1): 175–186.

Pew Research Center. 2012. *The Global Religious Landscape*. https://www.pewresearch.org/religion/2012/12/18/global-religious-landscape-exec/. Accessed 1 Aug 2022.

Ward, Suzanne F. 2016. The art and science of conscious leadership. *AORN Journal* 104 (5): 383–386.

Index

P. S. Satsangi et al. (eds.), *Consciousness Studies in Sciences and Humanities: Eastern and Western Perspectives*, Studies in Neuroscience, Consciousness and Spirituality 8, https://doi.org/10.1007/978-3-031-13920-8

MIX
Papier aus verantwortungsvollen Quellen
Paper from responsible sources
FSC® C105338

If you have any concerns about our products,
you can contact us on
ProductSafety@springernature.com

In case Publisher is established outside the EU,
the EU authorized representative is:
Springer Nature Customer Service Center GmbH
Europaplatz 3, 69115 Heidelberg, Germany

Printed by Libri Plureos GmbH
in Hamburg, Germany